Sustaining Biodiversity and Ecosystem Functions

Economic Issues

Clement A. Tisdell

Professor Emeritus, School of Economics, The University of Queensland, Australia

Edward Elgar
PUBLISHING

Cheltenham, UK • Northampton, MA, USA

Published by
Edward Elgar Publishing Limited
The Lypiatts
15 Lansdown Road
Cheltenham
Glos GL50 2JA
UK

Edward Elgar Publishing, Inc.
William Pratt House
9 Dewey Court
Northampton
Massachusetts 01060
USA

A catalogue record for this book
is available from the British Library

Library of Congress Control Number: 2015943016

This book is available electronically in the Elgaronline
Economics subject collection
DOI 10.4337/9781785362682

MIX
Paper from
responsible sources
FSC® C013056

ISBN 978 1 78254 439 5 (cased)
ISBN 978 1 78536 268 2 (eBook)

Typeset by Servis Filmsetting Ltd, Stockport, Cheshire

Printed and bound in Great Britain by TJ International Ltd, Padstow

Contents

Preface

I was pleased when Edward Elgar Publishing agreed to accept my proposal to publish the manuscript for this book. It gave me an opportunity to combine in one volume many of my views about two major aspects of conservation economics, namely the economics of developing, conserving and changing the stock of human developed germplasm and ecosystems, as well as alterations to wild genetic diversity and natural ecosystems. Biological capital consists basically of two parts: human developed biological capital and natural biological capital, the stock of which is interdependent to some extent. A significant portion of human developed biological capital is a form of heritage capital. In my view, many contemporary studies of the economics are unbalanced because they focus almost exclusively on the importance of natural biological capital as a contributor to human well-being and neglect the importance of the human developed component of human developed biological capital in this regard. A comprehensive and balanced approach to conservation economics requires that both of these components (and their interdependence) be taken into account. This is what I have tried to do in this book. It is a start, but there is more to be done.

While writing the manuscript for this book, I became increasingly aware of the desirability of putting the issues covered into their long-term historical context. This is reflected in parts of this book. My awareness of the need for this was reinforced as a result of joint research on the economic development and diversity of ancient economics which I began in 2014 with Serge Svizzero of the Université de La Réunion, France. Thanks, Serge, for this cooperation.

It was a great encouragement to me (while writing the manuscript for this book) to receive the following email on 15 September, 2014 from Kenneth Arrow:

> Hi Clem,
> I have been following your work on *ResearchGate*. I am amazed by its amount and quantity. I just wanted to say, hello.
> As ever,
> Ken

Thanks, Ken. It is rare to get such positive feedback from such an eminent scholar as yourself. I'll treasure your communication along with a letter I received from Oskar Morgenstern (many years ago) letting me know that he was dying of cancer and urging me 'not to hide my light under a barrel'. A little bit of encouragement goes a long way.

The School of Economics at The University of Queensland has continued to support my research work and I am grateful to the present Head, Rodney Strachan, and the previous Head for doing this. Evelyn Smart did the word processing for the manuscript, helped me find references and assisted with proofreading. Once again, thanks Evelyn for the help and for successfully deciphering (mostly) my handwriting.

Again, two little dogs (Rani and Sparky) were my companions as I rose early most mornings to complete the manuscript, after which each morning I took them for a walk. Rani, however, is no longer around because she died of a heart attack while the manuscript was being completed.

I especially want to thank my wife, Mariel, for her understanding and support for my continuing interest in writing books. This has meant less shared time than otherwise, and my reluctance to 'tidy up'. Without Mariel's support, continuing my research and writing (over many years) would have been impossible. Thanks once again, Mariel.

Clem Tisdell
Brisbane, Australia

PART I

Biological conservation: basic economic issues

1. An introduction to the economic issues involved in sustaining biodiversity and ecosystem functions

1.1 WHY THIS SUBJECT IS IMPORTANT

Both the stock of the world's biological diversity and the state of its ecosystems are major influences on the ability of humankind to sustain and increase its well-being. They are major determinants of the availability of commodities essential to human life, such as food and water, and they are also, to a large extent, interdependent. However, it is not only the supply of necessities which is affected by the state of biodiversity and ecosystems. Also the availability of many non-essentials that add to the quality of human life is directly or indirectly dependent on the state of biodiversity and ecosystems. Human societies are embedded in ecological systems which they have been altering for many millennia.

Even before agriculture commenced about 9000 years ago, humans were both altering natural ecosystems to increase their well-being and adapting their economic activities to varied local and environmental conditions. These processes gathered considerable momentum following the commencement of agriculture and its geographical spread. Major acceleration occurred in humankind's ability to control ('and conquer') nature following the Industrial Revolution, which began in Europe: processes set in motion by this revolution (mainly increasing reliance on the application of advances in science and technology and associated capital accumulation for achieving economic growth) have (and continue to have) major impact on the state of the globe's biodiversity and ecosystems.

A feature of this long-term development process has been the increased decoupling of human economic activity from its surrounding local environmental conditions. This has not only been made possible by advances in science and technology and capital accumulation but also by the development of market systems and their extension. This 'increased decoupling' is a major cause of losses of pre-existing biodiversity and the reduced presence of ecosystems that once existed. In effect, humankind now has to adapt its economic activities less to local environments than in earlier times. This is because it alters local environments or it is able to a significant extent to insulate its activities from them, on a scale which could never have been imagined by its forebears. Although this has added greatly to human well-being, it is unclear for how long this type of development process can continue to add to human well-being, given global population growth and increasing demands on the earth's biological resources and ecosystems.

The world's human population is expected to increase by a large amount between now and 2060. At the same time, the number of people with higher incomes demanding more food, especially more animal protein, is predicted to rise considerably. Even if the globe's level of population stabilizes around 2060, increased human demands on the earth's biological resources (as well as its non-reproducing, non-renewed resources) may well increase. Consequently, continuing global economic growth presents increased challenges for conserving wild biodiversity and natural ecosystems.

However, it is not only the conservation of wild biodiversity and natural ecosystems which are being threatened by continuing economic growth and its nature. A further casualty is the loss of heritage germplasm (developed in the past by humans) for the production of food, fibres and other biological products. As well, previously available ecosystems developed by humans for biological production are being lost. While keeping all of these heritage-types of capital is unlikely to be rational, their potential economic value should not be overlooked.

They are not a part of natural capital but they have many of the qualities associated with biological types of natural capital. For example, it is likely to be impossible (or virtually impossible) to replace most heritage germplasm once it is lost. Many contemporary studies of the economics of biodiversity conservation and ecosystems fail to pay attention (or adequate attention) to the importance of biological heritage capital for the maintenance of human well-being and focus entirely (or almost so) on natural capital. This is, in my view, an erroneous approach. Therefore, in this book considerable attention (but not exclusive attention) is given to the conservation of biological heritage capital as a contribution to human well-being and its sustainability. Attention is also given to reasons why this resource is being lost, and like natural capital, is being lost in ways that reduce human welfare or prospects for its sustainability.

It might be noted that heritage biological capital is just one type of heritage capital. Another is the stock of human knowledge. It is possible for at least parts of this knowledge to be irretrievably lost, with potential negative consequences for the future welfare of humanity.

This book grapples with all the above-mentioned issues. Part I (which includes this chapter) provides a general conceptual backdrop to contemporary debate about these issues. Part II concentrates on changes in human developed germplasm and sustainability issues involving human developed ecosystems. The focus in Part III is on the economics of conserving wild biodiversity and assessing the economics of the utilization of natural ecosystems and the supply of wanted ecosystem services. More detailed information about the coverage of this book follows.

1.2 AN OVERVIEW OF THE REMAINDER OF PART I: BASIC ISSUES

In order to assess the economic value of sustaining the existing stock of biodiversity and ecosystems, it is important to classify appropriately this stock and changes to it. One of the main objectives of Chapter 2 is to do this. It is argued in this chapter that most recent economic studies of this subject have failed to pay enough attention to the nature of changes in the basic composition of biodiversity and in the attributes of ecosystems. They have failed to give enough attention to human developed germplasm and ecosystems because they have concentrated almost exclusively on the value of natural biodiversity and natural ecosystems as a part of natural capital. As a result, insufficient attention has been given to the economics of trade-offs between conserving wild biodiversity and natural ecosystems and the impacts of human development of new germplasm and the overall consequences of human alterations to ecosystems. Consequently, there are major deficiencies in much contemporary economic analysis of the net benefits of conserving wild biodiversity and natural ecosystems.

Furthermore, there appears to be a tendency in much of the recent economic literature on the valuation of biodiversity and ecosystems to take account only of the positive contributions of wild biodiversity and natural ecosystems to human well-being. While this may be justified from the political perspective that most members of the general public undervalue the services provided by wild biodiversity and natural ecosystems, it is not a balanced perspective from a scientific point of view. It is pointed out in Chapter 2 that natural ecosystems (or some of their attributes) can be either assets or liabilities from a human perspective and that some wild organisms may be pests or assets. Balanced evaluation of changes to biodiversity and natural ecosystems requires all these aspects to be taken into account.

In assessing changes to biodiversity and ecosystems, we need to give attention not only to losses of wild biodiversity and natural ecosystems but also to losses in heritage (human developed) germplasm and human developed ecosystems. The latter losses are neglected in some recent influential studies (for example, Kumar, 2010) of the economics of biodiversity and ecosystem services. Socio-economic developments are the main drivers of these losses in the modern era. Socio-economic processes involved in transforming the genetic stock and altering ecosystems are identified in Chapter 3. It provides a general introduction to this subject.

In Chapter 3, the geographical extension of human contact and human migration are identified as important sources of alterations to the global genetic stock and to ecosystems. Their influence has been present for millennia. However, following the European Age of Discovery, these influences became a global phenomenon. In this chapter also ways in which capital accumulation and advances in science and technology have brought about changes in biodiversity and ecosystems are outlined. Furthermore, it is indicated how the development of market systems and their extension have been significant drivers of biodiversity loss and changes in ecosystem. Other institutional influences which are considered are the preference of supermarkets and fast food outlets for standardized products, the granting of property rights in new genetic material and the operations of political structures and special interest groups. Additional factors which are taken into account include increased labour mobility, the growth of large organizations, such as public companies, government failure to reduce open access to natural resources in a timely manner and path dependence in the human development of germplasm and ecosystems. In later chapters, more detailed consideration is given to most of the above relationships.

The above influences on changes in biodiversity and in the nature of ecosystems have implications for the sustainability of economic development. Sustainable development and changes in the genetic stock and ecosystems is the subject of Chapter 4. First, several general concepts of economic development are identified. Several of these contemporary criteria are inconsistent in the types of development paths that they favour. All these sustainable criteria are normative and in some cases they result in indeterminacy and in choices that are Paretian inferior. An alternative possible criterion based on safety first is suggested. Current welfare criteria for achieving sustainable development are very hypothetical. It is argued that it would be worthwhile giving greater attention empirically to determining the degree of concern which current generations show (as well as express) for the welfare of future generations. This would give empirical substance to an idea raised by David Pearce (1998, p. 71) and also considered by Herman Daly (1999, pp. 115–16). This is likely to be more productive than concentrating only on the theoretical consideration of this subject.

There is no guarantee that any of the welfare criteria suggested for sustainable development will be able to be met. Humanity may not be willing to make sufficient economic sacrifices and to curb its population growth to achieve any of these objectives. Gowdy and Krall (2013), for example, warn that modern societies are in danger of becoming trapped in a system of ultrasociality in which they engage in mindless economic accumulation and population growth, like some species of ants.

Different views have been expressed about the conditions which need to be satisfied for achieving sustainable economic development. On the one hand, probably most mainstream economists believe that only weak conditions need to be imposed on the utilization of natural resources, including biological resources, to achieve sustainable economic development. Politically this is a popular view. On the other hand, most ecological economists suggest that strong conditions on the use of natural resources would need to be imposed to achieve sustainable economic development. These opposing views are discussed.

Uncertainty about the future makes it difficult to plan for sustainable development. Some of the problems that arise are discussed and particular attention is given to the precautionary principle. It is shown that, contrary to common belief, this principle does not always favour the conservation of natural resources, including the conservation of biodiversity.

A more specific sustainability issue, which is discussed from a general perspective in this chapter, is the sustainability of agroecosystems. Its focus is on ideas which were originally suggested by Gordon Conway (1985; 1987). This analysis is given further attention in Chapter 8, which examines the economic value of human developed and modified ecosystems from a general perspective. However, all the chapters in Part II raise issues which are relevant to agricultural sustainability and to other bio-industries.

1.3 AN OVERVIEW OF PART II – HUMAN DEVELOPED GERMPLASM AND ECOSYSTEMS – ECONOMIC ANALYSIS OF ISSUES INVOLVED IN BIOLOGICAL CONSERVATION

Chapter 5 discusses socio-economic factors that have influenced the development of crop varieties (mostly food crops) and their consequential impacts on biodiversity loss and change. Socio-economic factors which have resulted in the loss of food crop germplasm include changes in the demand for different types of food, such as increased demand for prepared and ready-to-eat food, and alterations in the nature of the food chain. This chain has become longer and institutional developments have transformed it, for example, the evolution of supermarket chains and large food processors. An additional influence on biodiversity change has been an increase in the number of end uses of particular types of crops. For example, some crops are used now not only to feed humans but to feed livestock and produce biofuel, as well as other end products.

On the supply side, the arrival of new species or varieties of food crops often results in loss of biodiversity present in existing crops. This is discussed and examples are given. Additional influences on biodiversity loss in crops include improved methods and reduced costs of transporting produce (as well as easier availability of off-farm inputs). New methods for harvesting crops and the development of new post-harvest technologies have also had an influence. Various technological and market developments have also resulted in the monoculture of crops replacing polyculture, for economic reasons. This is discussed. The increased adoption of monoculture contributes to the loss of pre-existing crop biodiversity.

Market extension (made possible both by technological and scientific advances and by institutional changes) has been a major force for altering the global stock of biodiversity. It has altered the geographical pattern of the composition of biodiversity and has contributed to biodiversity loss. This is a result of increased regional economic specialization in production. As pointed out in Chapter 5, this is not only due to increased regional specialization in the supply of end products but is also facilitated by greater international and inter-regional trade in agricultural inputs. It is found that the overall impact of increased (economic) globalization is to reduce global biodiversity, particularly the diversity of existing germplasm.

Because of the socially dominating structure of modern (market) economies (their ultrasociality?) the political dictates of economic growth make it difficult (virtually impossible) to avoid continuing loss of global biodiversity for reasons which are discussed. Furthermore, as a result of the large anticipated increase in global population and rising per capita income in this century (for example, the continuing emergence of a sizeable middle class in Asia), global demand for food (especially animal protein) is expected to increase substantially. It is unlikely that this demand will be met without further biodiversity loss. In the light of this prospect, two divergent views have emerged about how to respond to it in order to minimize environmental costs, including the loss of biodiversity. One view favours a land-saving (land-sparing) approach involving the intensification of production by bio-industries, including agriculture. The other favours a land-sharing (mixed land-use) approach which would result in extension of land use by agriculture and other bio-industries. These views are discussed. They also receive further attention later in this book.

The economics of conserving the germplasm of crops has been given little attention by economists probably because it is a difficult topic requiring considerable interdisciplinary cooperation. Some economic factors which are relevant to deciding on the desirability of conserving this germplasm are explored. These factors include the nature of demand for different types of food crops (the level and elasticities of demand for them), the extent to which livelihoods and income depend on particular crops, and anticipated changes in environmental conditions, including climate change. Consideration is given to the economics of conserving food crops by adopting different methods, such as their conservation in seed banks, *in situ* and *ex situ* or on state farms and research stations.

Chapter 6 concentrates on the identification of socio-economic factors that have contributed to the loss of genetic diversity of domesticated animals, particularly livestock. It complements the discussion in Chapter 5 because most of the socio-economic factors that have contributed to biodiversity loss in food crops have also contributed to loss of livestock biodiversity. It provides additional economic analysis which is relevant not only to understanding the causes of biodiversity loss in livestock but which throws further light on the reasons for a decline in crop biodiversity. In addition, it examines several other issues related to livestock husbandry such as its environmental consequences and its impact on food security.

In this chapter, the proximate causes of the loss of livestock breeds and the various socio-economic (and other) processes that have contributed to this are identified and analysed. Particular attention is given to the Swanson dominance-effect, the consequences of market extension and economic globalization, loss of multi-purpose breeds of livestock, and the impact on the diversity of breeds of the increased decoupling of animal husbandry from local environments. The disappearance of local breeds of pigs in Vietnam provides a short illustrative case study. This case study is followed by a discussion of the consequences of the development of livestock husbandry for the loss of wild biodiversity. Taking Australia as an example, it is pointed out that some Australian wild species have benefited from pastoralism but others have been endangered or have become extinct as a result of it. The net effect of pastoral activities has been to reduce global wild biodiversity. The impacts of livestock husbandry on food security are discussed. It is argued that, in various circumstances, the keeping of livestock adds to human food security and food supplies, even though there are situations in which it reduces the amount of food available for human consumption. It is usually the latter case that obtains coverage in the literature discussing this matter, rather than the former possibility. Both aspects are considered in this chapter.

In recent decades, genetic engineering has opened up new possibilities for altering biodiversity and for changing the way in which ecosystems are used. Species from several biological kingdoms (for example, bacteria, plants and animals) have been genetically modified. However, Chapter 7 concentrates mainly on the genetic modifications of crops (plants) and examines its ecological and economic consequences. Both are considered because economic assessments of the value of transgenic crops require their ecological effects to be taken into account. Various types of attributes which had been imparted to crops by genetic engineering are identified and then consideration is given to the ecological processes involved in biodiversity loss as a result of the use of genetically modified (GM) crops.

Subsequently, it is pointed out that trade-off of attributes and erosion of biological fitness problems usually occur after a GMO (genetically modified organism) comes into use (and some examples are given). When a particular organism is genetically modified to increase its biological fitness for a particular purpose, some of its other desirable attributes may be weakened or may become negative. It is even possible for the overall fitness of genetically modified organisms to decline both from a biological and an economic point of view. Furthermore, when crops are engineered to resist particular pests, this may provide a niche for populations of non-targeted pests to increase. The effectiveness of the genetic modification in controlling pests of the genetically modified crops can be expected to decline with the passage of time. Continuing effectiveness is likely to require increasingly more 'powerful' genetic modification. Some measures which can be adopted to counteract biological erosion in the biological fitness of a genetically modified crop are discussed.

The release of GMOs involves considerable environmental risk and uncertainty. Therefore, as discussed, consideration has to be given to how much caution should be exercised before releasing a GMO, the appropriate social criterion to apply when considering the release of a GMO, and the legal liability that ought to be placed on suppliers and on adopters of GMOs for any unwanted consequences of its use. These matters are discussed.

The economics of developing and marketing GMOs is an important subject. It influences the type of GMO developed and market structures. These and related aspects are also discussed in Chapter 7. Other socio-economic issues covered include patent protection and property rights in GMOs, the lack of biological and social co-evolution and different types of social conflicts arising from the use of GMOs, as well as problems posed by asymmetric information about GMOs. Before concluding Chapter 7, further assessment is undertaken of legal liability as a means of responding to any adverse consequences of the use of GMOs, such as negative environmental spillovers.

Chapter 8 explores the economic value of human developed and modified ecosystems. In the recent economic literature, the economic value of these has been neglected compared to consideration of the values of natural ecosystems. Reasons why this has happened are suggested. The various concepts of an ecosystem are outlined and their relevance to human developed ecosystems is examined. In addition, the dynamic properties of ecosystems are investigated, such as the nature of their resilience and their resistance to change.

Economic methods of valuing human developed and modified ecosystems are outlined and evaluated. The general environmental consequences of ecosystems altered by humans for cultivation purposes are identified, paying attention to onsite and offsite biotic and abiotic changes. A key component of this chapter is its assessment of the comparative economic value of alternative types of agroecosystems, as well as different forms of aquaculture and silviculture. Particular attention is given to evaluating high external resource-using agroecosystems and to the concept of sustainable agricultural intensification. The latter potential development is currently touted by many scholars as the ideal means to increase global food production at minimum (or even zero) economic cost and without loss of environmental quality. This view is examined critically.

In the following discussion, the prospects of utilizing increased intensification of bio-industries to increase food production and reduce their environmental footprint are investigated analytically. It is argued that although this win–win outcome may be possible, political and market failures can result in this outcome not being achieved. Furthermore, the evaluation of environmental change is usually ambiguous because it is normally the case that both positive and negative environmental changes occur when ecosystems are transformed. In most cases, it seems likely that intensification of ecosystems will have both positive and negative environmental effects and so their comparative net values will need to take this into account. Even changes which seem overall to be environmentally friendly are likely to have some negative environmental consequences.

In Chapter 7, the subject of the reduced biological fitness of organisms that have been genetically modified to resist pests and diseases is given attention. Chapter 9 further contributes to this subject by exploring the basic economics of evolutionary resistance to control techniques of pests and diseases. This resistance can develop for the use of a wide range of control techniques. These include controls based on the use of antibiotics, chemicals (such as herbicides), genetic modification to organisms, and even in the case of classical biological methods of pest control, as for example happened in Australia with the release of the myxomatosis virus to control Australia's population of rabbits.

A significant economic feature of the use of many techniques (which decline in their biological effectiveness for controlling pests or diseases as the frequency or duration of their use increases) is the failure of users to take into account the collective user cost of their use. This is the main subject investigated in Chapter 9.

It is shown that if the supply of this type of control method is available under conditions of pure competition, or is made available at the cost of supply by government (as has happened in some lower-income countries), this promotes its socially excessive (economic) use. The possibility that a monopoly would result in the efficient supply and use of such a technique is also investigated. It is shown that this is possible under special conditions. However, a monopolist substantially reduces the economic surplus of consumers (users) of the control technique. This can be excessively so from a social economic welfare point of view. On the other hand, this might be counterbalanced (in a dynamic setting) by a faster rate of technological progress under monopoly, if the theories of Clark (1940) and Schumpeter (1942) apply.

As part of the analysis in this chapter, the types of users that ought to be given priority in their access to techniques that decline in effectiveness with use are identified. The value judgement underlying this is that it ought to be based on willingness to pay. However, I would like to emphasize that this is a contentious assumption (but a standard one in economics) and is unlikely to have widespread community support in cases where the health and survival of individual human beings is at stake. Communal pressures are likely to exist to make these types of treatments available to (at least) some needy groups of individuals who are unable to pay for the cost of their provision, for example, by means of public health schemes subsidizing health treatment. Moreover, strong social pressures are likely to exist to allow greater current use of these techniques than is strictly optimal from an economics point of view. This is a fertile subject for further research.

From Chapter 9, it is apparent that property rights in much human developed genetic material play an important role in their development and in their sustainable use. Moreover, they can also be a significant influence on the conservation of existing human developed germplasm. Chapter 10 examines the economics of property rights in human developed genetic material, giving particular attention to relatively new institutional developments, such as plant variety rights and biopatenting. Furthermore, the consequences of contemporary legal developments, particularly international ones, are scrutinized in relation to this topic. These agreements include the International Convention on the Protection of New Variety of Plants (UPOV), the Convention on Biological Diversity, the Agreement on Trade-Related Aspects of Intellectual Property Rights (TRIPS) and the International Treaty on Plant and Genetic Resources for Food and Agriculture (ITPGFA). Some important inconsistencies between these agreements are identified, some of which can be attributed to what is described by Feindt (2012) as 'layering'. In addition, it is contended that these inconsistencies, as well as the vagueness of some clauses in these agreements, reflect international political and social conflicts.

1.4 OVERVIEW OF PART III – WILD BIODIVERSITY AND NATURAL ECOSYSTEMS: MORE ECONOMIC ANALYSIS OF ECONOMIC ISSUES INVOLVED IN BIOLOGICAL CONSERVATION

Whereas the main focus in Part II is on the economics of the conservation and loss of human developed germplasm and human modified ecosystems, the focus in Part III shifts to the economics of the conservation and the loss of wild biodiversity and natural ecosystems. Chapter 11 provides a general introduction to this part. It begins by pointing out the shortcomings of many economic studies which purport to analyse the economics of biodiversity conservation and loss. They actually fail to value diversity as such and consider only the economic value of particular natural objects, not their variety and the extent of their differences.

The economic valuation of the non-material benefits of the conservation of wild biodiversity is discussed and some empirical information is given about this. This is followed by consideration of the economics of the potential material benefits of wild biodiversity conservation, for example, its potential value for supplying pharmaceuticals.

It is important to take into account the total economic value of wild biodiversity and natural ecosystems. Some of the basic and little discussed problems (both empirical and philosophical) which arise in doing this are identified. Several divergent views exist about the comparative economic value of natural and modified ecosystems. For example, scholars differ in their views about the extent to which natural ecosystems can continue to be modified without reducing human well-being, and about relationships between the modifications to particular ecosystems and their impact on human welfare. These views are considered in Chapter 11.

Several important points are made in Chapter 11. For example, the importance of recognizing the relative nature of economic valuation and cost-effectiveness analysis as a basis for economic decision-making is emphasized. It shows that the economic value ascribed to a natural object may increase even though its degradation occurs. Moreover, the extent to which cost-effectiveness analysis can obviate the need for economic valuation is questioned.

Socio-economic reasons for socially detrimental alterations to natural ecosystems and biodiversity are identified. These are wide ranging in nature and include market failure, political problems, and shortcomings in public administration and in decision-making by NGOs. These are all discussed and analysed.

Several policy issues are raised by the coverage of the topic in Chapter 11. These are addressed in Section 11.10. It is pointed out that a wide range of different public policies are possible for influencing the conservation of biodiversity and the supply of ecosystem services. Payment for the supply of ecosystem services is just one possibility and may not always be the socially most desirable policy option. It is important to compare the social desirability of alternative policies and their sustainability, as becomes evident from some of the subsequent chapters, for instance, Chapter 16.

The main purpose of Chapter 12 is to demonstrate the importance of *relative* values in making decisions about land use in order to minimize the opportunity cost of conserving wild species in their natural habitats. Several authors (mostly ecologists) make the mistake that such decisions should be based on absolute values. For example, cases occur in the literature recommending that land areas which give the highest (absolute) returns to commercial use should be allocated to this purpose rather than to nature conservation, or that areas able to support the highest densities of the species to be conserved ought to be used for this purpose.

The conservation of the orangutan is taken as an example and both the private and social opportunity costs of its conservation are considered. It is assumed that a land-saving strategy rather than a land-sharing conservation strategy is adopted. However, in Chapter 17, the optimality of land-sparing versus a land-sharing conservation strategy is considered. Once again, the conservation of the orangutan is used to illustrate this analysis.

Chapter 13 pays particular attention to the role which private property rights in wildlife can play (or fail to do so) in their conservation. It also discusses features of international agreements which affect trade in wildlife species or their parts, and consequently, the conservation of individual species of wildlife.

In examining the prospects of use of wildlife for conservation purposes while on a private landholding or on areas controlled by an NGO, the conservation consequences are considered of the landholder being able to extract non-consumptive use value, or alternatively, consumptive use value from this presence of wildlife. This is shown to be influenced by the mobility patterns of different wildlife species. Also in the case of highly mobile wildlife harvested for consumptive use, it is found that the private property solution results in a different outcome from the open-access one. This has not previously been made clear in the relevant literature.

The conservation consequences of often conflicting global rights to trade in wildlife and their parts are examined. Particular consideration is given to the differing approaches adopted in the Convention on International Trade in Endangered Species (CITES) and in the Convention on Biological Diversity (CBD).

Chapter 14 extends the discussion in Chapter 13. It does this by giving further attention to the conservation consequences of granting property rights in wildlife species to landholders when these species are present on their land. It also considers the effects of other economic incentives to conserve wildlife species on private lands. Its coverage includes economic analysis, policy issues and examples. After considering private property rights and free markets as a means for conserving wildlife and pointing out some limitations of this approach, it identifies a range of other economic policies which may be adopted by governments for conserving wildlife on private land. These alternative economic policies are assessed, and this is followed by further examination of some actual government scheme to provide economic incentives for the conservation of wildlife on private land.

The relatively radical thesis that there is no regular relationship between the level of the rate of interest and loss of wild species (also loss of natural capital) is developed in Chapter 15. The usual view presented in the economic literature is that an increase in the rate of interest increases the likelihood of extirpation of wild species. However, this is shown to be only true for some microeconomic situations. Furthermore, when account is taken of macroeconomic relationships, it is demonstrated that the abovementioned hypothesis does not hold. Taking into account macroeconomic relationships, increased loss of natural capital (and, therefore, greater loss of wild biodiversity) can occur, in some circumstances when the rate of interest falls, and in other cases when it rises.

This is because increases in the level of aggregate economic activity, and in the level of aggregate investment are the prime drivers of the loss of natural capital. These levels can be associated with either a rise or a fall in the rate of interest. In macroeconomic contexts, the rate of interest is a dependent variable not an independent one. Therefore, both on the basis of microeconomic and macroeconomic analysis, the definite conclusion is reached in Chapter 15 that there is no regular relationship between the loss of natural capital (and the decline in wild biodiversity) and the level of the rate of interest.

Processes involved in the economic valuation of natural ecosystems and their sustainability are the focus of Chapter 16. It begins with a discussion of the role of natural capital in influencing human well-being. It is pointed out that some recent models (for example, Costanza et al., 2014), which include natural capital as a contributor to human well-being, are too restrictive in their identification of contributors to human well-being. This is followed by critical observations on recent estimates of the global value of ecosystem services and a general discussion of alternative economic methods of valuing ecosystem services. The implications of some of these estimates are disturbing. Furthermore, most rely on a hodgepodge of data collected using different methods, some of which have been applied in a dubious manner, for example, the cost of replacement method. An important and interesting issue discussed is macroeconomic accounting for the value of ecosystem services versus microeconomic welfare accounting for their value.

Particular attention is given to the limitations of adopting cost-side methods of economic valuation of ecosystem services compared to demand-side valuation. Situations in which cost-side replacement valuation is an accurate indicator of economic value (and others in which it is not) are identified. Usually, both demand-side and cost-side factors need to be taken into account in economic valuation.

There is a danger when cost of replacement methods are used that existing ecosystem services will be overvalued. The cost of replacement methods never undervalue the loss of ecosystem services but very often overvalue them, as is demonstrated. This is followed by a discussion of the role of economics in the valuation of changes in ecosystems along with some illustrations involving the introduction of invasive (pests) species to new geographical regions, and the adoption of inappropriate methods of ecosystem 'restoration'. Taking China's Grain for Green Project as an example, for illustrative purposes, there is additional discussion (beyond that provided in Chapters 13 and 14) on the policy of paying landholders for the supply of ecosystem services. Attention is given to the potential consequences of the lack of permanency of such payments and to principal–agent problems associated with such payments.

The supply of ecosystem services and the economics of their provision depend on the spatial use of the biosphere by human beings. The economics of the spatial use of ecosystems are considered in Chapter 17 which, for analytical purposes, introduces a model which examines the economics of using a forest for wildlife conservation and for the supply of logs. The discussion is relevant to the debate about the desirability of land sharing versus land sparing as alternative means of conserving wild biodiversity and simultaneously meeting human demands on ecosystems (land use) for the supply of provisioning of services (such as timber and food). An aim of this chapter is to determine the optimal (economic) pattern of the use of a forested area that conserves a minimum viable population of a focal wildlife species and which (subject to this constraint) maximizes profit from logging. The model can also be applied if it is desired to conserve a higher targeted population of the focal species, thereby increasing the chances of its survival. Furthermore, it can be extended to include the conservation of more than a single species. However, when this is done, it usually imposes increased restrictions on forest operations (logging must be more selective) when the land-sharing option is considered. As a result, mixed land-use (land sharing) can be expected to become less profitable and so the probability of land sparing (spatial separation of the provision of ecosystems services) is likely to become more attractive.

The modelling is applied to the conservation of the orangutan. It therefore adds to the discussion in Chapter 13 about the conservation of the orangutan. Some dynamic spatial aspects of the recovery of the population of a wild species after its habitat has been disturbed are discussed. For example, the retention of nuclei of the wild population in small, fully conserved cells from which they can disperse as a forest recovers from logging.

The debate about the desirability of land sharing versus land sparing has mostly been conducted by ecologists. Most, however, do not specify the valuation functions that they have in mind. They are usually not made explicit. Therefore, it is not surprising that much of this debate is inconclusive. At the same time, it is clear from the analysis in Chapter 17 of this book (and elsewhere in this book) that social optimality of land sharing versus land sparing varies according to areal variations in economic and ecological (environmental) conditions. In addition, because there are many different forms of land sharing, it is also apparent that the optimal choice between land sharing and land sparing requires many different alternatives to be considered, not just two.

1.5 CONCLUDING COMMENTS

This book contains a constructively critical discussion of contemporary views about the economics of biodiversity conservation and the supply of ecosystem services. It also throws doubt on some propositions and policies for resource conservation that have been proposed by ecologists. It is important that the role of biodiversity conservation and the provision of ecosystem services for human well-being and for sustainable development be assessed from varied points of view, and that rational scepticism be exercised in doing so. Therefore, I present several perspectives which are at odds with mainstream contemporary views about biological conservation, where this seems to be warranted.

REFERENCES

Clark, J.M. (1940), 'Towards a concept of workable competition', *American Economic Review*, **30**, 241–56.

Conway, G.R. (1985), 'Agroecosystem analysis', *Agricultural Administration*, **20**, 31–55.

Conway, G.R. (1987), 'The properties of agroecosystems', *Agricultural Systems*, **24**, 95–117.

Costanza, R., R. De Groot, P. Sutton, S. Van der Ploeg, S.J. Anderson, I. Kubiszewski, S. Farber and R.K. Turner (2014), 'Changes in the global value of ecosystem services', *Global Environmental Change*, **26**, 152–8.

Daly, H.E. (1999), *Ecological Economics and the Ecology of Economics: Essays in Criticism*, Cheltenham, UK and Northampton, MA, USA: Edward Elgar Publishing.

Feindt, P.H. (2012), 'The politics of biopatents in food and agriculture, 1950–2010: Value conflict, competing paradigms and contested institutionalisation in multi-level governance', *Policy and Society*, **31**, 281–93.

Gowdy, J. and L. Krall (2013), 'The ultrasocial origin of the Anthropocene', *Ecological Economics*, **95**, 137–47.

Kumar, P. (ed.) (2010), *The Economics of Ecosystems and Biodiversity: Ecological and Economic Foundations*, London and Washington, DC: Earthscan.

Pearce, D.W. (1998), *Economics and Environment: Essays on Ecological Economics and Sustainable Development*, Cheltenham, UK and Northampton, MA, USA: Edward Elgar Publishing.

Schumpeter, J. (1942), *Capitalism, Socialism and Democracy*, 2nd edn, New York: Harper & Brothers.

2. Classifying the stock of genetic resources and ecosystems: their economic nature and patterns of biodiversity change

2.1 INTRODUCTION

Although biological taxonomies of living organisms are useful, other types of classification can be of greater help when considering the nature and economics of biodiversity change. Simple classifications are outlined in this chapter which make it clear that our current stock of genetic resources and ecosystems only consist partly of natural capital. Some (probably most) of our highly valued genetic resources and ecosystems are primarily the result of human effort and investment. This is not to suggest that natural genetic resources and ecosystems cannot be very valuable from an economics point of view. In fact, they usually have greater total economic value than is commonly realized. For example, natural wetlands and swamps are frequently seen by members of the general public as being of little or no economic value, but as pointed out in the Report on *The Economics of Ecosystems and Biodiversity* (TEEB) (de Groot et al., 2010, pp. 379–83), they often prove to be very valuable when all (or many of) their ecosystem services are accounted for.

Although some transformations of natural ecosystems by humans have reduced their total economic value, other transformations have added considerably to the economic value of biosphere use, despite these transformations not always being ideal.

The above perspective is not intended to downplay the importance of naturally occurring germplasm. Natural germplasm has provided the initial basis for the human development of germplasm for many bio-industries such as in agriculture, forestry, aquaculture and for some biological applications in medicine. Thrall et al. (2010, p. 405), for example, outline how

> evolution driven both by conscious and unconscious selection by humans has shaped the development of modern agriculture. Agriculture has been a crucible

> of evolutionary change ever since its inception thousands of years ago, and this change permeates agricultural endeavours at all levels of biological organization, ranging from the individual gene through to whole communities.

Given the above perspective, the assertion that 'from an economic point of view, biodiversity (and ecosystems) can broadly be seen as a part of natural capital' (Pascual et al., 2010, p. 187) is misleading. This is because many valuable ecosystems and genetic resources are human developed: they are a result of human investments, effort and continuing interventions with nature.

Both human developed ecosystems and genetic resources (as well as natural ones) can be assets (forms of capital), and sometimes elements of both can be economic liabilities. Obviously, some organisms are pests or liabilities as far as humankind is concerned, and there is little point in pretending otherwise. Balanced economic analysis of biodiversity conservation requires the whole spectrum of these phenomena (the human developed and natural attributes of biodiversity and the positive and negative characteristics of living things and ecosystems from an anthropocentric point of view) to be taken into account. I have argued elsewhere (Tisdell, 2014a) that a shortcoming of a major TEEB report (Kumar, 2010) is its failure to assess this full spectrum. For example, it gives insufficient attention to the assessment of the economic value of human developed ecosystems, (such as agroecosystems) and genetic resources developed by humans, such as varieties of cultivated crops and breeds of domesticated animals.

It is sometimes assumed that biodiversity loss implies that the total stock of biodiversity is declining, that is the total stock of genetic material is becoming smaller, with no additions being made to that existing stock. However, at the same time as some genetic resources are being lost, human beings are adding to this stock by their selection of genetic resources (especially the selection of favoured types of germplasm) and more recently, by means of genetic engineering. Nevertheless, the number of basic elements in the genetic stock has been falling rapidly due to human activity and because of human development of new genetic resources. This process is easy to grasp from the simple taxonomy (a representation) outlined below.

In developing this chapter initially, a simple taxonomy of genetic resources is outlined and discussed. Then the dynamics of changes in the genetic stock are examined, paying particular attention to the general economic processes generating these changes. Subsequently, the nature, classification and value of ecosystems are discussed and the problem of determining whether organisms and ecosystems are assets or liabilities is considered.

2.2 A TAXONOMY OF GENETIC RESOURCES INTENDED TO ASSIST ECONOMIC ANALYSIS

As a first approximation, the existing set of genetic resources (germplasm) can be divided into three sets:

1. The set that exists in the wild. Designate this as set A.
2. The set that exists under human management, for example, in crops under cultivation and in domesticated livestock. This set owes its existence primarily to genetic selection by human beings. Designate this as set B.
3. The set that has been produced by genetic engineering, set C.

These three sets are represented conceptually in Figure 2.1. Let set B plus set C be designated as set D. Set D owes its existence primarily to human effort.

Each element in the combined sets A, B and C, that is, within the complete set of existing genetic resources, is assumed to a genome (unit of germplasm) but the sets are conceptual and do not portray the actual number of genomes in each of the sets. Nevertheless, currently, set A seems to be the largest by far, set B is much smaller and set C is the smallest. Many of the genomes in set A are as yet unknown (particularly in the case of small organisms) and even many in set B are not known to the scientific community. For example, according to a report prepared for the FAO on the state of plant genetic resources for food and agriculture in Papua New Guinea, there is a serious lack of information and data on the existence of these resources in farmers' fields and in wild habitats throughout the country even though much biodiversity exists (FAO, 2009,

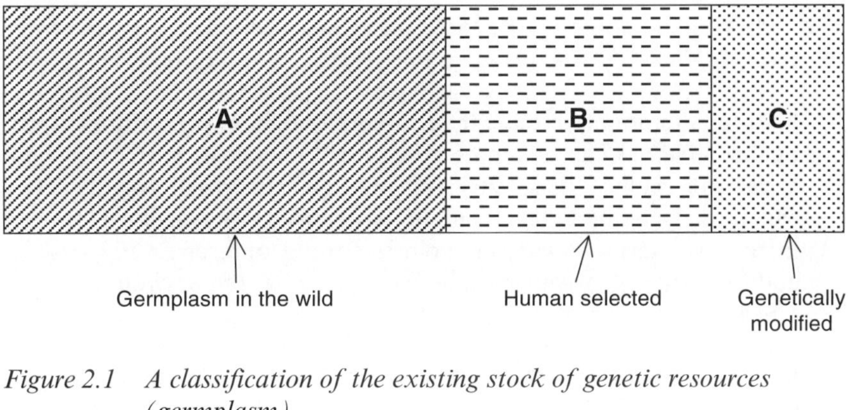

Figure 2.1 A classification of the existing stock of genetic resources (germplasm)

Section 2.1). Furthermore, some overlap occurs between set A and set B. Some organisms exist in the wild as well as under human cultivation or husbandry. For example, some species of mushroom are cultivated and also grow in the wild. Some previously domesticated animals have become feral, for example, pigs in Australia and in the United States (Tisdell, 1982). Furthermore, human selection of organisms in the wild has occurred in the past (Gilmore, 1934) and still continues, for example, in some natural forest stands as a result of selective removal (thinning) of trees. Evolution of species in the wild has been influenced for millennia by human practices. For instance, the systematic use of fire to burn vegetation (mosaic burning) by Australian Aborigines altered landscapes, thereby favouring the survival of particular wild species (Gammage, 2012; Flannery, 1995). Therefore, the classification illustrated by Figure 2.1 is somewhat 'fuzzy'. Nevertheless, as will be demonstrated, it helps with the understanding of important features of the processes involved in changes in genetic diversity (the stock of germplasm) and the influence of economic factors on that change.

2.3 DETERMINANTS OF CHANGES IN THE GENETIC STOCK

Let us consider in turn factors that result in changes in the stock of germplasm, focusing first on set A and then on sets B and C but also bearing in mind that a variation in one of these sets can alter the stock contained in the other sets.

Germplasm in the Wild (Changes in Set A)

The following are some of the factors that may eliminate germplasm in set A.

1. Natural alterations in habitat or in environmental conditions in particular localities can eliminate species or varieties of species in these areas, especially if these environmental changes occur rapidly. They can initially reduce the global stock of wild biodiversity. However, in the long term, biological evolution could once again give rise to additions of new organisms adapted to the changed environments.
2. Unsustainable harvesting or hunting of wild species can result in their extinction. For example, the Maoris in New Zealand are believed to have hunted moas (large flightless birds) to extinction. See, for example, Anon (2014a).

3. Habitat loss resulting from human activity is a threat to the continuing existence of many wild species. Reasons for this include the conversion of their existing habitats for use in agriculture, aquaculture and forest plantations, as well as urban expansion and the building of infrastructures to support human settlements. The process of habitat conversion is influenced by capital investment and new technologies (including improved varieties of domesticates), population growth and the seemingly insatiable demands of human beings for a greater quantity of produced commodities and/or those of superior quality.
4. The introduction of exotic species to new areas by humans (sometimes deliberately, and sometimes accidentally) often results in a global reduction in the biodiversity. The deliberate introduction of the cane toad (*Rhinella marina*, formerly *Bufo marinus*) to control insects in sugar cane crops threatens the survival of several Australian native species (Anon, 2014b). This toad contains a powerful poison which kills animals that ingest it. The introduction of the European carp (*Cyprinus carpio*) to some freshwater rivers in Australia is a threat to the survival of several native fish species (Koehn, 2004). There has also been accidental spread of pathogens, for example, *Phytophthora* species. These fungi infect the roots of plants that are not adapted to their presence and weaken or kill them (see Anon, 2014c). *Phytophthora* can adversely affect wild species of plants as well as cultivated ones.

The natural evolution of germplasm does not stand still. Species continue to evolve as habitats change, and as a result set A alters. While natural evolution is a slow process for species that reproduce infrequently, for those that reproduce rapidly, such as most microbes and insects, it is a relatively speedy process. In the latter case, mutants and variants in the population of the species, which prove to be fitter under changed environmental conditions (or even in stationary conditions) than others, reproduce rapidly. Furthermore, the frequency of mutations per unit of time in rapidly reproducing species may be higher than in slowly reproducing ones. For example, new strains of influenza evolve with a high relative frequency with the passage of time. Pathogens and pests that resist chemical and biological controls seem to evolve fairly rapidly. This also seems to be so for fungal diseases (such as rusts) that attack several types of crops (Thrall et al., 2010).

The development of improved germplasm in sets B and C (that is, set D) stimulates the growth of bio-industries, such as agriculture, and often leads to their extension as well as their intensification on land (or in water, in the case of aquaculture) already used for these activities. This usually results in habitat loss for wild species and tends to reduce the size of set A.

Furthermore, as a result of the development of bio-industries, some wild species become pests and attempts may be made to eliminate these. This can also reduce the wild gene pool. In addition, genetic resources developed for use in bio-industries can change the composition of wild genetic resources in set A in other ways: an organism developed for use in a bio-industry may become feral or cross-fertilization may occur with the wild stock. Thus, there are several ways in which the development of the genetic stock in sets B or C can alter the genetic stock contained in set A.

However, alterations in the stock of germplasm contained in sets B and C may not only alter set A but can also result in changes in the elements of sets B or C. Improved varieties of crops or breeds of animals added to set B or C may displace less economic ones already in set B and C. This displacement is probably more likely if the new varieties in sets B or C are accompanied by a profitable package of technologies which tend to decouple (to divorce) the use of these varieties from local eco-geographic conditions. Many 'Green Revolution' technologies employed in agriculture have these attributes (Alauddin and Tisdell, 1991). However, the increased decoupling of bio-industries from surrounding environmental conditions can also occur in the absence of the development of new varieties of domesticates. For example, the extension of irrigation reduces the dependence of crops on local rainfall, artificial fertilizers may compensate for local soil deficiencies, and pest controls may isolate domesticates from local pestilence. These types of technologies are a powerful force for enabling bio-industries to rely on a reduced set of varieties of domesticates for their production.

As a rule, the above-mentioned types of development tend to reduce the number of existing elements in set B, but some exceptions can occur. As before, let D represent set B plus C. When a new variety or breed is developed in set D, then initially the elements in set D increase because it takes time for the use of the new germplasm to diffuse. If in some locations, the use of the new germplasm is more profitable than existing germplasm, then replacement will occur. However, this might be confined to only part of the area where the existing germplasm is used and all the existing germplasm may continue to be employed in other locations. Consequently, there could be a global net addition to biodiversity. However, in practice, this seems to be the exception rather than the rule.

Socio-economic developments also play a major role in altering the size and composition of the genetic stock. Market extension and globalization have, for example, resulted in the loss of many varieties of domesticated organisms. The FAO (n.d.) reports that the globalization of the food system, and market extension and marketing developments have resulted in the cultivation of fewer varieties of crops and a reduction in the number

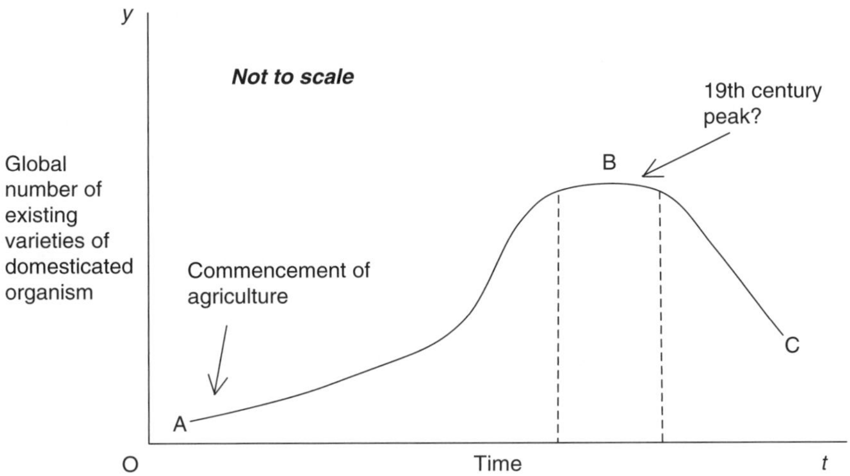

Figure 2.2 Hypothetical time-pattern for the global number of varieties of domesticated organisms present in agriculture

of breeds of animals used for economic production. In addition, growing urbanization (accompanied by changed lifestyles) has favoured varieties of agricultural produce which are amenable to transportation and easy to store, which can be standardized, and which are suitable for the production of convenience foods. These processes will be given attention in Chapter 3 and specific examples will be discussed in Chapters 5 and 6. These losses of biodiversity are often associated with the replacement of local varieties of crops (and breeds of domesticated animals) by improved or exotic varieties and species (FAO, 2004).

As a result of all these developments, the global number of varieties of domesticated organisms has probably followed a pattern like that indicated in Figure 2.2, at least as far as agriculture is concerned. Following the commencement of agriculture early in the Holocene area in the Levant, and its independent beginnings in other parts of the globe several thousand years ago (Renfrew, 2007), the number of domesticated species would have increased and numerous new varieties would have been developed to suit local eco-geographic conditions. This number may have reached its maximum at some time in the nineteenth century. Since then, it has declined. The FAO (2004) reports that in the twentieth century 'some 75 per cent of plant genetic diversity has been lost as farmers worldwide have left their multiple local varieties and landraces for genetically uniform, high yielding varieties'. In addition, the FAO (n.d.) estimates that in the twentieth century 1000 domestic animal breeds were lost. Furthermore, it estimates that around a third of the 1335 domestic animal breeds present in this century are threatened with extinction.

Note that the pattern shown in Figure 2.2 may not apply to aquaculture, which has had a late spurt of development. It is possible that the number of varieties of organism in use in aquaculture is still increasing but at the same time, the composition of varieties is altering (Tisdell, 2009). Furthermore, it is not clear when in the Holocene era the number of existing wild species was at its peak, but that may also have occurred in the nineteenth century, or even earlier.

2.4 THE NATURE AND CLASSIFICATION AND VALUE OF ECOSYSTEMS

The Millennium Economic Assessment (2003; 2005) stressed the importance of conserving ecosystems (particularly natural ones) because they make a very valuable contribution to the level and maintenance of human well-being. Their contribution is often poorly known or appreciated by members of the general public. This emphasis on the contribution of ecosystems to human well-being is also reflected in reports prepared as a result of the TEEB (The Economics of Ecosystems and Biodiversity) project (see for example, Kumar, 2010). Kumar (2010) provides an overview of economic estimates of the value of ecosystems and some of the economic methods used to estimate these values. Kumar (2010) reveals significant gaps and shortcomings in the current economic valuation of ecosystems and biodiversity (Tisdell, 2014a), many of which will be difficult to eliminate. TEEB gives most of its attention to the economic value of 'natural' ecosystems (assessed on an extremely broad basis) and pays inadequate attention to the comparative economic value of human modified or altered ecosystems, such as alternative agroecosystems and diverse patterns or mosaics of transformed natural ecosystems (Tisdell, 2014a; Hof and Bevers, 1998; 2002). This is a serious shortcoming because it fails to take adequate account of many of our most valuable ecosystems and varied patterns of biosphere use, developed as a result of human effort and investment. That is not, however, to suggest that transformed ecosystems are always more valuable from an economics point of view than are the natural systems they have replaced, nor that this replacement has always occurred in an economically optimal manner.

Types of Ecosystem Services Identified in the Millennium Ecosystem Assessment

The Millennium Ecosystem Assessment (2003; 2005) focused on identifying services provided by ecosystems which are valuable for humans. Its approach is anthropocentric, as is that of economics. It emphasizes that the conservation of biodiversity is intimately bound up with the conservation of ecosystems, and this approach also dominates TEEB reports; for example, Kumar (2010). Whether or not this is an entirely satisfactory approach to assessing the economics of biodiversity conservation will be discussed later.

The Millennium Ecosystem Assessment divides the services provided by ecosystems into four main categories:

1. *Supporting services*. Those which are essential for supporting the supply of other services. These are primary services and are essential for the supply of the services listed below.
2. *Provisioning services*. They include the supply of provisions, such as food, for human consumption.
3. *Regulatory services*. These encompass the role of ecosystems in regulating environmental conditions, such as regulating floods and weather conditions.
4. *Cultural services*. These are rather intangible services compared to the more tangible provisioning services. They include the aesthetic and spiritual values generated by ecosystem services.

Table 2.1 provides examples (identified in the Millennium Ecosystem Assessment) of these services.

Whether this Millennium Ecosystem Assessment classification is the most satisfactory possible classification of ecosystem services, especially from the point of view of completing economic valuations, is unclear. In any case, the classification was not devised to facilitate economic measurement of the value of these different services. Furthermore, the value of some ecosystem services, particularly cultural ones (such as aesthetic and spiritual ones), are hardly amenable to accurate economic measurement, especially if the values of individuals conflict. On the other hand, more reliable (but not perfect) economic methods exist for measuring the recreational value of ecosystems, or (at least) portions of these systems. Whether or not economic methods for measuring non-use values (such as contingent valuation) would provide a satisfactory measure of aesthetic and spiritual values is debatable. In some cases, they might provide a guide to such values but they would still be subject to serious limitations.

Table 2.1 Main categories of ecosystem services identified in the Millennium Ecosystem Assessment, with examples

Supporting	Regulatory
Nutrient cycling	Climate regulation
Soil formation	Flood control
Primary production	Disease and pest regulation
Others	Water purification
Provisioning	**Cultural**
Food	Aesthetic
Fibre and wood	Spiritual
Genetic resources	Educational
Biochemicals	Recreational
Ornamentals	
Fresh water	
Fruit	

Source: Extracted from information in Millennium Ecosystem Assessment (2005, pp. 19, 33–7).

TEEB and the Economic Valuation of Ecosystems

The difficulty of completing comprehensive and reliable valuations of the services provided by different ecosystems is highlighted by the estimates given in Kumar (2010, Appendix 3). These estimates were compiled by relying on results from previous economic studies of the values of ecosystems. Most of these studies did not include the valuations of all the services of ecosystems (such as those listed in Table 2.1) and most were completed only for restricted geographic locations within ecosystems.

The ranges of per hectare estimated economic values of various ecosystems (derived from these studies) are shown. The variation in the estimated ranges of economic values (per hectare) for each ecosystem is huge. This is partly explained by differences in the number of ecosystem services accounted for in each of the estimates, and the varied geographical locations and areas for which they were completed.

What can one conclude from the summary given in Table 2.2? First, on a per hectare basis, many ecosystems (in some locations) have a high economic value. Second, these values are likely to be greater (in many instances) than the general public realizes. For instance, many wetlands have a high economic value, particularly when the value of their supporting and regulatory services is taken into account.

Table 2.2 Monetary values[(a)] (US$/ha/year) of ecosystem services identified by TEEB for a selection of ecosystems, arranged in descending order of maximum values

Ecosystem	Range of values (US$)	No. of estimates used
1. Coral reefs	14 – 1 195 478	101
2. Coastal wetlands	1995 – 215 349	102
3. Coastal systems	248 – 78 580	112
4. Inland wetlands	198 – 44 597	86
5. Tropical forests	91 – 23 222	140
6. Lakes and rivers	1779 – 13 488	12
7. Temperate and boreal forests	30 – 4863	40
8. Grasslands	297 – 3091	25
9. Woodlands	16 – 1950	18
10. Open ocean	13 – 84	6
11. Polar and high mountain systems	Discussed but no estimates	
12. Cultivated systems	No estimates given	
13. Urban systems	No estimates given	

Note: (a) These are 2007 international US dollar values.

Source: Extracted from Appendix 3 of Kumar (2010).

Many valuable ecosystem services (such as those generated by wetlands) are unmarketed (this is because some are not amenable to being marketed) or they give rise to environmental externalities. Furthermore, these services frequently lack visibility. Usually, all these factors result in the general public undervaluing these services.

A 'problem' with the valuations given in Table 2.2 is that the values do not depend solely on the inherent traits of ecosystems. They depend also on human-related phenomena. For example, the economic value of coral reefs for recreation reflects the ease of access to them and their proximity to human settlements and so on. Consequently, reefs of low ecological quality in prime geographical locations (for instance, close to Honolulu) will be assigned a higher economic value than ecologically pristine reefs in remote locations.

When the range of services included in the valuations given in Table 2.2 are examined, it is found that very few include an allowance for the economic value of genetic resources preserved by the various ecosystems assessed. This may reflect the difficulties of making such estimates. This is worrying because it shows a major gap in economic knowledge. Furthermore, the Millennium Ecosystem Assessment stresses that conserving ecosystems is very important because this is essential for conserving biodiversity. However, the economic value of doing this in order to conserve biodiversity is poorly known.

This raises two issues: (1) is ecosystem analysis the appropriate form of reference for assessing the economic valuation of biodiversity conservation? (2) Has adequate attention been given to the comparative economic values of transformed (in some cases replaced) ecosystems relative to 'natural' ones? Transformed systems include partially altered natural ecosystems, for example mosaic patterns of transformation within natural systems.

First, it is questionable whether ecosystem-based analysis should be the sole or main focus for the economic valuation of environmental alternatives. For one thing, the geographic boundaries of an ecosystem are indefinite and the designated boundaries vary with the judgement and perspective of those framing these (Tansley, 1935). Nevertheless, despite this shortcoming, all ecosystem analysis focuses on relationships between complexes of organisms and their environments, including the physical and chemical nature of those environments. Sometimes, the boundaries of ecosystems are drawn to cover quite a small geographical area but on other occasions they are assessed on a macro-scale of the type portrayed in Table 2.2. However, at such a macro-scale, there is an extraordinary degree of variation within ecosystems, such as the types of habitats they provide, their surrounding physical environments and so on. Therefore, it is doubtful (from a pragmatic point of view) whether such macro-analysis can give helpful policy guidance. The sole value of this type of analysis may be to create increased awareness about the possible economic value of alternative ecosystems. However, that is not to suggest that all valuations involving sets of organisms and their relationships with their surrounding and supporting environments are of little use.

A major shortcoming of the TEEB analysis reported in Kumar (2010) is its failure to consider adequately comparative economic returns from transformed (human altered) natural ecosystems or their virtual replacement by human engineered and managed systems, for example, agroecosystems. This shortcoming is highlighted by the fact that no economic estimates are provided by TEEB for the value of cultivated systems (see Table 2.2) and no trade-offs are reported for the replacement of natural environments by cultivation. It is worth briefly considering at this stage some of the general issues involved. Further analysis of the subject will be presented in several later chapters.

Economic Trade-offs Involved in the Transformation of Ecosystems

TEEB fails to give sufficient weight to the economic value of transforming existing natural ecosystems and environments in order to better satisfy human wants. This may be because it wishes to stress that natural systems have been altered in ways that result in human deprivation. While this is sometimes true, in many cases human transformations of natural ecosystems and environments have resulted in substantial benefits to humankind. Some simple models can be used to consider the benefits or disbenefits to humans of such transformations.

Let y represent the quantity of an ecosystem service (for instance, a hydrological service) provided by a natural ecosystem existing within a defined geographical area of z_m km^2 in size. Assume that this natural environment can be transformed to support agriculture. Let z represent the area of land allocated to agriculture of a particular type or mix where the transformation occurs according to a specific geographical progression. For instance, the more productive land may be converted to agriculture first, and subsequently land of decreasing agricultural productivity may be used for agriculture. Where x represents the quantity of agriculture production, the production function:

$$x = g(z) \qquad z \leq z_m \tag{2.1}$$

where $g' > 0$ and $g'' < 0$ may apply. In other words, agricultural production may increase at a decreasing rate as a larger amount of the natural area is converted to agriculture.

The corresponding trade-off frontier between the environmental service provided by the natural environment and agricultural production might be like that shown by curve ABCDE in Figure 2.3. This indicates that as more land is allocated to agriculture, agriculture production rises at the expense of the supply of the relevant natural ecosystem service, y. If a Bergson (1938) social welfare function is adopted to portray society's preference for the environmental service and agriculture production, social welfare indifference curves of the type represented by W_0W_0 and W_1W_1 may emerge. Assuming that higher social welfare indifference curves designate preferred combinations of the commodities, x and y, the combination at B is socially optimal. Society's welfare will not be maximized if the original natural ecosystem is not transformed. Society will be better off for any transformation which enables it to reach a combination above the curve W_0W_0; for example, the combination at C would be better than that at A, but not ideal. However, a combination anywhere on the segment of the frontier between D and E would be inferior.

It is likely that the level of agricultural production does not only depend

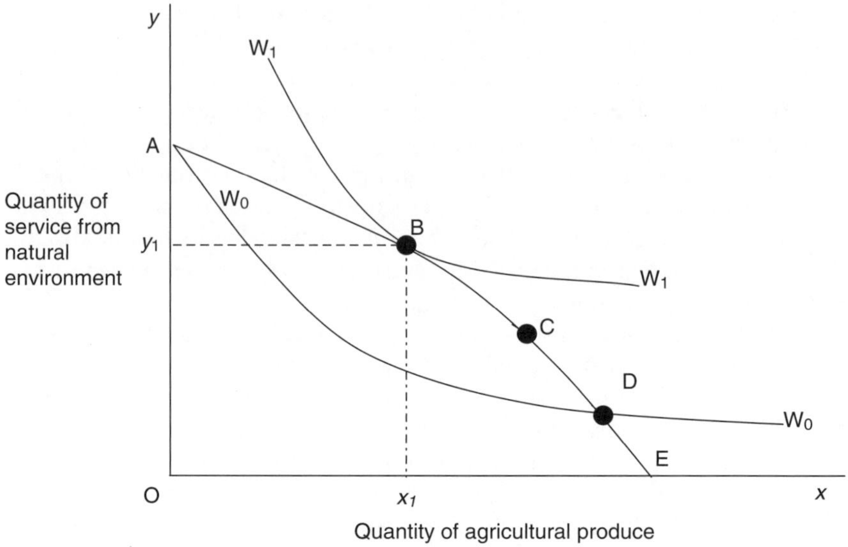

Figure 2.3 A case in which transforming a natural ecosystem can add to social welfare, but for which only a limited range of transformations are able to increase social welfare

on the area of land devoted to agriculture, z, but also on the amount of the environmental service, y, available. Hence, a more complex functional relationship may apply, such as:

$$x = h(z, y) \tag{2.2}$$

where y is also a function of z. Given this type of relationship, the set of possible levels of the supplied ecosystem service, y, and of agricultural production may be re-entrant. This is because once a large enough proportion of the territory z_m is converted to agriculture production, the reduction in the availability of the ecosystem service impacts negatively on the level of agricultural production. Consequently, a trade-off function of the type shown in Figure 2.4 by ABCDE may emerge.

If both the ecosystem service and the level of agricultural production are valued, no combination on the segment CDE of the frontier shown in Figure 2.4 can be socially optimal. This will also be so if the ecosystem service has no value independently of its contribution to agricultural production. Unfortunately, however, cases probably occur (for example, due to market or political failures) that result in economic systems operating at a point, such as D, on the segment CDE of the possibility frontier.

Figure 2.4 A re-entrant set of possibilities for agricultural production and the supply of an ecosystem service

A third possibility also exists. In a few cases, the transformation of natural systems may enhance the provision of environmental resources. For example, it has been claimed that rice production based on terracing improves water flow. In such cases, the frontiers shown in the previous figures may actually rise over a part of their range initially.

Applied situations are much more complex than the above illustrations indicate, for example, they are multidimensional in nature. Furthermore, different groups in society often subscribe to different social welfare functions and social conflict emerges about the extent to which natural ecosystems or existing environments should be transformed. This complicates social economic choice.

2.5 THE CLASSIFICATION OF ECOSYSTEMS, THEIR DESIGNATION AS NATURAL CAPITAL AND ORGANISMS AS PESTS AND ASSETS

This section discusses possible general classifications of ecosystems (whether natural or managed) or more broadly, possible partitions of the biosphere, considers the categorization of species as assets or pests, qualifies the claim that all ecosystems are a part of natural capital, and explores the likelihood that portions of ecosystems may be liabilities rather than assets and that, as well, organisms may be pests rather than assets.

Classifying Ecosystems and the Determination of the Status of Partitions of the Biosphere

Classifying ecosystems and the status of parts of the biosphere (on a grand scale) is a daunting task because of the diversity of traits or attributes which can be used for this categorization. One possibility is to use biomes as the basis of such classification, as seems to have been done for the categorization in Table 2.2. However, this is a very coarse approach and fails to capture the extent of biological physical and environmental variation within biomes and ecosystems. This is because they are identified at a macro-level. In particular, this approach does not provide a detailed enough characterization of human managed or transformed ecosystems or the various possible partitions of the biosphere.

Sometimes parts of the biosphere are classified according to the perceived extent to which they have deviated from pre-existing natural ecosystems. Some alterations to natural ecosystems involve little change whereas others involve a high degree of transformation, and in extreme cases they almost entirely replace natural ecosystems. But such an approach is relatively subjective.

One can also consider (as a basis for their categorization) the extent to which ecosystems or portions of the biosphere (such as those used for agriculture) are self-sustaining and independent of other parts of the biosphere. For instance, the degree to which agroecosystems depend on human effort and external inputs can vary greatly. Most modern agroecosystems rely on a high degree of human effort and the use of a considerable amount of off-farm (external) inputs, such as chemical fertilizers, purchased seed, chemicals for pest control, electricity and fuels, such as petrol or diesel, and so on. In the case of crops, modern cultivation is dominated by monoculture based on improved plant varieties, and farmers tend to specialize in the products they supply. This contrasts with traditional small-scale farming practices involving mixed farming and a low degree of reliance on external inputs. Technological and scientific advances in agriculture coupled with the extension of markets have made it possible to increasingly decouple (delink) agricultural production from its local environment and available regional resources. For example, the practice of hydroponics (under glasshouse conditions) largely isolates the growing of plants from the local environment and from the availability of naturally occurring *in situ* resources. As is discussed later in this book, this 'decoupling' contributes to a loss of the traditional varieties of crops and breeds of domesticated animals.

Ecosystems as Natural Capital

It is sometimes assumed that all ecosystems are a form of natural capital. For example, Pascual et al. (2010, p. 187) state that 'from an economic point of view, biodiversity (and ecosystems) can broadly be seen as a part of natural capital and the flow of ecosystem services is the interest on that capital (Costanza and Daly, 1992)'. While this may be true of natural ecosystems, it does not apply to human generated ecosystems (such as agroecosystems, aquaculture systems and silviculture of many kinds). These systems usually include elements of natural capital but they are primarily a result of human effort and investment. In that regard, they consist of a large amount of man-made capital, much of which can be regarded as heritage capital. The same is true of the stock of a considerable amount of genetic resources, as should be clear from the discussion in section 2.2 of this chapter.

Furthermore, it is not necessarily the case that all natural or extant ecosystems can be regarded as assets from a human perspective. Some desert ecosystems may be considered to be liabilities and their spread can have negative economic consequences. Furthermore, some community members may consider wetlands consisting of swamps where malaria-carrying mosquitoes breed to be liabilities. These members might support their drainage and conversion to agriculture. Whether or not that conversion would result in a net economic gain if the Kaldor–Hicks criterion (also designated the potential Pareto improvement criterion) is adopted, depends on the specific case. In many cases, social choice about whether existing ecosystems should be transformed to a different state involves social conflict. This is because it is often the case that some members of the community gain whereas others lose as a result of this transformation. Consequently, difficult income distribution issues can arise.

Organisms as Assets or Pests (Liabilities)

Just as some parts of the biosphere can be assets or liabilities from a human perspective, the same is true of organisms. It is widely agreed that some organisms have a negative value from a human perspective, for instance, organisms causing diseases such as smallpox, poliomyelitis or tuberculosis. The same is true of some organisms causing diseases in crops or in livestock. However, the assessment of the economic value of many organisms is complicated when some members of the community regard them as pests whereas others see them as assets. Given the possibility of such conflict, how should organisms or species be valued from an economics point of view?

One possibility is to adopt the Kaldor–Hicks approach to such valuation. If this approach is adopted, the net social value of a species is equal to the difference between the economic value placed on its existence by those who regard it as an asset and the negative value placed on it by those who regard the species as a pest. However, these values are likely to depend on several independent variables, one of which could be the level of population of a species being assessed. Suppose, for simplicity, that this is the only independent variable of consequence. Then the asset-like status of a species is liable to alter with its population, as was previously examined in Tisdell (2014b, Ch. 2), using a somewhat different exposition from the one adopted here.

Figure 2.5 illustrates this possibility for a selected species. For example, it may apply to the population of red kangaroos (*Macropus rufus*) in Australia. When they are present in large numbers, most graziers consider

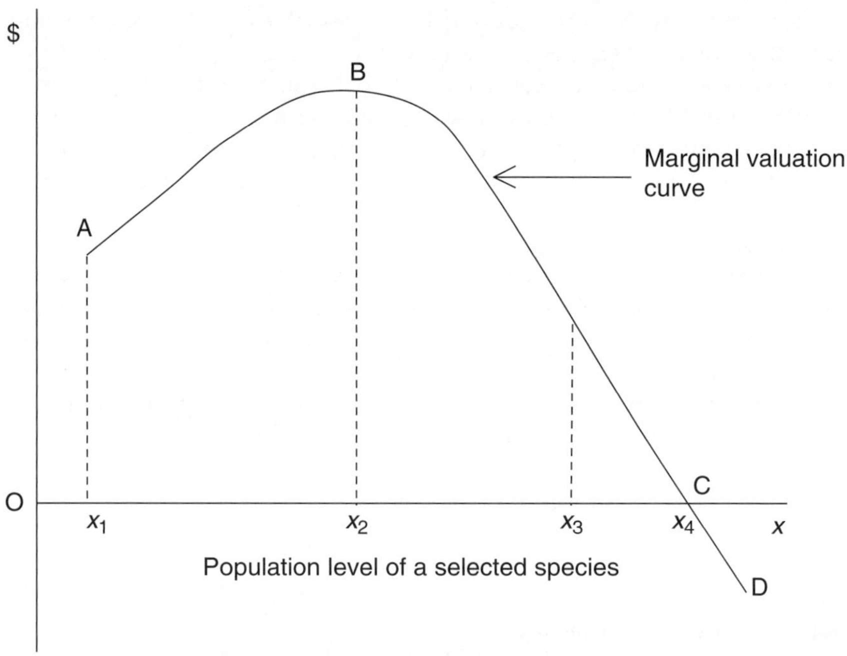

Figure 2.5 A case in which the economic status of a species as an asset or a pest varies with the level of its population

red kangaroos to be pests, but at the same time, the general public values their existence. Let x represent the level of population of the wild species under consideration. Then the marginal net (aggregate) valuation curve, ABCD, may apply. It assumes that when the population of the species drops below x_1, the species will become extinct in a relatively short time, and that the probability of its extinction (within a specified period) declines as x increases. The total net value of the species increases with its increasing population until $x = x_4$. Beyond that level of population of the species, the aggregate economic value of the species declines, and results in a reduction in human well-being when this is measured using the Kaldor–Hicks method. Observe that even for some values of x greater than x_4, the total aggregate economic value of the species could still remain positive, but when x becomes very large, it is increasingly likely to become negative.

Figure 2.5 implies that the optimal level of population of the focal species is x_4 if the economic well-being of the community is to be maximized on the basis of the Kaldor–Hicks principle. However, this does not imply that it is necessarily economic for the community to try to alter the level of the population of the species so that it reaches x_4. The economics of altering the population level of the species depends not only on the extra economic value obtained from this but also on the extra cost required to bring about this change.

In Figure 2.6, curve ABC illustrates a case in which the marginal economic value of a species (based on the Kaldor–Hicks principle) is negative for all possible viable populations of the species. The focal species may be negatively valued by all individuals, or by the aggregation of the values of those who consider it to be an asset and others who rate it as a pest. The latter case seems to typify the valuation of the feral pig in Australia (Tisdell, 1982). The loss in economic value caused to agricultural industries by this pig and to natural ecosystems seems to far outweigh its value for amateur hunting and other types of hunting.

In practice, it is not always easy to determine truly whether the status of a species is a pest or asset even if the Kaldor–Hicks principle is adopted. Reliance on the stated preferences of individuals can be misleading. For example, respondents may be poorly informed about the function of a species in an ecosystem or its role in the web of interconnected life-forms and cultural bias may influence their perceptions (Tisdell, 2014b). Furthermore, the valuation by individuals of species can vary with the passage of time (Tisdell, 2014b, Ch. 2). Nevertheless, from a human perspective, it is clear that not all species are assets, nor are they all a part of natural capital.

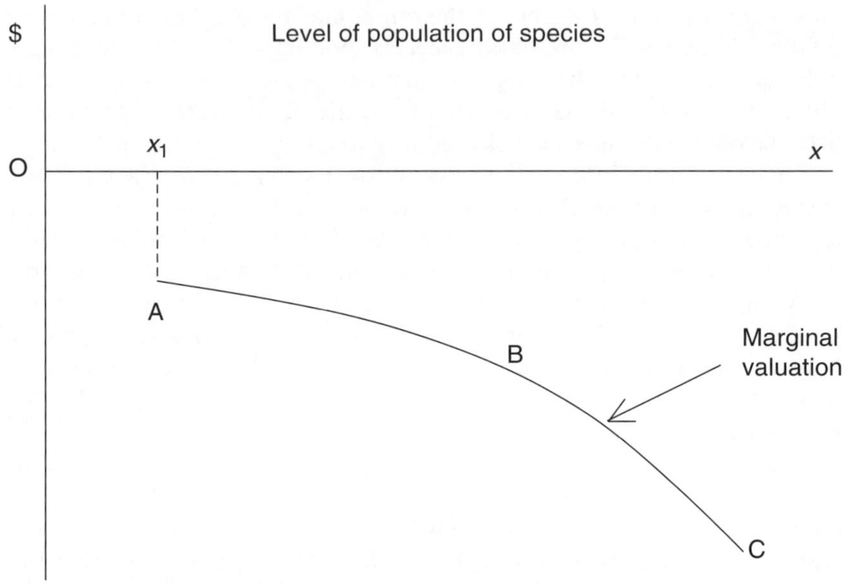

Figure 2.6 *A case in which all the viable levels of populations of a species yield a negative economic value: it is a 'net' pest from an aggregative point of view*

2.6 CONCLUSION

Consideration of the composition of the stock of genetic resources (that is, the stock of germplasm), as well as the range of existing ecosystems and the contents of portions of the biosphere highlight the fact that they do not all constitute natural capital. Many are, to a significant extent, a result of human effort and investment. Numerous human-induced transformations of living natural resources increase human well-being, whereas others reduce human welfare or are not performed in ways which maximize human benefits. As was explored in this chapter (in a general way), economic trade-offs are involved.

Economic assessment of the value of species (as well as the status of parts of the biosphere) can be complicated by conflicting individual valuations of these things. Some individuals may regard these as assets whereas for others, they are liabilities. While the Kaldor–Hicks principle provides one method of allowing for the discrepancies, its application is limited as a method of conflict resolution, and it can have unwelcome consequences for the distribution of income. Furthermore, the valuation of species and ecosystems (or parts of these) can alter as some variables on which this depends change. For example, the status of a species as a pest or asset is liable to vary, in many cases, with the level of its population. Furthermore, the valuations of portions of an ecosystem usually vary with their location relative to human settlements.

It is also apparent from the discussion and analysis (in this chapter) of the classification of genetic resources and the categorization of ecosystems, that anthropocentric alterations to these can result in both subtractions and additions to the stock of genetic resources. In modern societies, anthropocentric influences on the availability of genetic resources can exert themselves in a relatively short period of time and are primarily a consequence of economic forces. The role of these forces in influencing biological evolution (or change) is explored in a general manner in the next chapter, and is given more detailed consideration in several of the subsequent chapters.

REFERENCES

Alauddin, M. and C.A. Tisdell (1991), *The 'Green Revolution' and Economic Development: The Process and its Impact in Bangladesh*, London: Macmillan.

Anon (2014a), 'Moa', Wikipedia. Available at: http://en.wikipedia.org/wiki/Moa. (Accessed 10 April, 2014).

Anon (2014b), 'Cane toads in Australia', Wikipedia. Available at: http://en.wikipedia.org/wiki/cane_toads_in_Australia. (Accessed 10 April, 2014).

Anon (2014c), 'Phytophthora', Wikipedia. Available at: http://en.wikipedia.org/wiki/Phytophthora. (Accessed 10 April, 2014).

Bergson, A. (1938), 'A reformulation of certain aspects of welfare economics', *Quarterly Journal of Economics*, **52**, 310–34.

Costanza, R. and H.E. Daly (1992), 'Natural capital and sustainable development', *Conservation Biology*, **6**, 37–46.

De Groot, R., P. Kumar, S. Van den Ploeg and P. Sukhdev (2010), 'Estimates of the monetary value of ecosystem services', in P. Kumar (ed.), *The Economics of Ecosystems and Biodiversity: Ecological and Economic Foundations*, London and Washington, DC: Earthscan, pp. 367–401.

FAO (2004), 'What is agrobiodiversity?', *Building on Gender, Agrobiodiversity and Local Knowledge* [Online] Rome: FAO. Gender and Development Service, Sustainable Development Department. Available at: http://www.fao.org/docrep/007/y5609e/y5609e01.htm. (Accessed 21 February, 2014).

FAO (2009), *The State of Plant Genetic Resources for Food and Agriculture: Country Report of Papua New Guinea*, Rome: FAO.

FAO (n.d.), 'Food security and the environment', [Online] Rome: FAO. Available at: http://www.google.com.au/url?sa=t&rct=j&q=&esrc=s&source=web&cd=1&ved=0CCoQFjAA&url=http%3A%2F%2Fwww.fao.org%2Fworldfoodsummit%2Fenglish%2Ffsheets%2Fenvironment.pdf&ei=ocswU56UCIifkAWr3YH

wBQ&usg=AFQjCNFslMSYNGJU5O2hs8jWi85p0o-Fdw&sig2=W2000eZyj6emp2AS2ywYVg&bvm=bv.63587204,d.dGI. (Accessed 25 March, 2014).
Flannery, T. (1995), *The Future Eaters: An Ecological History of the Australian Lands and People*, Sydney: Reed Books.
Gammage, B. (2012), *The Biggest Estate on Earth: How Aborigines Made Australia*, Sydney: Allen and Unwin.
Gilmore, M. (1934), *Old Days, Old Ways*, Sydney: Angus and Robertson.
Hof, J.G. and M. Bevers (1998), *Spatial Optimization for Managed Ecosystems*, New York: Columbia University Press.
Hof, J.G. and M. Bevers (2002), *Spatial Optimization in Ecological Applications*, New York: Columbia University Press.
Koehn, J.D. (2004), 'Carp (*Cyprinus carpio*) as a powerful invader in Australian waterways', *Freshwater Biology*, **49**, 882–94.
Kumar, P. (ed.) (2010), *The Economics of Ecosystems and Biodiversity: Ecological and Economic Foundations*, London and Washington, DC: Earthscan.
Millennium Ecosystem Assessment (2003), *Ecosystems and Human Well-being: A Framework for Assessment*, Washington, DC: Island Press.
Millennium Ecosystem Assessment (2005), *Ecosystems and Human Well-being: A Synthesis*, Washington, DC: Island Press.
Pascual, U., R. Muradian, L. Brander, E. Gómez-Baggethun, B. Martin-López and M. Verma (2010), 'The economics of valuing ecosystems services and biodiversity', in P. Kumar (ed.), *The Economics of Ecosystems and Biodiversity: Ecological and Economic Foundations*, London and Washington, DC: Earthscan, pp. 184–256.
Renfrew, C. (2007), *Prehistory: The Making of the Human Mind*, London: Weidenfeld and Nicolson.
Tansley, A.G. (1935), 'The use and abuse of vegetational terms and concepts', *Ecology*, **16** (3), 268–71.
Thrall, P.H., J.D. Bever and J.J. Burdon (2010), 'Evolutionary change and agriculture: The past, present and future', *Evolutionary Applications*, **13**, 405–408.
Tisdell, C.A. (1982), *Wild Pigs: Environmental Pest or Economic Resource?*, Sydney, Oxford, New York: Pergamon Press.
Tisdell, C.A. (2009), 'The economics of fish biodiversity: Linkages between aquaculture and fisheries – some perspectives', in K. Ninan (ed.), *Conserving and Valuing Ecosystem Services and Biodiversity: Economic Institutional and Social Challenges*, London, UK and Sterling, VA, USA: Earthscan, pp. 47–57.
Tisdell, C.A. (2014a), 'Ecosystem functions and genetic diversity: TEEB raises challenges for the economics discipline', *Economic Analysis and Policy*, **44** (1), 14–20.
Tisdell, C.A. (2014b), *Human Values and Biodiversity Conservation: The Survival of Wild Species*, Cheltenham, UK and Northampton, MA, USA: Edward Elgar Publishing.

3. Socio-economic processes transforming the genetic stock and altering ecosystems

3.1 INTRODUCTION

Homo sapiens has become the major source of global environmental changes as a result of growing populations of humans, increasing economic production and greater man-made capital accumulation, as well as new techniques of production associated with the expansion of knowledge. While environmental changes due to natural phenomena still occur independently of human action, they pale in significance compared to the impact of humans on the natural environment. Such impact arises as a direct result of human actions (for example, the removal of natural vegetation to grow crops) and indirectly (as in the case of global warming and climate change caused by increased greenhouse gas emissions resulting from human activity). Particularly, since the beginning of agriculture in the Holocene era, humans have increasingly changed the global genetic stock and altered ecosystems.

Genetic material has been selected and developed by human beings in order to increase the economic value of the genetic stock. Natural ecosystems have been transformed to enhance their economic productivity from a human point of view. In addition to that, ecosystems and environments previously developed by humans for economic production have themselves been subject to alterations with the passage of time. For example, several hundred years ago, shifting agriculture (swidden agriculture) was practised in Europe but was eventually replaced by the more permanent use of cultivated fields for the growing of crops. Crop rotation and the application of fertilizers helped to maintain yields.

Human beings usually decide to alter the genetic stock and ecosystems in order to foster their own self-interest, or that of a small group of persons to which they belong. For example, the selection of wheat varieties which do not shatter and scatter their seeds once they reach maturity was clearly desirable from the point of view of early cultivators of wheat in the Levant (see Thrall et al., 2010, p. 405). Today, this still remains a desirable property of cultivated wheat and other cereals. Therefore, human selection of varieties of wheat and other cereals has favoured the conservation of varieties with this attribute as well as with other properties desired by human beings. Furthermore, in order to grow crops successfully, they need to be protected from biological competition, and this results in changes to existing natural ecosystems and, frequently, loss of wild genetic material. For example, competitive plants need to be removed from (or controlled in) soils where crops are to be grown. Pastures for the grazing of livestock may be made more suitable for that purpose by removing vegetation that reduces the availability of suitable pasture for livestock. This can require tree cover to be reduced. Consequently, environmental change occurs which alters pre-existing ecosystems and in some cases results in the destruction of natural ecosystems and the loss of wild biodiversity.

However, the pursuit of individual self-interest (or that of small groups) in altering the genetic stock and changing ecosystems does not always promote collective welfare, nor does it always result in the outcomes anticipated by those acting in accordance with their perceived self-interest. It may have adverse long-term economic consequences, which may or may not be anticipated by those pursuing their own self-interest. The likely occurrence and the magnitude of these effects can be expected to vary with the institutional structures of societies. These structures tend to alter with economic development, and other factors, as time passes. Therefore, changes in genetic diversity and ecosystems (particularly, in the Holocene era) need to be studied in their historical contexts.

While this chapter focuses mainly on the consequences for biodiversity of contemporary economic systems (and their relationship to political systems), historical developments are not ignored. For example, the consequences for global biodiversity of the geographical extension of human contact which increased global access to new genetic resources is given particular attention as is the growing extension of markets (resulting in greater inter-regional trade). Both processes are associated with growing globalization.

In developing this chapter, the consequences for biodiversity of the geographical extension of human contacts (such as those made possible by the European 'Age of Discovery') are discussed first and then this is followed by consideration of the consequences for biodiversity of human capital accumulation and advances in scientific and technological knowledge. Subsequently, the role of market systems and the geographical (including international) extension of trade in altering biodiversity are examined from a general point of view. Connections between market systems and political processes are given attention because of their implications for biodiversity conservation. Some related modern institutional developments are covered in the penultimate section. These include the developments of property rights in genetic material, the growth of larger businesses or productive units (such as companies trading on the stock exchange) and the evolution of supermarket chains and other retail chains. These institutional developments all have consequences for the way in which biodiversity is altered.

3.2 THE GEOGRAPHICAL EXTENSION OF HUMAN CONTACTS AND MIGRATION

Human beings have been and continue to be important vectors for the translocation of genetic material. They have transported this material both deliberately and accidentally, and this has resulted in major changes in the composition of the global genetic stock. These movements have been a significant factor in the reduction of biodiversity in the wild (for example, as a result of the extension of cultivation) and have contributed to the global loss of traditional varieties of cultivated crops, for example, as a result of the substitution of introduced species of crops for existing ones.

Due to the extension of the geographical range of human contact (often accompanied by migration), individuals in all parts of the world have been able to access a wider range of germplasm. They have been able to enjoy an increased suite of services or benefits as a result of this availability, but this has been achieved at the expense of a reduction in stock of previously existing biodiversity. Consequently, although this process has had economic benefits, it has been achieved at an environmental cost. While in some cases, the social economic benefits would have outweighed the social cost of the deliberate translocation of genetic material, in other cases, the social net benefit has been negative. For example, the deliberate introduction of the cane toad (*Rhinella marina*) to Australia to control insect pests in sugar cane was not an effective biological control and threatens the survival of several native Australian wildlife species (Anon, 2014b). The social net benefit of its introduction has been negative.

Considerable changes in biodiversity are associated with human migration, increased human social contact and trade. According to some scholars, the diffusion of agriculture (following its commencement in the early Holocene period) was (at least in Eurasia) mainly a result of migration by those who had already adopted agriculture as their main source of livelihood (Ammerman and Cavelli-Sforza, 1984). This has been named the demic theory of diffusion. However, the spread of agricultural methods of production was probably also a result of imitation facilitated by social contact and trade (Svizzero and Tisdell, 2014). As a result of the ongoing replacement of hunting and gathering activities by agriculture and other means of material production, major transformations of natural ecosystems occurred. Although many hunter-gatherer communities had already had some impacts on the survival of species (see, for example, Bulte et al., 2006) and had brought about some changes in natural ecosystems, after the commencement of agriculture and subsequent economic developments, human activity eventually became the major force shaping the nature of ecosystems and the composition of the genetic stock.

In modern times, the European 'Age of Discovery' began a series of events which (combined with continuing technological and scientific progress, capital accumulation and economic growth) had major (and to some extent, irreversible) impacts on the stock of global biodiversity. This age began towards the end of the fifteenth century with the journeys of Christopher Columbus to the Americas and the rounding of Africa's Cape of Good Hope by Vasco da Gama. These events resulted eventually in massive geographical movements of germplasm, both into and out of newly discovered territories. These movements were accentuated by the European colonization of (settlements in) many of the newly discovered lands, such as the Americas, Australia, New Zealand and parts of Southern Africa, and by the establishment of plantations in South-East Asia and elsewhere. Furthermore, agricultural techniques were introduced from Europe to newly discovered territories and further developed. This resulted in major changes to natural ecosystems (or near natural ones) in newly discovered territories. In addition, the availability of germplasm from newly found territories brought about changes in ecosystems and the composition of agricultural production in geographical areas already familiar to Europeans as well as new territories. Consider some examples.

Wheat and grapes, cattle, horses, pigs and sheep were introduced to the Americas, Australia and New Zealand. The American Prairies were converted from natural grasslands to the growing of wheat, corn (maize) and other crops, and several wild species disappeared. More recently, the introduction of soya bean (soybean) (*Glycine max*) cultivation to the Americas from Asia has increased the land area used for cultivation. It has, for instance, been one of the agricultural developments associated with the continuing clearing of the Amazon's rainforest. The growing of soya beans had its origin in East Asia but today the Americas account for the major portion of their global supply.

Significant movements of germplasm out of the Americas also occurred and from places other than those mentioned above. The growing of maize (corn) (*Zea mays*), potatoes (*Solanum tuberosum*), tomatoes (*Solanum lycopersicum*) and capsicums (*Capsicum* spp.) spread from the Americas to elsewhere in the world. Now, the potato is a major global food crop, and today China grows more potatoes than any other country. The growing of cacao trees (*Theobroma cacao*) spread from Central America to the west coast of Africa and propagules for growing rubber (*Hevea brasiliensis*) were obtained from Brazil to establish rubber plantations in South-East Asia and elsewhere. Oil palm (*Elaeis guineensis*) was introduced from West Africa to South-East Asia and elsewhere. Often plantations of these three crops replaced areas of tropical rainforest. Moreover, oil palm plantations are being established on peatland. This expansion adds to greenhouse gas emissions and is a threat to biodiversity in the wild. For example, the expansion of oil palm cultivation is a serious threat to the survival of the orangutan (Swarna Nantha and Tisdell, 2009).

Access of humans to additional stocks of valuable genetic material following the commencement of the European Age of Discovery provided an economic bonanza by expanding global possibilities for (commercial) biological production. This resulted in an enormous expansion in the scope for increased production by bio-industries. It also increased the varieties of food available to humans in most places. However, at the same time, it contributed to a decline in global biodiversity, especially biodiversity in the wild. Now that global geographical knowledge of the stock of germplasm is so substantial (but not complete), less scope exists for expanding the production of food, fibre and timber by the discovery and transfer of genetic material from one region of the world to another. Therefore, advances in the productivity of bio-industries are likely to increasingly depend on the selection and genetic manipulation of existing germplasm, and on scientific and technological advances in production methods.

3.3 CAPITAL ACCUMULATION AND ADVANCES IN SCIENCE AND TECHNOLOGY

Particularly in the last 300 years or so, capital accumulation combined with scientific and technological progress has facilitated the geographical extension and intensification of agriculture and other bio-industries. For example, the contribution of animal power and human energy to environmental change has been increasingly replaced by that of machines dependent on internal combustion engines or electric motors. This has made it easier to alter existing ecosystems to make them more suitable for agriculture, silviculture and similar industries.

New crop varieties and the availability of chemical fertilizers and pesticides have resulted in changes in agroecosystems and other ecosystems. For example, they have facilitated multiple annual cropping in many areas. Consequently, the length of time for which cropland is fallowed is being reduced. New high-yielding varieties of crops (such as rice and wheat) can significantly raise annual crop yields provided the necessary complementary package of inputs (such as fertilizer, water, pesticides and so on) is available and carefully controlled.

The supply of many of the inputs required for this input package depends on markets and trade. For example, chemical fertilizers and pesticides are often imported or some of the components for their manufacture may be imported. A major portion of the inputs needed to successfully grow new high-yielding variety of crops cannot be economically produced on farms but need to be purchased in the marketplace by growers. Advances in technology and the extension of the market have both made it possible for those engaged in bio-industries to increase their control of (local) environmental conditions in which organisms are cultured or husbanded. This has resulted in the loss of many varieties of cultivated crops which were previously selected by growers because of their suitability for cultivation under local ecological conditions using available local inputs.

Economic developments have tended to decouple (delink) production by bio-industries from local environmental conditions and from the availability of local supplies of inputs. While these processes have been a strong force for increasing the yields of bio-industries, questions have been raised about the sustainability of the increased production achieved by these processes. These developments have, among other things, resulted in a reduction in the genetic stock existing prior to their occurrence (see the previous chapter). The implications for economic sustainability of those developments will be discussed in the next chapter.

In addition to the above, advances in soil science have enabled agricultural production to be extended to areas which were previously unsuitable for agriculture. For instance, the addition of trace elements to some soils has enabled them to be used for agriculture. This has contributed to the loss of biodiversity in the wild because it has extended the area of agriculture. Furthermore, the use of standard chemical fertilizers has helped to extend the area used for agriculture. For example, by using chemical fertilizers, it is (privately) economic to cultivate oil palm on peatland in Indonesia. Consequently, as indicated above, natural areas of peatland (which, for example, provide suitable habitat for orangutans) are being converted to agriculture. This conversion is having significant negative environmental consequences: it is contributing to a loss of pre-existing wild biodiversity on peatland and releasing greenhouse gases. The latter contributes to global warming and to climate change.

Capital accumulation and advances in science and technology are not only associated with biodiversity loss in industries that culture organisms. They are also factors in biodiversity loss in industries (such as fishing and forestry) that involve the harvesting of natural resources (Tisdell, 2011).

Biodiversity losses are likely to be marked if there is open access to natural resources and the cost of exploiting these resources falls due to technological progress, other things being held constant. Even in cases where a natural biological resource is private property, a reduction in the cost of harvesting it can result in its elimination. This, for example, can be expected if the land on which the natural biological resource exists would give a higher economic return if used for an alternative purpose (Tisdell, 2004). A reduction in the cost of harvesting or eliminating the natural biological resources makes it more likely that conversion to the alternative land use is economic. The same is true when the natural species occupying their land are regarded by landholders as pests.

3.4 MARKET SYSTEMS, THEIR EXTENSION AND BIODIVERSITY LOSS

First, it needs to be made clear that biodiversity loss and change due to anthropogenic factors are not exclusively due to the presence of market systems. They have occurred in hunter-and-gatherer societies (in which market exchange was virtually absent) and in centrally controlled command economies, such as that associated with the former Soviet Union (USSR). However, here the focus will be on market systems because nowadays they are globally pervasive.

Many economists have argued that market systems are a socially ideal means of economic organization. The reasons given vary. For example, one line of argument is that competitive markets result in allocative efficiency: a situation in which no individual can be made better off without making another worse off. This view has its genesis in the theory of Vilfredo Pareto (1927). Another point of view associated with the theories of Joseph Schumpeter (1954), but also inherent in the theory of workable competition proposed by John Maurice Clark (1940), is that although market systems may not result in allocative economic efficiency, they promote technological progress, risk-taking and capital accumulation, which in turn stimulates economic growth. This theory, in fact, postulates that intense market competition retards economic growth and increases in economic welfare. There are other points of view as well. For example, it is sometimes argued that the forces of economic selection result in only the most efficient producers surviving (Hirshleifer, 1977), and economic selection therefore promotes economic efficiency. These different points of view are discussed in depth in Tisdell (2013). However, none of these views pay much attention to issues involving the conservation of natural resources and alterations in the stock of biodiversity. This is also typical of many mainstream theories of economic growth which focus on the role of man-made capital, labour and technological progress as contributors to economic growth (see, for example, Romer, 1990; and Tisdell, 1990, Ch. 3, for a critique). Natural resources and the stock of biodiversity are absent from these theories of economic growth.

Those who argue that market systems do not result in ideal economic outcomes usually ascribe this to several types of market failure. Among other things, market failures normally result in human failures to conserve and develop an optimal stock of biological resources, subject to the natural constraints on this process.

Two important sources of market failure which affect the stock of genetic resources and the provision of ecosystem services are the occurrence of environmental externalities (spillovers) and the presence of goods and services that possess public good characteristics. In the case of positive environmental spillovers, individuals do not appropriate the full economic benefits of their actions. If negative spillovers occur, individuals do not bear the full social costs of their actions; costs are imposed on others for which they are not compensated. When individuals are motivated in their actions only by their own private gain, market failures occur in these cases. Furthermore, when commodities have the attributes of public goods, any individual providing these can appropriate only a small fraction of their economic benefit. Consequently, they tend not to be supplied at all or to be under-supplied from a social economic point of view unless collective action is taken to supply them. A pure public good has the following attributes: (1) once it is supplied, it is available to all (exclusion is impossible or uneconomic); and (2) its consumption by any single individual or group does not reduce its availability to others (its consumption or enjoyment is not competitive).

There are many examples of environmental spillovers arising from changes in the genetic stock and in ecosystems, or more generally anthropocentric alterations to the biosphere. For example, in the absence of government controls, individuals may introduce exotic germplasm to a country which benefits them but causes economic damage to others. The growing of genetically modified crops in some regions may result in cross-breeding with crops that are not genetically modified and cause economic damage to farmers growing these crops. The removal of tree-cover in river catchment areas can have negative spillover consequences for water flow and water quality. Individuals downstream may experience greater variability in water flows (more frequent floods, and reduced water availability in dry periods) and the water may be of poorer quality and have to be treated at extra cost to make it suitable for human consumption. The extra turbidity and silt in the water may also have negative impacts on aquatic life.

Many biological resources have the attributes of public goods, at least to some extent. In the latter case, they are mixed goods involving a combination of the characteristics of private and public goods. Most non-use values of desired ecosystems and wanted wildlife species display the attributes of public goods. These include the existence value and bequest value of these things. The non-use value of a wildlife species is often the major source of its economic value. There is strong evidence that this is so, for example, in the case of the Asian elephant (*Elephas maximus*) see Bandara and Tisdell (2003) or Tisdell (2014, Ch. 5) as well as for less well known species such as Australia's tree kangaroos and its northern hairy-nosed wombat (Tisdell, 2014, Chs 5 and 7).

It is also possible for biological resources to be regarded as public bads. For example, if a natural ecosystem is invaded by an exotic botanical pest, the existence of this transformed ecosystem may be viewed negatively. Furthermore, the existence of parasites or microbes that threaten the health of humans can constitute a public bad.

High transaction costs involved in the protection of private property can also result in market failure. They may result in natural resources being subject to open access. Hence their over-exploitation of biological resources from an economic point of view can be expected and this may result in extinction of species. This, however, does not imply that the economic establishment and maintenance of private property rights will ensure the conservation of all existing germplasm and ecosystems (see, for example, Tisdell, 2004).

Uncertainty about future demands is an additional possible impediment to the conservation of biodiversity. If individuals are uncertain, for example, about the future demand for an existing variety of crop which is in little current demand, they are likely to have little incentive to conserve it, especially if their property rights in this existing variety of the crop are uncertain or if they are costly to enforce. In such circumstances, there may be an economic case for conservation of the germplasm by collective (government) intervention. The cost of its conservation can be spread more widely as a result of collective action, and greater consideration can be given to the application of the Precautionary Principle. Conservation may take place either *in situ* or *ex situ*, depending on the species involved. Furthermore, if a species (or an ecosystem) has a high non-use value (or a substantial public good component), this strengthens the case (because of market failure) for government intervention to conserve the species (or ecosystem).

Market extension, including freer and greater international trade, can be a strong force for reducing the global diversity of biological resources. It has been already noted that market extension tends to increase the range and nature of inputs available to bio-industries in most locations. Consequently, farmers often find it profitable to replace landrace varieties of crops or livestock adapted to local conditions by other varieties that are more productive given the increased range of inputs available. Consequently, some varieties may become extinct.

Secondly, the extension of markets tends to result in regional specialization in production, in accordance with the law of comparative economic advantage. As a result of this extension, the composition of agricultural industries present in a region is liable to change. Some local agricultural industries may disappear and varieties of crops and breeds of livestock that are unique to these industries can also disappear. A case of this nature is reported for Southern Nigeria in Chapter 6 of this book.

Thirdly, the extension of markets and increased possibilities for international trade may lead to greater demands on natural or wild biological resources in some regions and countries. For example, demand may increase for timber from tropical forests and this can be expected to reduce wild genetic resources and alter natural ecosystems in these areas. Or, because of access to a wider market, the demand for a particular cultivated crop may expand in some regions and the area of its cultivation may increase at the expense of natural ecosystems. For example, the global demand for palm oil has resulted in the expansion of palm oil plantations in Indonesia and Malaysia (and elsewhere) at the expense of natural ecosystems. On the other hand, countries that expand their demand for imported timber may reduce the logging of their own forests, and this may reduce their rate of wild biodiversity loss.

3.5 INSTITUTIONAL AND OTHER CONSIDERATIONS

Several institutional and other factors (additional to those mentioned above) influence the conservation and development of the stock of biological resources. While a considerable amount of existing economic theory can help to explain the effects on biodiversity of the various institutions considered below, they are given little attention in economic analysis. Relevant institutional developments include the growth of chains of supermarkets and food outlets, the granting of property rights to developers of some new genetic material, and the occurrence of political and administrative structures which enable special interest groups to engage in activities resulting in biodiversity loss contrary to the general interest. Further relevant factors influencing changes in biodiversity include increased work and geographical mobility of individuals with the development of market systems and the evolution of large businesses and organizations with geographically scattered operations. Government failure in relation to the management of open-access resources is also relevant. Each of these aspects (and some additional ones) will now be considered in turn.

Growth in Chains of Supermarkets and Fast Food Outlets

The managers of chains of supermarkets and fast food outlets seem to prefer standardized products that can be supplied regularly and in adequate volumes to stock their multiple outlets. This most likely is an economic advantage for advertising products (in a wide geographical area), for packaging, and for the standard management of goods in-house. Furthermore, consumers can more easily judge the attributes of standard products compared to those displaying considerable variation. Many manufacturers also find it less costly to process biological products that are relatively uniform in nature rather than those that display considerable variation. Even at the farm level, uniformity in biological products may be preferred.

These preferences favour the conservation and development of varieties of crops (and other organisms) and the adoption of production methods able to satisfy these requirements. In addition, products that transport and store well are likely to be favoured. This can result in the loss of local varieties of crops if they lack these characteristics; for example, those that do not transport or store well and those which vary considerably in their attributes.

Property Rights in New Genetic Material

The practice of granting exclusive property rights (for a limited period of time) to developers of new germplasm has become more common. For example, plant variety rights are common and in some cases, genetically modified organism can be patented. These grants give the developer (for a limited period of time) a monopoly in new germplasm. Depending on economic conditions, these rights enable developers of new germplasm to obtain an economic reward from their contribution. This system provides an economic incentive for the development of new germplasm of *commercial value*. However, this new germplasm may displace part of the stock of existing germplasm (see Chapter 2 of this book).

Given the availability of these rights, factors which can influence private decisions to develop new germplasm include:

1. the prospects for and likely cost of developing new germplasm of commercial value;

2. the size of the market for this new germplasm and the nature of the demand curve for it, including its elasticity;
3. the cost of supplying the germplasm to customers;
4. the costs and ability of the developer to enforce his/her property rights; and
5. the anticipated speed at which competitive germplasm is expected to be developed by rivals. This will affect the commercial life of the new germplasm.

Other things being held constant, the larger the market for the new germplasm and the more inelastic the demand for it, the higher will be the profit of the developer of the new germplasm, if he or she acts as a monopolist. This can be illustrated, using comparative statics, by applying the theory of the behaviour of a profit-maximizing business.

Three alternative situations are illustrated by Figure 3.1. Here *X* represents the quantity of marketed seed incorporating new germplasm. Line AS represents the cost per unit of supplying this seed to the market. Suppose that D_1D_1 represents the demand curve for the seed if the market is large and that D_2D_2 represents this if the market is small. The monopolist will be able to extract a higher profit in the former case than in the latter case. Now, consider the possibility that the demand for *X* is of greater elasticity than in the cases just mentioned, elasticity being measured at the point where the demand curve intersects the per-unit cost curve of supplying the seed. For example, the elasticity of demand for the seed is higher if the

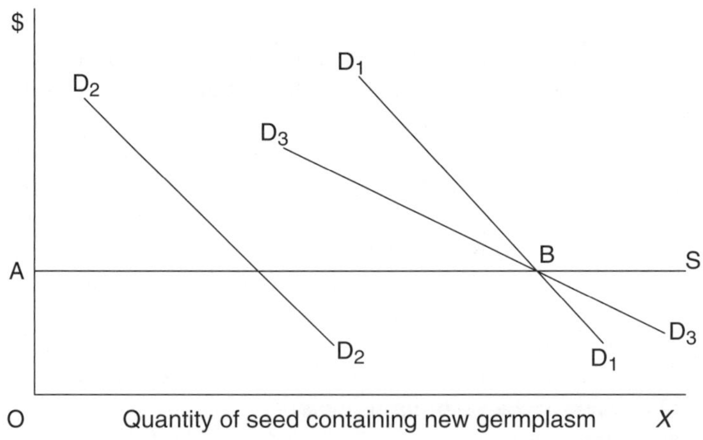

Figure 3.1 An illustration of different market situations that alter economic incentives for developing new germplasm

demand curve is as shown by line D_3D_3 rather than D_1D_1. In this case, the maximum profit of the monopolist will be lower. Each of the propositions can be proven by applying the economic theory of monopolistic behaviour, but that will not be done here.

Given the type of market alternatives illustrated in Figure 3.1, it will be more profitable (other things being held constant) to develop new germplasm for markets that are relatively large and/or for which demand is relatively inelastic. Markets in higher income countries seem more likely to satisfy those conditions than markets in lower income countries. This shapes the stock of new germplasm developed (given the institutional conditions just considered) and also has potential consequences for the loss of existing germplasm.

Large companies and businesses may be in a better position than smaller ones to benefit from systems that grant them temporary monopoly rights in new genetic material. First, economies of scale and scope may occur in the development of this material and the subsequent innovation process. Secondly, large businesses are likely to be in a stronger economic position than smaller ones to enforce their property rights; for example, to detect infringements on their rights and bear the cost of litigation to enforce their rights.

The extent to which the granting of private property in new germplasm promotes the general public interest is subject to some controversy. On the one hand, its monopoly element limits the level of social economic benefit obtained once the innovation has occurred. On the other hand, it provides economic incentives for discoveries and innovations that otherwise would not occur or which would be less likely to occur in the absence of this policy. Furthermore, doubts may be raised about whether this policy results in the best additions or socially desirable additions to the germplasm. This is evident, for example, in the debate about the desirability of adding genetically modified organisms (GMOs) to the genetic stock (see Chapter 7 in this book).

This raises the question of whether public bodies (including international research bodies associated with the United Nations) should have a role in developing new germplasm and if they do have a socially desirable role, what that role should be. In other words, should the role of creating new germplasm and promoting its use be left solely to the private sector? This matter will be discussed in Chapter 10.

Political Structures and Special Interest Groups

Many important decisions affecting the use and conservation of biological resources depend on government decisions. These in turn, depend on the structures of political systems and mechanisms available for individuals and groups of individuals to influence these decisions. The mechanisms for influencing these decisions can vary considerably. Apart from different voting systems and other formal mechanisms for making public decisions, lobbying of politicians, administrators and members of the general public are common. Both pro-conservation groups and developers engage in these activities.

In some cases, clandestine activities are engaged in by special interest groups or individuals wanting to pursue their own private self-interest. Bribes and other inducements may be given to politicians and public servants, for example, to allow natural areas to be used for commercial development. There is evidence of this occurring in making land available for the growing of oil palm, for instance in Indonesia (see McCarthy et al., 2012; Swarna Nantha, 2014), but it is not confined to this case. As a result, biodiversity is lost, and the social interest is mostly ignored. Because of free-riding and transaction costs, and because proponents of the general interest can only appropriate a small fraction of the public benefits from opposing special interests, in many cases, special interests are likely to prevail at the expense of the public interest. The political activities of special interest groups often contribute to unfavourable changes in the stock of biological resources when assessed from a collective point of view.

Increased Labour Mobility as a Result of the Expansion of the Market System

The expansion of the market system is associated with increased scope for labour mobility between occupations and geographical areas. This may reduce the motivation of individuals to conserve local resources, including local biological resources, because once these resources are exhausted, they may anticipate obtaining adequate employment elsewhere. This can also contribute to biodiversity loss. The matter is discussed in more depth in Tisdell (1990, Ch. 4)

The Growth of Large Organizations in which Responsibility for Decision-making is Shared Between Several Individuals

Many large organizations have evolved with operations in widely dispersed areas or covering large geographical areas. This includes many business organizations, government bodies and NGOs. Often decisions made by them are the responsibility of a small group of individuals: it is a shared responsibility. This may reduce their sense of personal moral responsibility for the decisions made. For example, their decisions may result in significant loss of biodiversity, which is not in the general interest, or may have undesirable consequences for the poor.

Furthermore, if the organization is large, those in the head office may not be fully aware of the consequences of decisions affecting the operations of the organization in distant areas. This can contribute to biodiversity loss. In addition, individuals and local communities are often at a disadvantage in trying to counteract the decisions of such large organizations that impact on them and their biological resources adversely. They are often not able to afford the cost of mounting an effective challenge to such developments, especially if the large organizations have government support.

In the case of large public companies, there is an additional issue because there is normally separation of their ownership and management. Agency problems make it difficult for shareholders to influence the decisions of managers. So shareholders may feel little responsibility for their actions. They could sell their shares if they are dissatisfied with the social consequences of decisions by their company but if shares in the company provide a high return, others are likely to buy them. So shareholders (who are dissatisfied with the social consequences of decisions by the company's managers) may not adopt this action on pragmatic grounds, namely that by selling their shares, they will have little or no influence on the actions of the managers of the company. For example, some shareholders may object to a bank financing development schemes that result in deforestation and to biodiversity loss but still may not sell their shares in the bank. There is more discussion of these aspects in Tisdell (1990, Ch. 4).

Government Failure to Reduce Open Access to Resources in a Timely Manner

Open access to natural resources can result in their over-exploitation and, in some cases, results in biodiversity loss. Due to political considerations, governments are often slow in moving to reduce open access to natural resources. This results in increasing dependence of some communities on these resources which may eventually be overused to such an extent that subsequently, community members experience poverty. This makes it even more difficult politically for the government to reduce the access of individuals to these resources since it is likely to push communities dependent on them below the poverty line, unless community members can obtain alternative means of livelihood. The exploitation of natural resources by local communities in the Okavango Delta in Botswana may accord with this pattern. The underlying theory of this process is outlined in Tisdell (2009).

Path Dependence

Some path dependence seems likely in the development of new germplasm having commercial value. The genetic resources of species supplying products that command a large market may be favoured for development compared to those which have a small market. This effect is self-reinforcing. At least three factors may contribute to this effect. First, the number of productive units involved in the larger market may be greater than in the smaller one. Hence, the number of units trying to select for more desirable traits in a species may be higher and the chances of favourable selection higher. Secondly, if public bodies are involved in the process of improving genetic material, the larger industry may be more successful than a smaller industry in lobbying for the public body to concentrate on their industry. Thirdly, if those developing new germplasm can obtain property rights in it, this is likely to favour larger industries, as was highlighted above.

Possibly, another type of path dependence also exists. It is related to the length of time species have been domesticated. Those which have been domesticated earlier rather than later have had a longer period of time in which humans could improve their traits so as to increase their economic value. This may favour the use of long established domesticates.

Consider Australia's position. Prior to the commencement of European settlement in Australia, Australian Aborigines domesticated no plants and animals, except dingoes. Since the arrival of Europeans, several species of native plants have proven to be suitable for domestication but their cultivation has commenced relatively recently. There has not been much time to improve their traits and methods of cultivation, for example by means of learning-by-doing. Substantial lags in the cultivation and use of Australian native plants have to be overcome in order to make them commercially viable. Aspects of the Australian case will be discussed in Chapter 5.

3.6 CONCLUDING COMMENTS

The above is not an exhaustive list of socio-economic factors influencing the conservation and development of biological resources. It does, however, indicate that many complex socio-economic factors and processes are involved. Particularly since the start of the Holocene era and within that era, beginning with the commencement of agriculture in different parts of the world, humans have exerted an increasing influence on the stock of biological resources, including the nature of ecosystems. Some historical occurrences have been especially important in that regard. For example, as shown above, developments following the European Age of

Discovery had a substantial impact on global diversity. Many other factors were also of significance in altering the genetic stock and ecosystems, such as continuing (man-made) capital accumulation, advances in science and technology and the extension of market systems, including the expansion of international trade.

In addition, a variety of other institutional features were shown to be relevant to changes in the stock of biological resources. For example, the following were shown to be relevant: the development of chains of supermarkets and food outlets; the granting of property rights in new germplasm; political structures and special interest groups; increased labour mobility as a result of expansion of the market system; the emergence of large organizations with special features affecting the nature of decision-making and failures in government policy to reduce open access to natural resources and path dependence influenced by socio-economic effects. Many of these factors have consequences for the sustainability of economic development. This is the main focus of the next chapter.

REFERENCES

Ammerman, A.J. and L. Cavelli-Sforza (1984), *The Neolithic Innovation and the Genetics of Populations in Europe*, Princeton, NJ: Princeton University Press.

Anon (2014b), 'Cane toads in Australia', Wikipedia. Available at: http://en.wikipedia.org/wiki/cane_toads_in_Australia. (Accessed 10 April, 2014).

Bandara, R. and C.A. Tisdell (2003), 'Use and non-use values of wild Asian elephants', *Sri Lankan Economic Journal*, **4** (2), 3–30.

Bulte, E., R.D. Horan and J.F. Shogren (2006), 'Megafauna extinction: A paleoeconomic theory of human overkill in the Pleistocene', *Journal of Economic Behaviour and Organization*, **59**, 291–323.

Clark, J.M. (1940), 'Towards a concept of workable competition', *American Economic Review*, **30**, 241–56.

Hirshleifer, J. (1977), 'Economics from a biological viewpoint', *The Journal of Law and Economics*, **20**, 1–52.

McCarthy, J.F., J.A.C. Vel and S. Afiff (2012), 'Trajectories of land acquisition and enclosure: Development schemes, virtual land grabs and green acquisitions in Indonesia's Outer Islands', *The Journal of Peasant Studies*, **39**, 521–49.

Pareto, V. (1927), *Manuel d'Économie Politique*, 2nd edn, Paris: Giard.

Romer, P. (1990), 'Endogenous technological change', *Journal of Political Economy*, **98**, 245–75.

Schumpeter, J. (1954), *Capitalism, Socialism and Democracy*, 4th edn, London: Allen and Unwin.

Svizzero, S. and C.A. Tisdell (2014), 'The Neolithic Revolution and human societies: Diverse origins and development paths', *Working Papers on Economics, Ecology and the Environment*, no. 192. Brisbane: School of Economics, The University of Queensland.

Swarna Nantha, H. (2014), 'The financial and political opportunity costs of orangutan conservation in the face of oil-palm expansion', *Working Papers on Economics, Ecology and the Environment*, no. 195. Brisbane: School of Economics, The University of Queensland.

Swarna Nantha, H. and C.A. Tisdell (2009), 'The orangutan-oil palm conflict: Economic constraints and opportunities for conservation', *Biodiversity Conservation*, **18**, 487–502.

Thrall, P.H., J.D. Bever and J.J. Burdon (2010), 'Evolutionary change and agriculture: The past, present and future', *Evolutionary Applications*, **13**, 405–8.

Tisdell, C.A. (1990), *Natural Resources, Growth and Development*, New York, Westport, CT and London: Praeger.

Tisdell, C.A. (2004), 'Property rights in non-captive wildlife and biodiversity conservation', *International Journal of Global Environmental Issues*, **4** (4), 195–208.

Tisdell, C.A. (2009), 'Poverty, policy reforms for resource-use and economic efficiency: Neglected issues', *The Singapore Economic Review*, **54** (2), 155–66.

Tisdell, C.A. (2011), 'Biodiversity conservation, loss of natural capital and interest rates', *Ecological Economics*, **70**, 2511–15.

Tisdell, C.A. (2013), *Competition, Diversity and Economic Performance*, Cheltenham, UK and Northampton, MA, USA: Edward Elgar Publishing.

Tisdell, C.A. (2014), *Human Values and Biodiversity Conservation: The Survival of Wild Species*, Cheltenham, UK and Northampton, MA, USA: Edward Elgar Publishing.

4. Sustainable development and changes in the genetic stock and in ecosystems

4.1 INTRODUCTION

Without adequate and suitable biological resources, humankind could no longer exist. Human beings depend inextricably on other living things to supply them with food. In addition, they need water for their existence, and its supply and quality depends on the nature of ecosystems. Although human well-being has become increasingly dependent on the use of non-renewable, non-living resources (these resources, for example, now provide shelter, energy, means of transport and fibres), they are not as fundamental to human survival as biological resources. Furthermore, should non-renewable resources become depleted, biological resources could provide substitute commodities for many of those currently supplied by non-renewable resources. In addition, apart from supplying services essential to human existence, biological resources supply other services that add significantly to human well-being (Millennium Ecosystem Assessment, 2003; 2005). Therefore, it is appropriate to give particular attention to the economics of sustaining the supply of biological resources. There is no doubt that significant loss of biological resources can threaten the sustainability of human welfare and restrict economic development.

For the reasons outlined in the previous chapter, market and political failures (which adversely affect the conservation of biological resources and ecosystems) result in economic welfare being lower than it could be. These failures can also result in unsustainable economic development, for example, inability to maintain the future well-being of individuals and the level or growth of GDP. However, these failures are not the only possible source of unsustainable economic development. Actions and strong desires by existing individuals to increase their current incomes (for example, by utilizing and depleting natural resources) can result in the subsequent impoverishment of current and future generations. Time preferences (both stated and revealed) for resource use can affect the sustainability of economic development.[1]

In recent times, it has been widely accepted that sustainable development is desirable. While the sustainability attributes of different alternative development paths are relevant to choosing between the available alternatives, it is shown in this chapter that contemporary criteria for choosing between development paths can result in inconsistent choices and, in some cases, in irrational choices. It also becomes clear that sustainable development paths are not necessarily the most desirable ones. Furthermore, in some instances, commonly used criteria for sustainable development are unable to determine which one of the available alternative development paths is optimal. These and related conceptual issues are covered in section 4.2.

Requirements for achieving sustainable development, including the necessary conservation of the biological stock, are discussed in section 4.3. The existence of uncertainty poses major challenges for determining an optimal development path and for choosing the most desirable strategies for conserving biological resources. These challenges receive particular attention in Section 4.4 where the implications of the Precautionary Principle for the conservation of natural resources, especially the conservation of biological resources, are explored. It is shown that this principle (which has varied interpretations) does not always favour the conservation of existing biological resources.

Section 4.5 provides a preliminary examination of the sustainability of alternative agroecosystems. In doing so, it considers the criteria proposed by Gordon Conway (1985; 1987) for the evaluation of alternative agroecosystems. His comparisons between the properties of modern and traditional agrosystems, including differences in their sustainability, are discussed critically. Although Conway's early contribution to this analysis provides a useful starting point for comparing alternative agroecosystems, it is argued that it is incomplete. It does not, for example, give enough attention to biodiversity loss and to the increased dependence of modern agroecosystems on the supply of off-farm resources compared to traditional agroecosystems.

4.2 SUSTAINABILITY AND SUSTAINABLE DEVELOPMENT: CONCEPTS AND OBJECTIVES

In recent decades, it has been widely contended that sustainable development and the attribute of sustainability is desirable. However, these concepts are not straightforward and one finds different meanings and applications of these in the literature. Current views about these concepts need scrutiny and the different objectives posited as necessary for sustainable development require careful examination. Consider these matters.

Sustainability as a Concept

Sustainability is a possible attribute of an object. Whether the sustainability of an object is desirable or not depends on the object in question. For example, few would contend (today) that it is desirable to sustain slavery but most would agree that it is desirable to sustain the absence of smallpox. Or, to take another example, special interest groups that gain from restrictions on trade are likely to favour these restrictions being sustained whereas those who suffer an economic loss (as a result of such restrictions) will want them removed. Therefore, sustainability is not necessarily a desirable attribute of an object nor is there always agreement on whether it is a desirable trait of an object.

It has been claimed (for example, Barbier, 1987) that for economic activities to be sustainable, they should be:

1. economically sustainable (economically viable);
2. ecologically sustainable; and
3. socially acceptable.

These requirements do seem relevant but there can be differences of opinion and uncertainty about whether each of these conditions is satisfied in particular cases. Furthermore, the social acceptability of different types of economic activities is liable to alter with the passage of time because communal values (and values of individuals) often alter with the passage of time (Tisdell, 2014). These above criteria have been applied to assessing the sustainability of the farming of saltwater crocodiles (*Crocodylus porosus*) in Australia (principally in the Northern Territory) based (to a large extent) on the collection of their eggs from the wild (Tisdell, 2014, Ch.9). This case study illustrates the benefits and limitations of this type of analysis.

Sustainable Development: Alternative Concepts

Policies for sustainable development need to be economically and socially acceptable as well as ecologically sustainable. Otherwise, they will fail to last. For example, if policies to achieve sustainable development are not socially acceptable (which could partly be because they are not economically acceptable), they are unlikely to be adopted, or if adopted, they are unlikely to last.

The subject of sustainable economic development is complicated because there are several views about what constitutes sustainable development.

The following are four different views:

1. Sustainable development is development that enables the human race to exist for as long as God wills, or until some force (not the result of human actions) brings human existence to an end.
2. Sustainable economic development is development that does not restrict the capacity of future generations of human beings to meet their needs.
3. Sustainable economic development is development ensuring that the well-being of each future generation is not to be less than the well-being of its predecessor.
4. Sustainable economic development is development ensuring that no future generation is less well off than current generations.

Herman Daly (1980) originally proposed the first concept. The second concept was contained in the Brundtland Report (WCED World Commission on Environment and Development, 1987). It is, however, the third concept that is most commonly mentioned in the economic literature dealing with the concept of sustainable development but mostly it is not clearly distinguished from the fourth concept, (see, for example, Tietenberg, 2003, p. 94). Tietenberg states (ibid, p. 94): 'The *sustainability* criterion suggests that at a minimum, future generations should be left no worse off than current generations'. This corresponds to concept 4 above. However, on the same page, he also maintains: 'In essence, the sustainability criterion suggests that earlier generations are at liberty to use resources that would be denied to future generations as long as the well-being of future generations remains just as high as that of *all previous* [my italics] generations'. This corresponds to concept 3, and is more restrictive than concept 4. All of these concepts are anthropocentric.

The first mentioned concept puts maximum weight on the value of sustaining *Homo sapiens* for as long as possible. It assumes that this is the most desirable of all objectives. Moral underpinning for concepts 2 and 3 are sometimes based on Rawls' Principle of Justice (Rawls, 1971). For example, Tietenberg (2003, p. 94) relies on the moral relevance of this principle, but it is not necessarily compatible with concepts 3 and 4 above. This will become clear when Figures 4.2 and 4.3 are discussed. This principle assumes that any individual could be born into any generation and that individuals before being born do not know what generation they will be born into. Rawls supposes that in this uncertain state, all would opt for equality of well-being, unless inequality would result in a Paretian preferred outcome, that is, a situation in which some are better off than otherwise and none worse off than otherwise.[2]

The Maximin Gain Criterion and Sustainable Development

However, it is not clear that Rawls' Principle of Justice will be accepted by all. It implies an extremely risk-averse attitude, namely that individuals want to maximize their minimum possible gains or benefits given all the alternative available development paths, that is, they adopt a maximin gain strategy. If a maximin gain approach is adopted, it can result in choices that do not seem to be reasonable.

For example, suppose that two development strategies are available. If Strategy I is adopted, development path AC in Figure 4.1 can certainly be followed. If Strategy II is adopted, suppose that either path AB or path AD is possible, but that it is almost certain that AD will be the outcome. The maximin gain criterion will result in Strategy I being adopted, despite the potential economic disadvantage of adopting Strategy II being very small and the probability of a large gain being very high for this strategy.

Choice Based on Rawls' Principle Compared with the Application of Criterion 4

Now consider the alternative development paths shown in Figure 4.2. If development Strategy I is adopted path ACF is followed, and if Strategy II

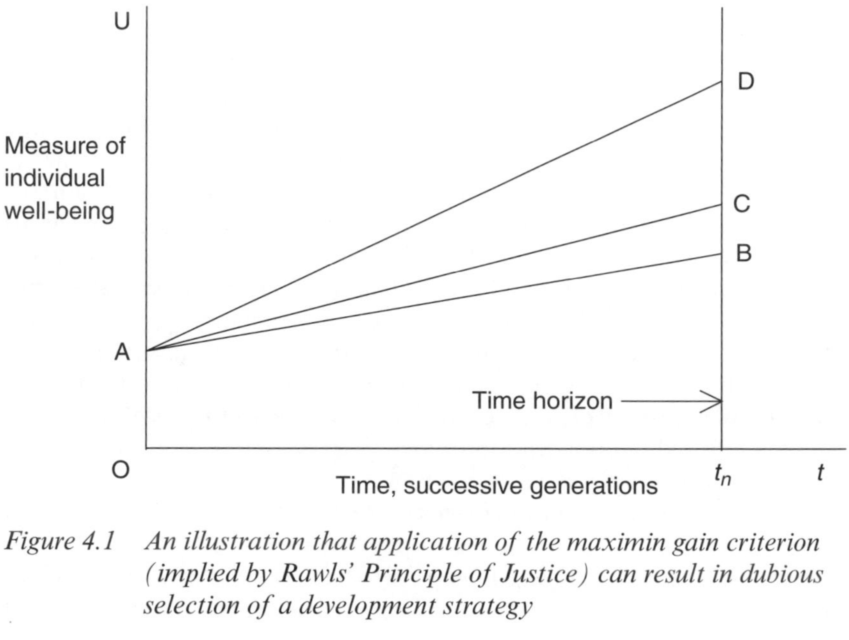

Figure 4.1 An illustration that application of the maximin gain criterion (implied by Rawls' Principle of Justice) can result in dubious selection of a development strategy

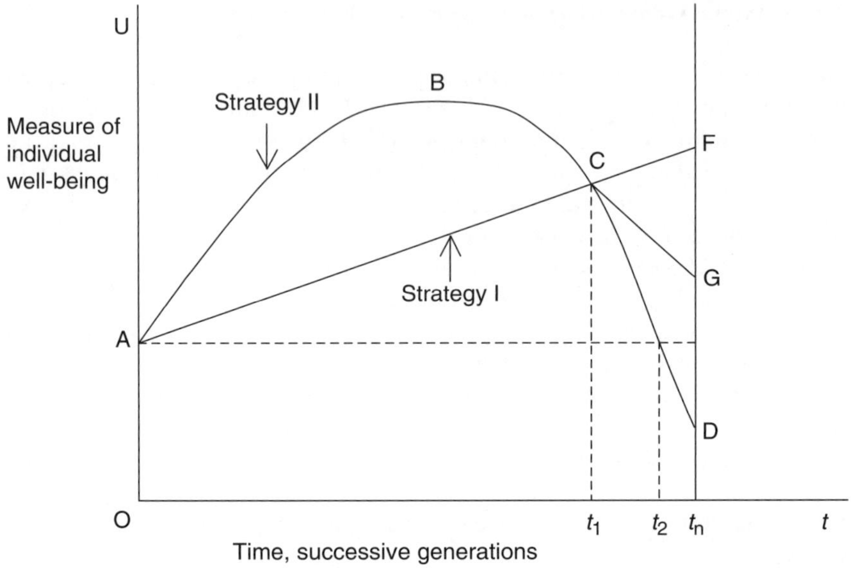

Figure 4.2 An assessment of some alternative development paths based on Rawls' Principle of Justice

is adopted, the outcome is path ABCD. Rawls recommends rejecting Strategy II and adopting Strategy I. In this particular case, some future generations (those existing beyond t_2) are worse off than current generations. However, this is not the basis of this rejection. If, for instance, Strategy II is adopted, and this happened to result in path ABCG being followed, this strategy would still be rejected by Rawls in favour of Strategy I, even though no future generations are worse off than the current ones if Strategy II is adopted. The rejection occurs because, in this case for $t > t_1$, all generations alive in this interval of time are worse off than if Strategy II had been adopted. Nevertheless, in cases of the latter type, the ethical grounds for rejecting a development strategy which leaves some generations less well off than some of their predecessors do not seem to be strong. Although some generations are worse off than their predecessors, all are better off than current generations. In this case, criterion 4 above is satisfied but not criterion 3.

Choice Involving Paretian Preference and Rawls' Principle

Figure 4.3 illustrates another interesting case. Three alternative development strategies are assumed to be available. If Strategy I is adopted, path AB

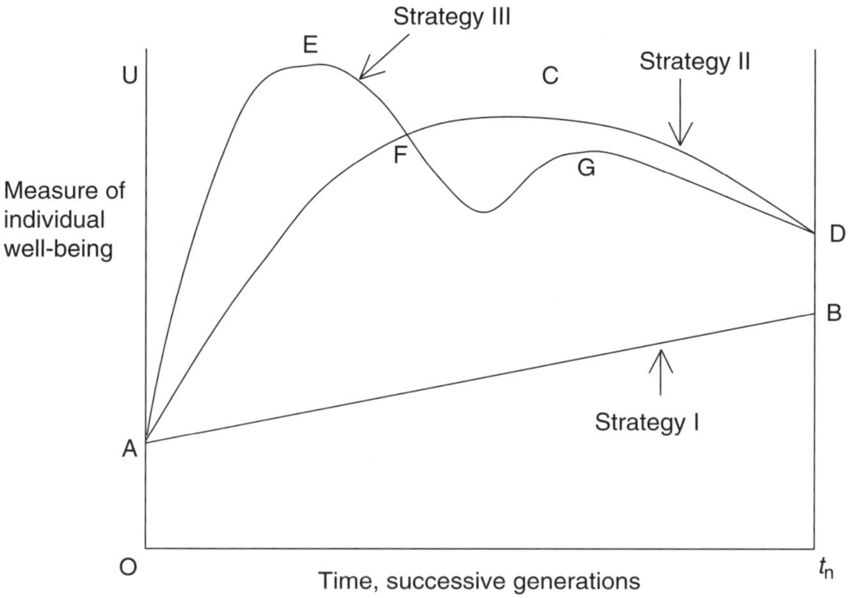

Figure 4.3 Paretian preferred development paths and Rawls' Principle of Justice: further limitations illustrated

emerges, and paths AFCD and AEFGD correspond to Strategies II and III respectively. Only path AB ensures that each succeeding generation is at least as well off as its predecessor. However, each of the other two paths are Paretian preferable to path AB because they will make some individuals better off than they otherwise would be, and none will be worse off. In this case, Rawls accepts that Strategies II and III are socially preferable to Strategy I even though they do not satisfy the economic criterion for sustainable development set out in point 3 above. Nevertheless, Rawls' Principle of Justice is unable to indicate whether Strategy II is preferable to Strategy III. It is inconclusive. This is because neither strategy is Paretian preferred to the other.

An Error about a Requirement for Sustainable Development, and Indeterminate Decisions

Despite indications to the contrary by Tietenberg (2003, pp. 93–94), Rawls' Principle of Justice does not imply that condition 3 above is essential for the achievement of sustainable development, nor does it imply that the satisfaction of condition 4 ensures sustainable development. This is clear from the discussion surrounding Figures 4.2 and 4.3. In the case of Figure 4.3, if Strategies I and II are the alternatives, Rawls would choose II even though it does not result in each succeeding generation being no worse off than its predecessor. In the case illustrated by Figure 4.2, Rawls would reject Strategy II in favour of Strategy I. However, if criterion 4 is applied, either Strategy I or Strategy II (*if* path ABCG is followed) would be chosen, but on its own, this criterion is unable to determine which should be adopted. It is inconclusive. Of course if the welfare of more distant generations is discounted at a sufficiently decreasing rate, then Strategy II would be selected in preference to Strategy I in this case.

Thus neither Rawls' principle nor criterion 4 are always able to determine the most desirable development path given the available alternatives. Moreover, if the trajectories of the well-being generated by two alternative development paths are both increasing and intersect, criterion 3 cannot be used to decide which is optimal. Therefore, an alternative to these criteria is worth considering.

A Safety-first Approach to Choosing Development Strategies

If individuals could hypothetically confer before being born in the circumstances proposed by Rawls, it is not obvious that they would adopt a maximin gain strategy.[3] They could differ in their preferences, or they might agree on a safety-first strategy. In the latter case, they may opt for maximizing their expected well-being subject to it not falling below a specified level. How this could affect their choice of development strategies can be illustrated by the example shown in Figure 4.4. Here, line MN represents the minimum level of well-being that the unborn are willing to accept. Consider three alternative available development strategies. Strategy I results in path AB, Strategy II corresponds to path AHDJ and Strategy III yields path ACDFG. Given this safety-first criterion, Strategy III would be rejected. Either Strategy II or I would be chosen, depending on which results in the highest expected level of well-being.

Model Limitations and an Assessment So Far of Proposed Objectives for Sustainable Development

Observe that the development models used in the above exposition rely on a simplistic assumption, namely that decisions by current generations about resource use determine fairly exactly future development paths. However, in reality, the influence of resource decisions by current generations on these paths is likely to be limited. This is because these paths also depend on decisions by future generations about their resource use and these

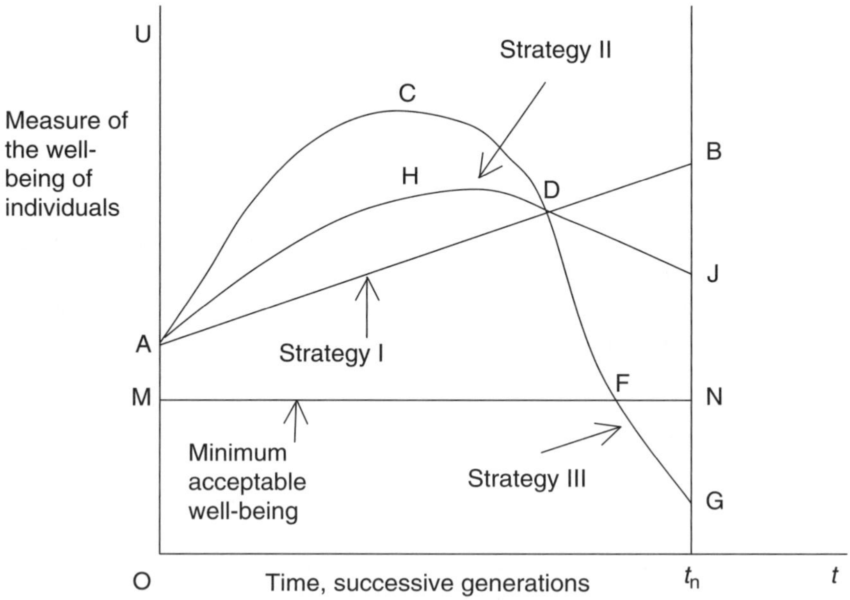

Figure 4.4 Illustration of a safety-first approach to choosing development strategies

decisions cannot be controlled by current generations. It is, for example, possible that some future generations could squander resources conserved by current generations thereby impoverishing their successors. The uncertainties about this, and about unknown possible exogenous events, complicate the choice of strategies intended to promote sustainable development. Furthermore, as has been demonstrated, some of the criteria for attaining sustainable economic development have significant shortcomings.

However, this does not imply that the quest for attaining desirable development paths should be abandoned. As shown, the view that sustainable development ethically requires the well-being of each successive future generation to be not less than that of its predecessor fails to satisfy Rawls' Principle of Justice in some circumstances. This was illustrated by the discussion based on Figure 4.3. Furthermore, it is not obvious that individuals are as risk-averse as assumed by Rawls. They might, for example, prefer a safety-first type of approach like that illustrated in Figure 4.4. If so, this does not imply that there would be no concern about the sustainability of development. Current generations would still desire to make sure that the well-being of future generations is maintained or sustained above a minimum acceptable level.

Most of the above objectives postulated for sustainable development are based on untested hypotheses about their social acceptability. There has been inadequate empirical investigation of the extent to which the above-mentioned objectives are accepted by current generations and about their concerns for the welfare of future generations. This is important because future generations are powerless to influence the resource-use decisions of preceding generations, including current generations. Both what current generations believe ought to be the objectives for assisting future generations and the sacrifices they are actually willing to make to achieve these objectives need to be empirically investigated. Although this agenda cannot be pursued here, some observations about the concerns of current generations for future generations are in order.

Coefficients of Concern for Future Generations

David Pearce (1998) has recognized that the *actual degree of concern* of current generations for future generations is relevant to debates about whether sustainable development can be achieved. Pearce (1998, p. 71) suggested that it would be desirable to determine coefficients indicating the extent to which individuals belonging to current generations are concerned about the welfare of future generations. He hypothesizes that each generation is, at the most, concerned about the welfare of three of four generations, namely their own, that of their children and grandchildren, and possibly that of their great grandchildren. Therefore, this effective span of concern of each generation may be less than 80 years, and could decline sharply with possibilities that are distant in time. Furthermore, individuals in each generation are likely to be more concerned about the welfare of their own offspring than the welfare of the offspring of others. This is because individuals are likely to have greater empathy with their offspring than with others, and if the hypothesis of Dawkins (1989) is correct, humans have a strong desire to protect their own genetic legacy. However, as I have argued elsewhere (Tisdell, 2013, Chapter 7), this legacy declines sharply with succeeding generations.

Why might one expect the relevant coefficients of concern to be declining the further one goes into the future? First, each individual is likely to have reduced empathy with more distant generations. Secondly, individuals currently alive may believe their decisions will have diminishing and eventually minute effects on the options available to distant generations and on their choices. Thirdly, the options available to distant generations may be viewed by current individuals to be so uncertain that no rational basis exists for influencing them. In addition, bounded rationality may dictate that there is little benefit in current individuals assessing such distant possibilities.

Furthermore, it seems likely that coefficients of concern of individuals for others will vary considerably between individuals and in relation to different individuals belonging to the different social groups. These coefficients probably also vary with an individual's age and other factors. Research is needed to determine the sizes of these coefficients and how they influence choices by individuals; for example, the nature of their bequests. However, it seems probable that the likely level of welfare of generations existing more than a few generations into the future will have little effect on choices by current generations.

Given this truncated vision, will the welfare of very distant generations be entirely neglected? It might be argued that their welfare will not be entirely neglected because each successive generation will care (to some extent) for a few following generations. Nevertheless, it is possible that *very long-term* effects on future generations of decisions of each generation will be disregarded by all generations. In addition, bequests of preceding generations for future generations may consist primarily of private goods given to their own kin. Due to political and market failures, the bequest of public goods of value to future generations may be significantly undersupplied, as has been suggested by Margolis (1957) and by Krutilla (1967).

Daly (1999, pp. 115–116) adopts the view that provision for distant generations is essentially a public good, and he is of the view that 'most people do not take any individual actions on behalf of their descendants beyond their grandchildren'. Therefore, collective rather than individual arrangements are needed if it is desired to increase economic assistance to future generations.

Uncertainty and the Concepts of Sustainability and Sustainable Development

Insufficient attention is usually given to the relationship between risk and uncertainty and the sustainability of an object, including the sustainability of development. One must consider how certain one can be about the extent to which welfare will be sustained if particular development strategies are chosen. To what extent should high-risk development strategies be avoided?

The type of decision-making problem involved can be illustrated by reference to Figure 4.5. Envisage two alternative development strategies. If Strategy I is adopted, either path CD or CF eventuates, depending on subsequent events. Those paths may, for example, have a subjective probability of 0.5 each. If Strategy II is adopted, path AB is followed with certainty.

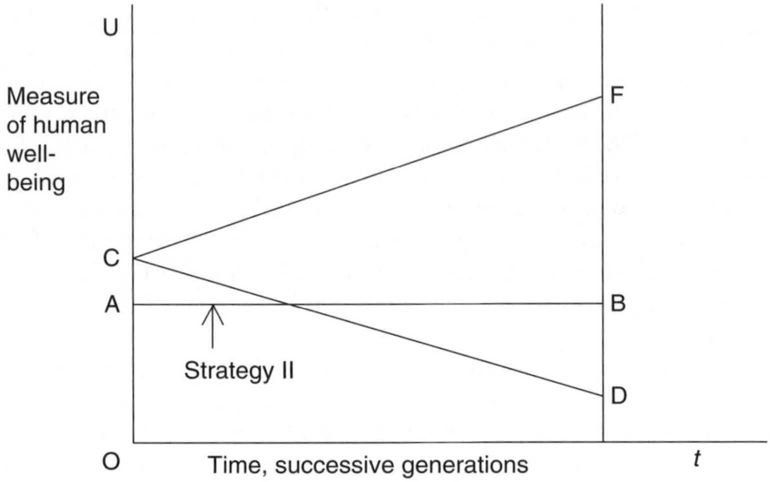

Figure 4.5 *An illustration of the relevance of uncertainty when choosing between alternative development strategies*

Consequently, Strategy II ensures greater sustainability of human well-being than does Strategy I.

The attitudes of individuals to risk-taking when choosing alternative development strategies need empirical investigation. For example, are current generations less willing to subject themselves and nearby succeeding generations to risk than distant future generations? To what extent do individuals believe that risk-avoidance is important in choosing between alternative development strategies? These types of issues involving risk-taking are integral to debates about the desirability of alternative strategies for economic development. Nevertheless, it is reasonable to suppose that there will be a strong desire to avoid unwanted deviations from desired development paths, particularly unpredictable deviations of this kind. Particular attention is given to uncertainty, sustainability and the precautionary principle in section 4.4.

4.3 REQUIREMENTS FOR ACHIEVING SUSTAINABLE DEVELOPMENT, INCLUDING THE NECESSARY CONSERVATION OF THE BIOLOGICAL STOCK

Opinions differ considerably about what is needed to achieve sustainable development, even when there is agreement about what sustainable development is and on what constitutes a desirable development path when intergenerational equity is accounted for. As a rule, the discussion of what is required to achieve sustainable development focuses on what economically valuable natural and other assets should be conserved or transformed to achieve this goal and the consequences of these actions for environmental change.

Alternative Views about Natural Resource Conservation and the Achievement of Sustainable Development

Of particular interest are the consequences for sustainable development of the utilization of natural resources to supply products for consumption and build physical capital. These processes result (to varying degrees) in the depletion of natural resources, some of which are renewable (if they are not over-exploited) and the use of others that are non-renewable and depletable. Most biological resources belong to the former group whereas minerals, for example, belong to the latter group.

At one end of the spectrum, some scholars (for example, Daly, 1999) believe that it is necessary to take particular care to conserve natural resources (including natural ecosystems and genetic material) if human welfare is to be sustained, including the welfare of future generations or if the longevity of existence of *Homo sapiens* is to be maximized. They believe that significant restrictions on the use of natural resources for economic production (for consumption and for the creation of human produced capital accumulation) are necessary. These 'strong' conditions for sustainable development contrast with the 'weak' conditions for sustainability favoured by many mainstream economists who take the view that significant government restrictions on natural resource use are unnecessary for achieving sustainable development.

The reasons given by those supporting weak conditions for sustainable development vary. However, often the following reasons are given:

1. Natural resources are very abundant. Their abundance is underestimated by those who favour strong limitations on their use.
2. If one or more natural resources become scarce, it is always possible to rely on other natural resources to help make up for the shortfall in their availability.
3. Investment in physical and human capital is often a better bequest for future generations than is the conservation of natural capital.
4. Provided there is sufficient investment in research and development, advances in science and technology will more than counteract any

adverse economic consequences that may stem from the reduced availability or quality of natural resources.

John Maynard Keynes was, for example, very optimistic about the prospects of humankind being able to maintain its well-being into the future. In this regard, he states 'the resources of Nature and man's desires are just as fertile and productive as they were' (Keynes, 1931, p. 135). This opinion was even more strongly voiced by Engels, when he rejected Malthus' theory that human population growth is a potential barrier to rising living standards. Engels declared that nothing is impossible to science, and that even if population were to increase as a geometric progression, the increase in economic production (made possible by scientific advance) would occur at an even faster geometric rate (Engels, 1959, p. 204).

Those who adopt a more cautious view about the prospects for sustaining human welfare (given the impacts of humans on natural resource availability) have the following concerns:

1. Many natural resources are already scarce, are becoming scarcer and are being depleted. Their increasing scarcity can be expected to reduce the quality and availability of inputs for economic production. Furthermore, the magnitude of the adverse environmental effects from the use of these resources is accelerating.
2. While natural resource substitutes are available for some natural resources, they are imperfect substitutes and are likely to become fewer in number as different types of natural resources are depleted. Moreover, the costs of using these substitutes could be high.
3. The extent to which future well-being can be sustained or raised by using natural capital to accumulate physical human-supplied capital and human capital is unclear. Nevertheless, substantially reducing natural capital for these purposes poses a threat to the sustainability of human well-being. The threat becomes more imminent once there has been a substantial reduction in the stock of natural capital (Tisdell, 2005, Ch. 12).
4. There is no guarantee that advances in science and technology will be able to offset the negative economic and other costs of natural resource loss and environmental deterioration brought about by reducing the stock of natural capital.

In addition, this group of concerned individuals often favour strong sustainability conditions because they believe that market and political failures of the type discussed in the previous chapter result in excessive use of natural resources from a sustainability point of view (see, for example,

Daly, 1999, Chs. 4 and 15). There are strong economic forces favouring the increased production of private goods because of the employment and private income implications involved. Given current economic systems, pro-economic growth sentiments are the dominant forces influencing economic policies (Tisdell, 1999, Ch. 6). The current global economic system (and the associated political system) pay little heed to the desirability of conserving natural capital, especially natural biological resources.

There is little doubt that continuing economic growth by many lower income countries (such as China and India) will place increasing pressure on global natural resources and the state of the global environment. With economic growth, both these nations (and others) have become increasingly dependent on imports of natural resources (and particularly in China's case, commodities used in agriculture, for example grains for livestock production and timber for fabrications), the supply of which can have significant negative global environmental consequences. Daly (1999, p. 126) warns that when national economic growth is fuelled by such imports, 'free trade enables the ecological burden to be spread more evenly across the globe, thereby buying time before facing up to limits, but at the cost of having to face the problem simultaneously and globally rather than sequentially and nationally.' If the average per capita income of all the population of the world were to reach the level of that of the population of the highest income countries today, the drain on the globe's natural resources and the impact on its environment would be staggering, especially since the world's population is expected to increase to about 9 billion by the middle of this century.

Biological Conservation and Sustainable Development

The conservation of biological resources is an important consideration in relation to the achievement of sustainable development. However, it is necessary not only to consider the conservation of natural (wild) genetic material and ecosystems but also those that have been developed mainly by human effort and investment. The type of sustainability analysis outlined earlier in this section does not take account of the latter type of resources, for example heritage and recently developed germplasm (see Chapter 2).

The conservation of germplasm involves significant challenges because its ability to reproduce (its viability) must be maintained. Among other things, this usually requires the conservation of minimum viable populations of wild germplasm, and most human developed germplasm relies on continuing human management for its survival. The successful reproduction of human developed germplasm normally depends on the adoption of favourable cultivation and propagation methods. Propagative material of plants which are stored for a very long time (for example, in seed banks) may not remain viable. To maintain the viability of the reproduction of plants, their propagules usually have to be replanted periodically and new seeds or propagative material collected. The process is an ongoing one.

It is probably not economic (from a social point of view) to try to conserve all the global existing stock of germplasm and this may be unnecessary to ensure the attainment of a desirable development path. Nevertheless, risks of undesirable impacts on the level of human well-being increase as the size of the genetic stock is reduced. This applies to reduction both in the wild stock and in the human developed stock. There are several reasons why this is so.

First, a reduction in the global genetic stock usually reduces the ability of humans and ecological systems to adjust to changing environmental conditions. For example, with a reduction in the genetic stock, the amount of genetic stock adapted to different eco-geographic conditions shrinks. Consequently, there is less room to select genetic material suited to changes in these conditions and for selective breeding of organisms. Furthermore, the amount of genetic material available for recombining DNA is reduced. This can be expected to reduce the scope for creating new valuable organisms by means of genetic engineering. A consequence of this loss could be, for example, a reduction in the potential to adjust the earth's biological stock to respond to climate change.

Secondly, there are also debates about the extent to which genetic diversity at regional and local levels is desirable. It is, for example, often claimed that ecosystems that contain greater diversity of species (and other genetic material) are more sustainable. While the increasing elimination of species from an existing ecosystem may reduce its stability and robustness, it is not obvious that species-rich ecosystems possess greater sustainability than less diverse systems when subjected to environmental stresses of similar magnitude. One must take into account the environmental tolerances of the individual species in the ecosystem and their interdependence. It is possible, therefore, that species-rich natural tropical ecosystems are more vulnerable to environmental change than are those in temperate zones that are less species-rich but that contain many species displaying a high degree of environmental tolerance (Tisdell, 1999, Ch. 4).

As for human developed germplasm, considerable risks to sustainable development arise when food production becomes dependent on a very narrow range of plant varieties and breeds of animals. Narrowing of this type is happening both regionally and globally for the reasons which were outlined in the previous chapter. The economic vulnerability of economies that become reliant on a narrow genetic base for their food production is most concerning for staple crops that account for a large portion of food production (such as cereal crops and some root crops). For example, the welfare of the Irish was seriously reduced by the Irish potato blight in the mid-1800s and so was that of Samoa in 1993–94 as the result of the taro blight (Anon, 2014; Hunter, 2000). Possibly, the growing of a greater diversity of varieties of these crops, and less dependence for food on a single crop species, could have reduced the economic dislocation experienced as a result of these blights. However, it seems that all domesticated varieties of potatoes are susceptible to potato blight and that this blight needs to be controlled by adopting cultivation practices that reduce its likelihood of occurrence (Anon, 2014). On the other hand, not all varieties of taro are susceptible to the taro leaf blight caused by the pathogen *Phytophthora colocasiae*. Nevertheless, Samoa mainly depended on production from two susceptible taro varieties when its crop was devastated by taro leaf blight in 1993–94 (Hunter, 2000).

The extent to which economies are dislocated by unfavourable environmental changes in the above circumstances depends on their available 'fall-back' options and how easily and quickly these can be accessed. In the case of the occurrence of taro leaf blight in Samoa, taro varieties resistant to this blight were available fortunately in Micronesia, Palau and the Philippines and these could be used to develop resistant varieties in Samoa (Hunter, 2000). Although Nordhaus (1974) has considered back-up options in relation to the availability of energy resources, no similar exercise seems to have been undertaken for biological resources. Nevertheless, with a reduction in the size of the genetic stock, it is likely to become more difficult to find genetic options that will enable human well-being to be sustained as environmental conditions alter.

Assessing the environmental and ecological consequences of changes in ecosystems is complicated but these changes can also contribute to a loss of economic sustainability. For example, loss of forested areas as a result of their conversion to land use for agriculture (and other purposes) contributes to rising CO_2 levels and consequently to global warming. Global warming is expected to make it more difficult to sustain the welfare of future generations (Stern, 2007; Tisdell, 2009, Ch. 11). The loss of some human developed ecosystems could have negative effects on economic welfare. For example, rice cultivation based on terracing is believed to have beneficial effects on water flows as well as other environmental benefits compared to the abandonment of the land. It is predicted that a loss of this agroecosystem could result in a decrease in human well-being (Ninan and Inoue, 2013).

4.4 UNCERTAINTY, SUSTAINABILITY AND THE PRECAUTIONARY PRINCIPLE

As was pointed out in section 4.2, the presence of uncertainty about the impacts on future human welfare of alternative development strategies is a major consideration in selecting development policies. One needs to consider how much caution should be displayed in choosing between development strategies and to take account of the implications of uncertainty for biological conservation. One view is that if uncertainty exists about the future value of existing biological resources, their conservation should be favoured because this will help achieve sustainable development. This is one of the main inferences drawn from the Precautionary Principle. Although this principle has been given many different interpretations (Tisdell, 2010), its basic message is that given the presence of uncertainty about the future economic value of many existing biological resources, it is rational to err in favour of their conservation. Unfortunately, it is *not always* the case that this is so. Whether it is depends on attitudes to risk-taking and the structure of the choice problem. Moreover, even when risk-aversion is strong, it is not always rational to err in favour of conserving all existing valued biological resources – as the following examples will show. Furthermore, these examples also indicate that the desire to avoid the possibility of losing existing economic benefits as a result of failure to conserve biological resources is stronger than the desire to conserve biological resources in order to obtain *potentially* large economic gains. Consider first examples involving hypothetical games against nature and then cases involving the maximization of expected gains from conservation.

Hypothetical Games Against Nature: Precaution Does Not Always Favour Conservation

Ciriacy-Wantrup (1968) has argued that because the cost of conserving individual species is low relative to the uncertain but potentially large future economic benefits of their conservation, the conservation of a minimum viable population of (all?) species is desirable. His arguments have been extended by Bishop (1978; 1979). However, even if risk-aversion is strong, these circumstances do not necessarily make it rational to conserve a species or a class of organisms. For example, suppose that a maximin gain (minimax loss) strategy is adopted. This implies a high degree of aversion to risk-taking.

For simplicity, assume that the hypothetical situation illustrated in the matrix shown in Table 4.1 is applicable. This represents a hypothetical zero-sum game against nature. Two alternative conservation strategies

Table 4.1 *A hypothetical zero-sum game against nature in which the minimax loss (maximin gain) criterion is used to decide whether or not to conserve an endangered species or an endangered variety of an organism*

	States of nature s_1	States of nature s_2	
Conserve, α_1	−1	200	−1
Do not conserve, α_2	2	2	2*

are considered: α_1, conserve a minimum viable population of a particular species, and α_2, do not conserve this species at all. Two states of nature are possible. State of nature s_1 corresponds to a situation in which the conserved species fails to have any economic value in the future, and s_2 is a state of nature for which the conservation of the species would result in a very large economic benefit if it is conserved. The body of the matrix (Table 4.1) shows the hypothetical economic pay-offs corresponding to the different conservation strategies and possible states of nature. There is a pay-off of 2 if the species is not conserved, for example, if it is a wild species and its habitat is destroyed to make way for agriculture, or if it is a domesticated species (or crop variety) and it is replaced by another domesticated species (or crop variety). In this case, the minimax loss (maximin gain) strategy is α_2, that is, not to conserve the species, a possible loss of 1 is avoided and a gain of 2 is assured. Note that in this case, allocating sufficient resources to conserve a minimum viable population (quantity) of the endangered species (or variety of a species) involves a net cost. Nevertheless, provided the minimum possible net benefit from conserving a minimum viable population of the endangered species or variety is less than 2, the optimal minimax loss policy is not to conserve it.

Despite the precautionary principle, the strategy of conserving a minimum viable population of a species in the circumstances illustrated by Table 4.1 is irrational. On the other hand, this is not so if the objective happens to be to minimize the maximum possible level of economic regret. This is clear from Table 4.2, which shows the regret matrix corresponding to the matrix in Table 4.1. In this case, the minimax regret strategy is α_1, that is to conserve a minimum viable population of the species.

The values shown in Table 4.2 assume that there is a potential large benefit from conserving a minimum viable population of a species. On the other hand, what is the situation if it is assumed that there is the possibility of a large economic loss if a particular type of germplasm is not conserved? This may, for example, be a variety of crop which will disappear if special efforts are not made to conserve it because it will be replaced by another

Table 4.2 The regret matrix corresponding to that in Table 4.1 and the optimal minimax regret strategy

	States of nature s_1	States of nature s_2	
Conserve, α_1	3	0	3*
Do not conserve, α_2	0	188	188

Table 4.3 Another zero-sum game against nature: in this case, the minimax loss criterion favours biological conservation, unlike in Table 4.1

	States of nature s_1	States of nature s_2	
Conserve endangered variety, α_1	1	80	1*
Do not conserve this, α_2	200	0	0

crop variety. Possibly, its replacement may eventually succumb to a pathogen to which the endangered variety is resistant. As a result, the hypothetical economic pay-offs indicated by the matrix in Table 4.3 might apply.

In the matrix in Table 4.3, s_1 indicates a situation in which the new (substituted) variety of a crop is not affected by a potential pathogen. On the other hand, s_2 corresponds to a situation in which this variety is rendered unproductive by a pathogen to which the endangered variety is resistant. The endangered variety yields a smaller economic pay-off than the new variety in the absence of the pathogen. The economic pay-offs for all the possible events are shown in the body of the matrix (Table 4.3). If the pathogen does occur, conservation of the endangered variety will give a pay-off of 80 but this is lost if it is not conserved. The minimax loss strategy is to conserve the endangered variety. Precaution in this case favours conservation.

Further Limitations of Ciriacy-Wantrup's Approach to Conservation

Additional problems can arise in applying Ciriacy-Wantrup's safety-first approach to biological conservation. First, contrary to Ciriacy-Wantrup's assumption, the economic costs of conserving some species can be quite high. This is so, for example, for the orangutan, which requires a very large area of suitable habitat to support a minimum viable population. Singleton et al. (2004) estimated that a minimum viable population of orangutans is around 500. Given common densities of orangutans, this may require about 400 km^2 of suitable tropical habitat to be conserved. If much of this habitat is suitable for growing oil palm, substantial opportunity costs would be incurred in conserving the orangutan *in situ* (see, for example, Swarna Nantha and Tisdell, 2009). Furthermore, the potential economic benefit of conserving every species may not be high. Indeed, some may be pests with a negative economic value. Moreover, there is no level of population of a species which can absolutely ensure its long-term survival (Hohl and Tisdell, 1993). There is also the question of whether the safety-first approach to saving species should be applied to sub-species and varieties of species. Finally, if not all species or biological resources can be conserved at minimum viable population levels because of cost constraints, Ciriacy-Wantrup's approach does not indicate which ones should be chosen for survival (Tisdell, 1990).

Expected Values, the Status Quo and Conservation Preferences

Given diminishing marginal utility of increased income, it is usually rational to take greater precaution to avoid a possible decrease in income than to do likewise to achieve a similar possible increase in income. As can be seen from the example illustrated in Figure 4.6, a reduction in an individual's income from Y_2 to Y_1 results in that person's utility declining by $U_2 - U_1$, whereas the same increase in this individual's income from Y_2 to Y_3 brings about an increase in that person's utility of only $U_3 - U_2$. This is due to the strict concavity of the utility function. Furthermore, if the expected reduction in utility for a subjective probability distribution of reduced income is compared with the expected increase in utility for the same probability distribution of increased income, the absolute expected loss in utility will exceed the expected gain in utility. This adds further weight to the hypothesis that when individuals have to make a choice, they are likely to prefer conservation strategies that reduce the likelihood of income losses to those that promise the same probability of income increases. This hypothesis, however, requires empirical testing, for example, by the use of experimental methods.

The presence of the status quo effect (Kahneman and Tversky, 1979) tends to reinforce preferences for conservation measures intended to avoid income reduction compared to those offering similar prospects of increased income. Hysteresis (lack of reversibility) occurs in the utility relationship when the status quo effect is present. For example, in Figure 4.6, the utility curve HABC may apply for increases in income but once an income level is established, utility may not follow the same path if income is reduced. For instance, once an income level of Y_2 is established, utility may follow the path BFG if income is reduced, rather than BAH.

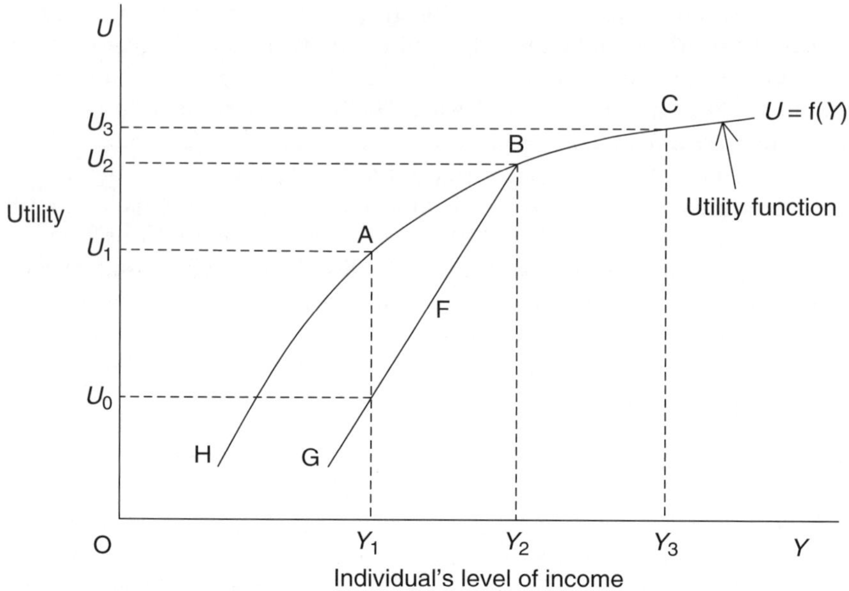

Figure 4.6 *An illustration of the proposition that conservation measures to avoid income reduction are likely to be preferred to conservation measures promising a similar increase in income*

This magnifies aversion to reductions in income. In this case, a reduction in income from Y_2 to Y_1 results in a reduction of utility of $U_2 - U_0$ rather than the one of $U_2 - U_1$. Bandara and Tisdell (2005) found evidence indicating that support for the conservation of the Asian elephant exhibits a status quo effect in relation to their abundance.

It is well known that even if there is a risk-neutral attitude towards economic returns, conserving a resource to keep options open (which would be lost otherwise) usually adds to economic benefits (Tisdell, 1970; 1971; Fisher et al., 1972; Arrow and Fisher, 1974). Nevertheless, even though economic value can normally be added by keeping options open, this does not necessarily imply that doing this is optimal. As Arrow and Fisher (1974, p. 319) state, 'the point is that the expected benefits of an irreversible decision should be adjusted to reflect the loss of options it entails'. They also mention on the same page: 'Just because an action is irreversible does not mean that it should not be undertaken. Rather the effect of irreversibility is to reduce the [comparative] benefits [of adopting an irreversible action].'

How much precaution should be taken in conserving resources depends upon one's attitude to risk-taking and the comparative costs and benefits of adopting a cautious approach compared to a less cautious one. In the case of the irreversible loss of biological resources, account should be taken of the cost of this due to its likely reduction of future economic options. Also additional flow-on effects of this loss need to be considered. Failure to take all these factors into account in decision-making can result in the irreversible loss of biological resources (and other natural resources) which should rationally be saved from extinction. However, the Precautionary Principle does not imply that irreversible loss of natural resources should always be avoided from an economics point of view.

4.5 PROPERTIES OF AGROECOSYSTEMS: CONWAY'S ANALYSIS

Taking into account their sustainability characteristics, Gordon Conway (1985; 1987) provided one of the earliest assessments of the properties of modern agroecosystems compared to traditional ones. In doing so, he considers their comparative economic ecological and social attributes. Although his analysis is subject to limitations, it provides a useful starting point for considering the sustainability and other properties of alternative agroecosystems, and similar systems (such as aquaculture systems), for the management or husbandry of domesticated organisms. A significant limitation of Conway's analysis is its failure to take account of the biodiversity loss resulting from the development of modern agroecosystems. This loss can imperil the sustainability of agricultural yields.

The following attributes of agroecosystems are taken into account in Conway's assessment of alternative agrosystems:

1. the level of crop yields or (in some interpretations) the level of economic returns obtained by farmers from crops;
2. the variability of the above;
3. the ecological sustainability of yields;
4. the consequences of an agroecosystem for income distribution of farmers.

Conway argues that modern agroecosystems provide higher yields and returns to farmers but that these are more variable than in the case of traditional agroecosystems. Furthermore, he hypothesizes that the yields of modern agroecosystems are less sustainable than those of traditional ones because they are more vulnerable to adverse ecological events which can permanently reduce their yields and in addition, their yields are more variable. Moreover, he claims that modern agroecosystems result in greater inequality of farm incomes than traditional agroecosystems and this can make them less socially acceptable than traditional ones.

On the whole, it does seem that modern varieties of crops and breeds of animals are capable of giving higher yields than traditional ones but the yields of modern varieties are very sensitive to environmental change. High yields are only obtained from modern varieties if their required complementary package of inputs is appropriately managed or controlled. In other words, most modern varieties of crops and breeds of domesticated animals show less biological tolerance to changed environmental conditions experienced by them than do traditional ones. This is illustrated in Figure 4.7.

Most modern agrosystems depend on a high proportion of inputs from sources outside the farm whereas traditional systems normally have a relatively low level of external inputs. The availability of external inputs helps farmers to manage the supply of inputs to their cultivated or husbanded organism so that the yields of modern varieties are higher and less variable

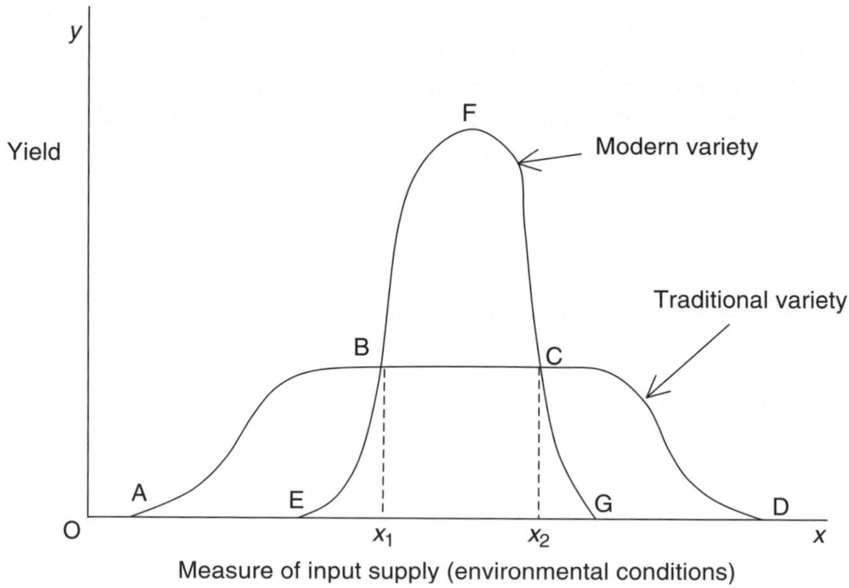

Figure 4.7 An illustration of the proposition that the yield of modern varieties of crops and breeds of domesticated animals usually show less tolerance to changed environmental conditions than do traditional varieties of crops and breeds of domesticated animals

than they otherwise would be in the absence of these external supplies. If this supply chain is not interrupted it is, in fact, possible for yields from modern varieties to be less variable than those from traditional varieties that are cultivated using a low external input system.[4]

The variability and the sustainability of yields of modern varieties are highly dependent on the nature of the supply chain. Furthermore, modern agroecosystems exhibit a high degree of dependence on the use of non-renewable natural resources, for example, fossil fuels and some chemical fertilizers. This and the nature of the supply chain must be taken into account in assessing the sustainability and the variability of returns from these systems.

Because of the high degree of dependence of modern agroecosystems on external inputs they are often not well suited to low-income countries. In such countries, many farmers may be unable to afford the purchase of the necessary complementary inputs and the supply chain of external inputs may be subject to impediments. For example, the road system and other infrastructures may be poor and the cost of bringing external inputs to the farm gate may be high.

Even though yields from modern varieties of crops and breeds of domesticated animals may be more vulnerable to external forces than traditional varieties of crops, they are not immune from these. For example, when a traditional variety has been developed in an environment devoid of pathogens occurring in another part of the world, it may succumb to these pathogens if they introduced to its environment, as happened in the case of the taro blight in Samoa (Hunter, 2000).

Further attention will be given to sustainability issues surrounding different ecosystems in the next chapter dealing with the conservation of crop varieties. For example, organic versus non-organic farming will be considered.

4.6 CONCLUDING COMMENTS

As was highlighted, sustainability is not in itself a desirable trait. Its desirability depends on what it is intended to sustain. As for the desirability of sustainable development, it was observed that a variety of views (some of which are incompatible) exist in the literature about what constitutes sustainable development. Furthermore, the application of some of the current criteria for sustainable development can lead to the irrational rejection of superior development strategies. For example, Paretian dominant strategies (in contravention of Rawls' Principle of Justice) can be rejected if the sustainability criterion is adopted that postulates that the well-being of every successive generation should not be less than that of its predecessor. Furthermore, it is not apparent that under the veil of ignorance, all would opt for the adoption of Rawls' principle. For example, as suggested, an alternative could be to adopt a safety-first approach to development.

Apart from suggesting normative criteria for sustainable economic development, there is a need: (1) to determine how socially acceptable these criteria are; (2) to determine the extent to which individuals are willing to take account of the well-being of future generations; and (3) to identify the actions they really adopt in an effort to sustain the welfare of future generations. Empirical studies of such relationships examining coefficients of concern for future generations are rare, to say the least. It was suggested that coefficients of concern fall rapidly (probably to zero after three future generations) and that they are heavily weighted in favour of individuals' own offspring. Even though each generation has some concern for a few generations succeeding it, this may not result in socially optimal development paths being followed.

Furthermore, it was pointed out that even when there is agreement about the objectives for sustainable economic development, opinions differ about what is required to achieve it. Some economists argue that the use of natural resources does not need to be limited (or at most, weakly so), whereas others believe that substantial conservation of these resources is needed in order to ensure that sustainable economic development is achieved.

The stock of biological resources bequeathed to future generations is important for the welfare of future generations. This applies both to the stock of natural (wild) germplasm and ecosystems as well as those developed by human effort and investment. Reductions in the stock can make it more difficult for future generations to sustain their welfare.

Uncertainty complicates planning for a sustainable economic future. The Precautionary Principle has been utilized by several authors to argue that, given the presence of uncertainty about the future, it is rational to err in favour of conserving existing biodiversity. Unfortunately, as shown by the examples given in this chapter, this is not always so. It depends on the structure of the decision-making problem. However, on the basis of the evidence provided in this chapter, the economic benefit of conserving biological resources to avoid a potential income reduction is stronger than doing that to obtain a comparable increase (or even somewhat greater increase) in income. In addition, the basis of Ciriacy-Wantrup's view that it is rational to conserve a minimum viable population of all species was questioned.

Human well-being today depends heavily not only on natural ecosystems but also on agroecosystems and similar human developed biotic systems. Both must be accounted for in discussing the sustainability of human well-being. Gordon Conway has stated that modern agroecosystems have some attributes that do not compare favourably with those of traditional agroecosystems. However, the situation is more complicated than envisaged by Conway. He does not, for example, identify the sources (such as loss of biodiversity, decreased fitness of new crop varieties or supply-chain factors) that might result in modern agricultural systems being less sustainable than traditional ones. Furthermore, if the off-farm supply chains of modern agroecosystems are not disrupted, the yields obtained using modern agroecosystems may be less variable than traditional systems. More attention will be given to the traits of alternative agroecosystems in the next chapter.

Note that while it has been emphasized in this chapter that biodiversity loss could lead to the impoverishment of future generations, ecological and environmental change is so rapid now that even the well-being of current generations can be undermined by such changes. During the average lifespan of humans, yields of new varieties of crops can become unsustainable because new pathogens evolve to which they succumb. This has occurred for some new varieties of barley because with the efflux of time, they have become increasingly subject to fungal infections. Or to give another example, some bacteria causing diseases (such as pneumonia) in humans have developed resistance to antibiotics in less than 50 years. Hence, the types of problems considered in this book are relevant to current generations as well as future ones.

The next part of this book (Part II) concentrates in more detail on biodiversity change associated with the socio-economics of variations in the stock of germplasm of domesticated organisms and alterations in the range of ecosystems developed by humans to use that germplasm. Part III deals mainly with socio-economic factors influencing the loss of wild biodiversity and selected policies to conserve wild biodiversity.

NOTES

1. Even if there are no political and market failures, myopia can result in unsustainable development. Correcting political and market failures does not ensure the achievement of sustainable development.
2. Rawls' assumption that all individuals who will be born could confer before birth and agree on the intergenerational distribution of achievable levels of well-being is extremely hypothetical.
3. Once again, the assumption that the unborn can confer and agree on anything is extremely hypothetical. However, individuals alive now might decide on behalf of the unborn. Given the presence of extreme uncertainty about when an unborn person might be born, it could be assumed that each unborn person (who is going to be born) is equally likely to be born at any time in the future. In this case, expected well-being of each person

who is going to be born could be calculated by relying on the Laplace criterion, also known as the principle of insufficient reason (Tisdell, 1968). Nevertheless, many uncertainties remain which limit this approach to determining the optimal intergenerational allocation of available resources.

4. There is some further discussion of Conway's model and the sustainability of modern agroecosystems in Tisdell (2015).

REFERENCES

Anon (2014), '*Phytophthera infestans*', *Wikipedia* [Online]. Available at http://en.wikipedia.org/wiki/Potato_blight. (Accessed 10 June, 2014).

Arrow, K.J. and A.C. Fisher (1974), 'Environmental preservation, uncertainty and irreversibility', *Quarterly Journal of Economics*, **88**, 312–19.

Bandara, R. and C. Tisdell (2005), 'Changing abundance of elephants and willingness to pay for their conservation', *Journal of Environmental Management*, **76**, 47–59.

Barbier, E.B. (1987), 'The concept of sustainable economic development', *Environmental Conservation*, **14**, 101–10.

Bishop, R.C. (1978), 'Endangered species, irreversibility and uncertainty: the economics of a safe minimum standard', *American Journal of Agricultural Economics*, **60**, 10–18.

Bishop, R.C. (1979), 'Endangered species, irreversibility and uncertainty: a reply', *American Journal of Agricultural Economics*, **51**, 376–79.

Ciriacy-Wantrup, S.V. (1968), *Resource Conservation: Economics and Policies*, 3rd edn, Berkely, CA: Division of Agricultural Science, University of California.

Conway, G.R. (1985), 'Agroecosystem analysis', *Agricultural Administration*, **20**, 31–55.

Conway, G.R. (1987), 'The properties of agroecosystems', *Agricultural Systems*, **24**, 95–117.

Daly, H.E. (1980), *Economics, Ecology, Ethics*, San Francisco: Freeman.

Daly, H.E. (1999), *Ecological Economics and the Ecology of Economics: Essays in Criticism*, Cheltenham, UK and Northampton, MA, USA: Edward Elgar Publishing.

Dawkins, R. (1989), *The Selfish Gene*, Oxford and New York: Oxford University Press.

Engels, F. (1959), 'Outlines of a critique of political economy', in K. Marx (ed.), *Economic and Philosophic Manuscripts of 1844*, Moscow: Foreign Languages Publishing House.

Fisher, A.C., J.V. Krutilla and C.J. Cicchetti (1972), 'The economics of environmental preservation: a theoretical and empirical analysis', *American Economic Review*, **62**, 605–19.

Hohl, A. and C.A. Tisdell (1993), 'How useful are environmental safety standards in economics? The example of safe minimum standards for protection of species', *Biodiversity and Conservation*, **2**, 168–81.

Hunter, D. (2000), 'Taro leaf blight in Samoa – tackling the problem as partners', [Online]. Available at http://aid.dfat.gov.au/publications/focus/0700/taroleaf.pdf. (Accessed 10 June, 2014).

Kahneman, D. and A. Tversky (1979), 'Prospect theory: an analysis of decision under risk', *Econometrica*, **47**, 263–92.

Keynes, J.M. (1931), *Essays in Persuasion*, London: Macmillan.
Krutilla, J.V. (1967), 'Conservation reconsidered', *American Economic Review*, **57**, 777–86.
Margolis, J. (1957), 'Secondary benefits, external economies, and justification of public investment', *The Review of Economics and Statistics*, **39**, 284–91. Reprinted with corrections in K.J. Arrow and T. Scitovsky (eds) (1969), *Readings in Welfare Economics*, London: George Allen and Unwin.
Millennium Ecosystem Assessment (2003), *Ecosystems and Human Well-being: A Framework for Assessment*, Washington, DC: Island Press.
Millennium Ecosystem Assessment (2005), *Ecosystems and Human Well-being: A Synthesis*, Washington, DC: World Resources Institute.
Ninan, K.N. and M. Inoue (2013), 'Valuing forest ecosystem services: What we know and what we don't', *Ecological Economics*, **93**, 137–49.
Nordhaus, W. (1974), 'Resources as a constraint or growth', *American Economic Review: Papers and Proceedings*, **64**, 22–6.
Pearce, D.W. (1998), *Economics and Environment: Essays on Ecological Economics and Sustainable Development*, Cheltenham, UK and Northampton, MA, USA: Edward Elgar Publishing.
Rawls, J.R. (1971), *A Theory of Justice*, Cambridge, MA: Harvard University Press.
Singleton, I., S.A. Wich, S.J. Husson, S.S.U. Atmoko, M. Leighton, N. Rosen, K. Traylor-Holzer, R. Lacy and O. Byers (2004), *Orangutan Population and Habitat Viability Assessment: Final Report*, Apple Valley, MN: IUCN/SSC Conservation Breeding Specialist Group.
Stern, N. (2007), *The Economics of Climate Change: The Stern Report*, Cambridge: Cambridge University Press.
Swarna Nantha, H. and C.A. Tisdell (2009), 'The orangutan-oil palm conflict: economic constraints and opportunities for conservation', *Biodiversity Conservation*, **18**, 487–502.
Tietenberg, T. (2003), *Environmental and Natural Resource Economics*, 6th edn, Boston: Addison Wesley.
Tisdell, C.A. (1968), *The Theory of Price Uncertainty, Production and Profit*, Princeton, NJ: Princeton University Press.
Tisdell, C.A. (1970), 'Implications of learning for economic planning', *Economic Planning*, **10** (3), 177–92. Reprinted in C.A. Tisdell (1996), *Bounded Rationality and Economic Evolution*, Cheltenham, UK and Brookfield, USA: Edward Elgar Publishing.
Tisdell, C.A. (1971), 'Economic policy, forecasting and flexibility', *Weltwirtschaftliches Archiv*, **106** (1), 34–54.
Tisdell, C.A. (1990), 'Economics and the debate about preservation of species, crop varieties and genetic diversity', *Ecological Economics*, **2**, 77–90.
Tisdell, C.A. (1999), *Biodiversity, Conservation and Sustainable Development*, Cheltenham, UK and Northampton, MA, USA: Edward Elgar Publishing.
Tisdell, C.A. (2005), *Economics of Environmental Conservation*, 2nd edn, Cheltenham, UK and Northampton, MA, USA: Edward Elgar Publishing.
Tisdell, C.A. (2009), *Resource and Environmental Economics: Modern Issues and Applications*, Singapore, Hackensack (New Jersey), London, Beijing: World Scientific.
Tisdell, C.A. (2010), 'The precautionary principle revisited: its interpretations and their conservation consequences', *Singapore Economic Review*, **55**, 335–52.

Tisdell, C.A. (2013), *Competition, Diversity and Economic Performance*, Cheltenham, UK and Northampton, MA, USA: Edward Elgar Publishing.
Tisdell, C.A. (2014), *Human Values and Biodiversity Conservation: The Survival of Wild Species*, Cheltenham, UK and Northampton, MA, USA: Edward Elgar Publishing.
Tisdell, C.A. (2015), 'Sustainable agricultural development and the environment: new perspectives and policy proposals – an assessment', in J.A. Daniels (ed.), *Advances in Environmental Research*, **40**, New York: Nova Science (in press).
WCED World Commission on Environment and Development (1987), *Our Common Future*, Oxford: Oxford University Press.

PART II

Human developed biological capital: germplasm and ecosystems

5. Crops and the socio-economics of biodiversity loss and change

5.1 INTRODUCTION

The development of agriculture and other bio-industries (such as aquaculture) has been, and continues to be, a major source of environmental change. Their development has resulted in additions to and large losses in the genetic stock. Moreover, natural ecosystems have been lost and transformed by cultivation systems supporting these industries.

After agriculture began to develop early in the Holocene era (about 9000 years ago),[1] it eventually produced more varied and greater quantities of food and other primary products. Increasing agricultural production has enabled the globe to sustain a massive increase in its human population which is expected to reach around 9.5 billion by 2060 (United Nations Department of Economic and Social Affairs, 2012).[2] Despite its past achievement, there are increasing concerns about whether agriculture and other bio-industries will meet the needs of this increased population mainly because of their environmental consequences and constraints. Undesirable consequences of agricultural expansion can include increasing genetic loss, rising CO_2 levels and, in many cases, unwanted changes to natural or near natural ecosystems. These combined changes can result in human welfare being lower than it need be, and can reduce prospects for continuing economic development.

The global stock of germplasm is continually changing. Changes in this stock have significant economic consequences and to a considerable and increasing extent are a product of socio-economic processes. In the next section of this chapter, important socio-economic factors that result in a loss of existing germplasm are identified and discussed. To some extent, these factors also influence the development of new germplasm by humans. This section extends the previous discussion in Chapter 3 about genetic changes specifically to cultivated food crops.

Decisions by individuals and organizations about resource use and conservation are partly a product of the economic and social system in which they are embedded. Therefore, in section 5.3, consideration is given to biodiversity loss caused by 'lock-in' to economic growth due mostly to macroeconomic factors. Structural features of market economies make it imperative that societies strive for economic growth, even though this may eventually prove to be unsustainable.

The successful cultivation of crops requires that agrosystems be developed that help crops thrive. However, the successful development of these systems can accelerate biodiversity loss. This aspect is considered briefly in this chapter because the main discussion of human modified or developed ecosystems is in Chapter 8. The discussion in this chapter is limited to the potential of intensification of crop production to reduce biodiversity loss.

A significant policy problem in biodiversity conservation is to decide what species and varieties of crops should have priority for their preservation and to determine the best (most economic) method of achieving their conservation, for example, by relying on seed banks or *in situ* options. The economics involved in these choices is discussed. Two short case studies follow before this chapter is concluded. One deals with the loss of food crop varieties in the Pacific Islands and the other briefly discusses the late commencement of the cultivation of Australian native plants for food.

5.2 SOCIO-ECONOMIC INFLUENCES ON THE LOSS OF EXISTING FOOD CROP GERMPLASM

Continuous socio-economic change results in alterations in the existing stock of germplasm, including that of food crops. Changes in market demand, in supply conditions and in economic frameworks are major drivers of losses in existing crop germplasm. This will now be substantiated. Note, however, that these influences often display interdependence. For example, in response to increased demand for a particular attribute in a crop, new varieties may be developed which prominently display this characteristic, thereby altering the nature of supply.

Changes in Demand and Genetic Loss

As a result of changes in the demand for different attributes of crops, some crop varieties may disappear. Shifts in demand for attributes of food supply may become evident at different points in the product supply chain, even though it is usually contended that consumers are sovereign, and therefore, changes in demand at any stage of the supply chain reflect alterations in the demands of end-purchasers, that is, derived demand. While this is basically so, it is rather simplistic. It can, for example, result in neglect of the role of institutional evolution (for instance, in marketing)

Figure 5.1 An example of one of many possible food supply chains: the number of intermediaries involved in these chains appears to increase as economies develop

in altering demand for different types of food. For example, as discussed below, the process of replacing independent retailers selling food in their shops by large supermarket chains alters demand for the attributes of food supplies. It can, however, be claimed that these institutional changes reflect alterations in what consumers want.

Food supply chains can be quite varied. As a result of increasing economic specialization and the extension of market exchange (both of which are consequences of economic development), food supply chains have become more varied and 'longer'. In higher income countries, there is little dependence on self-produced food and there is reduced use of local (municipal) markets selling fresh food to consumers. Dependence on processed food or semi-processed food supplied by supermarkets has increased. Therefore, the type of product supply chain shown in Figure 5.1 has become common. Evolution in food supply chains has altered the pattern of demand for different varieties of food crops, and in turn, this has resulted in the loss of some crop varieties and has favoured the conservation and development of other varieties for reasons which will now be outlined.

The socio-economic structure of communities alters as economies develop and this in turn changes the demand of consumers for different types of food and the places where it is purchased. With economic development, the demand for convenience foods which need little or no preparation by individuals prior to consumption has increased and this has favoured the patronage of supermarkets rather than purchases in municipal (local) markets.

This changed pattern of demand for food reflects growing urbanization and new work patterns. With economic development, there is greater participation of females in the workforce and they have less time for preparation of food. All adult members of a household are increasingly involved in working away from home. In these circumstances, the purchase of fresh unprocessed food from local markets becomes less attractive than the convenient purchase of mostly semi-processed or processed food from supermarkets. Therefore, the nature of the food supply chain alters. In turn, this alters the demand for different varieties of crops and results in fewer varieties being conserved.

It is widely acknowledged that supermarkets prefer to stock items that are uniform in nature or quality (see, for example, Goldman et al., 2002; Reardon et al., 2003). Basically, operators of supermarkets prefer to stock standardized products suitable for mass marketing. When they have multiple retail outlets (which is common), they also prefer to carry products that can be supplied in sufficiently large quantity to stock all these outlets, or at least, all those in a particular region. A preference may also exist for items that can be supplied for a longer length of time. This facilitates the achievement of economies of scale in promoting these products and their management in-house. Because a reduced set of varieties of individual food crops is likely to be demanded by supermarkets to achieve standardization and ensure sufficient supply of each variety, this is likely to reduce the demand for some varieties of food crops. Furthermore, those processing food for supply to supermarkets are likely to be under pressure to supply uniform products, and once again this is likely to favour some varieties of food crops at the expense of others.

It is also probable that many consumers prefer the supply of food products with standard attributes. This reduces the problem of checking on the attributes of the product. In some cases, it may also result in lower prices for the product than otherwise and an increase in consumer welfare (Tisdell, 2013, pp. 227–9).

Now consider a different influence on loss of biodiversity; namely an increase in the demand for a food crop generated by an expansion in the number of its end-uses. Examples include the use of grains and sugar-containing crops for producing biofuels and the growing demand for grains and soybeans to feed livestock.[3] Two possible consequences for biodiversity loss occur:

- Sometimes, the increase in the number of end-uses for a crop results in the extension of agriculture onto natural or near-natural areas with a loss of wild biodiversity.
- Other crops may be displaced by a crop for which the number of end-uses increases. This can result in a reduction in the number of varieties of the displaced crops.

In addition, unwelcome changes in social welfare can occur. For example, poor consumers may be forced to pay more to meet their basic food needs.

The potential effects of a proliferation of the end-uses of a food crop can be illustrated by Figure 5.2. The increased demand for grains to produce biofuel and livestock is taken as an example. Assume that the normal demand for grains for feeding humans is as indicated by the line marked D_1D_1 and that the market supply curve is as identified by line SS.

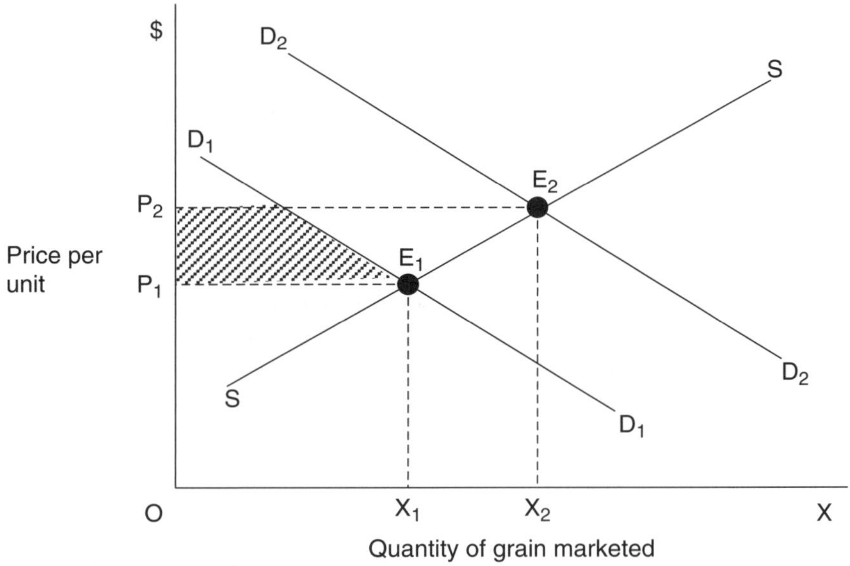

Figure 5.2 An example of the market consequences of increasing the end-uses of grain, for example, for biofuels and for feeding livestock: the price of grain rises, thereby reducing the surplus of consumers who purchase grain to eat, and biodiversity loss is likely to increase

The equilibrium price of grain is then P_1 and the quantity supplied is X_1. Now suppose that demand emerges for grain for supplying biofuels and for feeding livestock, other things remaining constant. The demand curve now shifts up to the line marked D_2D_2. The equilibrium price of grain rises to P_2 and the quantity supplied increases to X_2.

It can be seen from Figure 5.2 that the surplus of those consuming grain for food declines as the demand for grain is pushed up by increasing its end-uses. In the case illustrated, this reduction in consumer surplus is equal to the hatched area.[4] The increased supply of grain may be met by:

- an increase in grain yields on land already used for that purpose (intensification);
- displacement of other crops, that is, extension of grain-growing to land used for other crops;
- extension of grain growing to land not used for crops such as grazing land or onto natural or near natural areas; or
- a combination of all these responses.

Both the second and third responses are likely to reduce biodiversity. The second may reduce existing crop diversity and may also decrease wild biodiversity (see section 5.4). The third outcome may diminish wild biodiversity and/or livestock diversity. The fourth response can have all these effects. However, even the first response mentioned above can reduce wild biodiversity.

Supply-side Influences

The availability of new species of food crops and new varieties of existing food crops can result in displacement of existing ones if the new ones are preferred by consumers, or if they have economic advantages in cultivation and marketing. For example, hybrid taro varieties are replacing traditional varieties in some parts of the Pacific Islands because of their greater marketability and superior growing qualities. In parts of Africa and Asia, the consumption of millet (Anon, 2014a) and sorghum (Anon, 2014b) per head of population (FAO, 1995) has declined. A contributor has been the availability of corn (maize), the seed of which was introduced from the Americas. Many consumers prefer corn to millet and sorghum, even though the latter crops are more resistant to drought than corn (Anon, 2014a; 2014b).

Improved transport also results in regional populations not being so dependent on locally grown cereals. In China, for example, rice, wheat and other grains are more easily transported now than in early times to semi-arid regions where millet used to be a staple food crop, and their consumption has (to some extent) been substituted for the consumption of millet. Local farmers can then specialize in economic activities other than the growing of millet. As discussed below, the extension of trade can result in loss of genetic diversity of both cultivated and wild organisms.

Crops and varieties of crops preferred for cultivation alter as available cultivation methods and techniques change and the supply of inputs for their cultivation alters. However, the latter two factors are often interdependent. For example, the availability of irrigation water, chemical fertilizers and pesticides favour the growing of high yielding varieties (HYVs) of crops, at the expense of traditional varieties. On the other hand, the development of new HYVs may provide an incentive for adding to existing irrigation facilities. In addition, distribution systems may be improved to facilitate the supply of chemical fertilizers and pesticides to areas where the growing of HYVs becomes common.

Not only growing methods but harvesting methods (and supply requirements further down the food chain) influence the varieties of crop grown. For example, if mechanical harvesting is in use, crop varieties that are less susceptible to being damaged by this are likely to be favoured at the expense of those that are more prone to this damage. Nevertheless, economic trade-offs are often required. For instance, the taste of a variety of a tomato suitable for mechanical harvesting may be inferior to a variety that is unsuitable for this purpose. Despite this, the tomato variety with inferior taste may be supplied to the market because, as a result of mechanical harvesting, it can be supplied to the market at a sufficiently lower price to make it more attractive to consumers than a tomato variety with superior taste. The latter may have to be picked by hand at increased cost or, if mechanically harvested, the proportion of damaged tomatoes may be high. Both factors can increase the comparative cost of growing a flavoursome tomato. An additional consideration in the choice of a crop variety is how well it transports to end-users and how long it can be stored. Geographical lengthening of supply chains increases the importance of such considerations.

The costs of cultivating crops influences whether their polyculture or monoculture is practised. In more developed countries, relative costs favour monoculture because agricultural labour is scarcer in higher income countries than in nations with low incomes. The growing of a single variety of crop in a field enables it to benefit from cultivation methods specifically tailored to its requirements. Machines, fertilizer and pesticides can be adapted to the specific requirements of a monocultured crop. This is important when cultivation methods involving low intensity of labour must (for economic reasons) be substituted for those requiring a high intensity of labour. This occurs as economies experience economic growth and marginal agricultural labourers are attracted to other sectors of the economy (Clark, 1957). This trend in favour of monoculture (as economies develop) can contribute to loss of crop varieties. Increased mechanization of agriculture and reduced agricultural labour-intensity has other consequences. For instance, farms tend to increase in size to obtain economies of scale (for example, from their considerable investments in capital), and also specialization in production of agricultural products increases. Increased specialization contributes to loss of crop varieties.

Market Extension including Increased International Trade

Market extension involves changes both in demand and supply. The geographical extension of markets promotes growing geographical specialization in production. Regions tend to specialize in producing different products depending on their comparative economic advantage in doing this. As David Ricardo (1817) pointed out originally, this specialization can benefit all who participate in the extension of trade. However, this view also needs some qualifications, especially in dynamic situations. As well, account needs to be taken of externalities and the impacts on income distribution of the extension of trade. Furthermore, the consequences of the extension of trade for biodiversity conservation need consideration. Usually, this extension results in a reduction in the existing genetic stock, in particular that of domesticated organisms.

Because the extension of trade results in increased specialization in production, the production of some agricultural products in some regions is likely to cease. If their production is based on germplasm unique to these regions, this germplasm is lost. Consequently, there is a reduction in the existing global stock of human 'developed' germplasm.

Market extension probably also facilitates the loss of wild biodiversity. The situation in that regard is complex. Countries that specialize in non-agricultural production and import agricultural produce are able, to some extent, to conserve their wild biodiversity. On the other hand, nations from which they import agricultural produce may extend their frontiers of agriculture. This can result in considerable loss of wild biodiversity in nations specializing in the export of agricultural produce. For example China, by specializing in manufacturing and by importing agricultural produce and timber from countries such as Brazil, can increase its standard of living and conserve a larger natural area than otherwise. On the other hand, there is an incentive for Brazil to expand the area it uses for agricultural and timber production (for instance in the Amazon), thereby resulting in considerable loss of wild biodiversity.

Extension of trade in inputs used in agriculture results in farmers being less dependent on local environmental conditions when growing crops. Off-farm inputs can be used to compensate (to some extent) for shortcomings in on-farm inputs. Consequently, it may become less profitable to grow varieties of crops especially adapted to local environmental conditions. These varieties are likely to disappear.

The process of increased globalization is a wider concept than that of increased trade, although they usually go hand in hand. The overall impact of increased globalization appears to be to reduce global biodiversity (see Chapters 3 and 6 of this book).

Table 5.1 summarizes those crop-related economic factors discussed in this chapter which alter and usually reduce the global stock of existing germplasm. They are organized according to whether they are demand-related, supply-related or a consequence of the extension of trade. While it may appear from this table that demand and supply factors and the extension of trade exert independent influences on the loss of crop germplasm, in practice this is not the case.

Table 5.1 A list of crop-related economic factors affecting the loss of existing plant germplasm and relevant observations

Economic factors	Observations
Demand factors	
Changed consumer demand for crop attributes	Influenced by changed work patterns, urbanization and new technologies in homes.
Changed demand by market intermediaries	For example, the demand of supermarkets for standardized products that store well and look attractive.
Extension of the number of end-uses of crops	For example, the use of crops to produce biofuels or feed livestock as well as feeding humans.
Development of supermarkets and fast food outlets	Increased demand for standardized products supplied reliably in volume. Many modern food processors have similar requirements.
Supply factors	
Availability of new species of crops or varieties	These may have attributes preferred by consumers (e.g. maize versus sorghum) or to reduce the cost of supply, e.g. high-yielding varieties. They displace existing varieties of crops.
Increased agricultural labour costs	Associated with greater mechanization, capital investment in agriculture and larger sized farms. All tend to reduce existing biodiversity (see text).
New techniques for growing, harvesting and processing crops	Results in the selection of crops and variety of crops compatible with the use of these techniques.
Extension of trade	
Increased availability of off-farm inputs	Less dependence on local ecological conditions for growing crops. Reduced need to grow varieties adapted to local ecological conditions.
Increased specialization in production	Loss of crop varieties peculiar to some regions because these crops are no longer grown in these.
Geographically longer food chains	Increased demand for crops that transport and store well.

5.3 ECONOMIC GROWTH, MACROECONOMIC DICTATES AND BIODIVERSITY LOSS

Economic growth usually results in increasing loss of existing biodiversity and in general it is associated with an increase in demand for biological products, including agricultural products. The increased supply of agricultural products may be met by the intensification of agriculture or its extension, or both. In practice, historically both these responses have occurred.[5] Furthermore, when economic growth results in rising per capita incomes, the composition of demand for agricultural products usually alters. For example, there is a tendency for demand for higher protein food (especially for animal protein) to rise relative to agricultural products that are lower in protein content. This shift in the composition of demand is environmentally relevant and is a driver of agricultural extension and intensification.

Growth in GDP depends upon increases in population levels and rises in GDP (income) per head. Both can be expected to add to the demand for food. However, the demand for food is not only influenced by income per head but also depends on its distribution. It is, for example, expected that by around 2060, the world's population will increase by over 30 per cent compared to its level in 2013. If the average level of income were to remain unchanged and its distribution unaltered, then a 30 per cent increase in the demand for food can be expected, all other things being held constant. However, the increased demand for food will be even greater if per capita incomes continue to increase. Furthermore, if the percentage of the world's population on higher incomes increases, the demand for animal protein, for example, beef, is likely to rise substantially.[6] If this demand is met by feeding grains or other food suitable for human consumption to animals, the type of consequences illustrated by Figure 5.2 are likely to occur, and the type of biodiversity losses identified in the previous section can be expected. These losses are facilitated by the extension of international trade.

Despite it being generally agreed that the demand for food rises less than proportionately with increased income, the demand for food and agricultural produce does tend to increase with income. This contributes to loss of existing biodiversity. The pressures for increasing incomes are very strong in modern societies. This is so at the individual level and for macroeconomic reasons.

A macroeconomic problem in market economies is that the continual introduction of labour-saving technology needs to be offset by rising aggregate demand to ensure employment of the labour force at its existing level (or to employ an increasing labour force), particularly if wages are inflexible downwards. Conservation challenges are magnified because continual economic growth in market economies appears to be necessary for maintaining aggregate employment, and even faster economic growth is required to boost aggregate employment.[7] However, apart from this, economic growth is welcomed by most participants in market economies as a means for increasing their income levels. As a result, environmental conservation, including biodiversity conservation, is liable to attract limited political support. This may be especially so when the adverse environmental consequences of economic growth only occur (or become obvious) in the long term or when they are not very visible, as seems to be the case when the existing genetic stock is reduced.

5.4 OBSERVATIONS ON LAND-SAVING (LAND-SPARING) AGROECOSYSTEMS AND BIODIVERSITY LOSS

In section 5.3, it was noted that changed agricultural systems frequently result in loss of crop varieties. As was shown, this can happen, for example, as a result of the adoption of more capital-intensive agricultural techniques (in response to rising costs of agricultural labour) and the development of techniques specific to the cultivation and harvesting of particular crops. It is not intended here to discuss specific agroecosystems and their consequences for biodiversity loss because this is covered in Chapter 8. Instead attention will be given to whether agricultural techniques that increase yield per hectare (and, therefore, are said to be land-saving or land-sparing) are likely to result in reduced biodiversity loss compared to land sharing.

There is significant controversy among members of the scientific community about whether crop intensification (that is, increasing yields per hectare) is a favourable way of reducing (or eliminating) biodiversity loss when total agricultural production needs to be increased. One group of scientists (for example, Phalan et al., 2011a; 2011b) argue that it is favourable because it reduces the encroachment of agriculture onto natural habitats. On the other hand, scientists such as Tscharntke et al. (2012) are critical of this land-sparing approach as a way to reduce biodiversity loss. However, the situation is quite complex because biodiversity loss can be caused both by agricultural extension and by its intensification. The set of biological material lost (species or varieties of these) usually differs depending upon whether greater agricultural output is obtained by intensification or extension or a combination of both. Furthermore, differences can occur in the *nature* of intensification and extension of agriculture which are not adequately accounted for by simple dichotomous models of the biological and environmental consequences of these processes.

Substantial evidence exists of loss of biodiversity caused by agricultural intensification, and factors contributing to this have been identified (Bentham et al., 2003; Donald et al., 2006; Firbank et al., 2008). Most of these studies focus on loss of wild biodiversity as a result of agricultural intensification. However, agricultural intensification also results in loss of domesticated genetic material and changes in agroecosystems.

The following are some of the consequences of crop intensification for biodiversity loss:

- Crop intensification can be expected to favour crop varieties (HYVs) that are responsive in their returns to careful management and the use of off-farm inputs (such as chemical fertilizers and irrigation water). Consequently, less responsive crop varieties (traditional varieties?) tend to disappear.
- Often crop intensification is associated with increased mechanization and higher overhead costs on farms. These favour economies of scale. This tends to result in larger farms, larger-sized fields and removal of natural obstacles to enable the use of machinery in cropland. This can contribute to the loss both of some wildlife species and of crop varieties (Bentham et al., 2003; Tscharntke et al., 2012). These processes are believed to result in greater homogeneity of agricultural landscapes, thereby eliminating some habitats on which particular species of wildlife depend for their existence.
- The off-farm environmental externalities generated by increased agricultural intensification can have a negative impact on some wild species. For example, it may add to the concentration of nutrients (for example, nitrogen and phosphorus) in run-off water in agricultural areas selected for intensification. This can result in eutrophication of waterways receiving these nutrients and loss of species adversely affected by eutrophication. Biodiversity loss in some marine areas may also occur, for instance, loss of coral systems. In some areas, salting of soil and water can occur with negative effects on some wild species (for example, species of fish, amphibians and tortoises unable to survive in salty streams) and on the productivity of agricultural land.
- Economic activities designed to supply off-farm inputs to farms undergoing intensification can negatively impact on wild biodiversity. For example, irrigation dams can flood valuable wildlife habitat and alter water flows in streams and rivers, with negative consequences for the survival of wildlife. For example, some wild species rely on periodic flooding to initiate their breeding cycle and flooding may become less frequent or non-existent as a result of the building of dams. Several species of water birds are affected as well as other biota, such as red river gums in Australia. In addition, migrating aquatic animals such as salmon and eels may find that the presence of dams and weirs prevents their upstream movement and they are unable to complete their breeding cycle.

- Reduced availability of water in water bodies as a result of its diversion for use in increasing agriculture yields can have negative consequences for water-dependent wildlife species.

From the above, it is clear that agricultural intensification is a contributor to biodiversity loss. This also occurs as the result of agricultural extension. Therefore, if agricultural production is to be increased by either greater intensification or by extension (or a combination of both), it is necessary to choose between the alternative sets of biodiversity which are likely to be lost or conserved. A trade-off problem is involved and valuation of alternatives is needed to resolve it.

Economic factors will in any case determine the extent to which agricultural extension and intensification will occur if the demand for agricultural produce increases. In practice, both intensification and global spatial extension of agriculture are likely. However, because of market failures, this is unlikely to result in a socially optimal combination of intensification and extension, as is made clear in Chapter 8 of this book.

It should, nevertheless, be noted that increasing agricultural production is not the only available option for increasing food supply for humans. Tscharntke et al. (2012, p. 55) point out that other options are to reduce post-harvest food losses and waste, and to divert some food for consumption by humans which is currently used to feed livestock or to produce biofuel. Losses of harvested food intended for human consumption are high and amount to about one-third globally. Gustavsson et al. (2011) estimate that in developed countries about 40 per cent of food is lost (wasted) by retailers and consumers. Adopting these policy possibilities would increase the amount of food available to feed an increasing global population and would reduce the extent to which agriculture needs to be intensified or extended to increase food supplies and thereby diminish the loss of existing biodiversity. However, there is no guarantee that such policies to reduce these losses will be adopted. For example, consumers on higher incomes may not be supportive of policies designed to reduce the consumption of animal products, and they will undoubtedly be opposed by producers of these products.

5.5 PUBLIC POLICY AND ECONOMIC PRIORITIES FOR CONSERVING THE GERMPLASM OF CROPS

Public policies for conserving different types of crop germplasm should take into account the economic value of doing this. However, as is clear from section 4.4, this is not straightforward because of the presence of uncertainty about the future value of different types of germplasm and differing attitudes to the bearing of risks. Nevertheless, it is still important to explore the potential economic consequences of the loss of different types of germplasm because they are relevant to the conservation decisions which need to be made. In terms of game-against-nature scenarios, these potential consequences determine the pay-off values in the relevant matrices.

The economic costs and benefits of conserving different types of germplasm need to be considered. In this section, the focus is on the consideration of economic benefits as a guide to prioritizing crop germplasm for conservation. Some attention will be given to costs in the next section when economic aspects of the creation of gene banks are considered. Unfortunately, existing knowledge about the economics of alternative methods of conserving crop germplasm is very limited. Three aspects of the potential economic benefits of conserving different types of food crop germplasm will now be explored, namely (1) the implications of the nature of demand for the type of food supplied by different crops; (2) the extent to which economic livelihoods depend on particular crops; and (3) the consequences of relatively predictable changes in environmental conditions, such as some features of climate change.

The Nature of Demand and the Potential Economic Benefits of Conserving Crop Germplasm

The potential benefits to consumers of conserving the germplasm of a food crop depend upon the magnitude of the demand for it and how inelastic the demand for it is. Other things remaining constant, the potential loss in consumer surplus due to a fall in the yield of a crop (because, for example, its germplasm declines in biological fitness) tends to be larger the greater is the demand for the food produced by this crop and the more inelastic is that demand. The latter implies that consumers have few substitutes for the food produced by the crop, and the former indicates that it is a substantial item in the food budget of consumers. In some regions, staple crops appear to have these properties. The results suggest that priority should be given to conserving germplasm of crops for which demand has the above-mentioned attributes. Conserving this germplasm is likely to provide some insurance against a significant decline in the productivity of such crops due, for example, to the occurrence of a new disease or heat stress caused by climate change.

Figure 5.3 illustrates the above-mentioned theory. Suppose that the costs of supplying an agricultural commodity, X, for example, wheat, increases

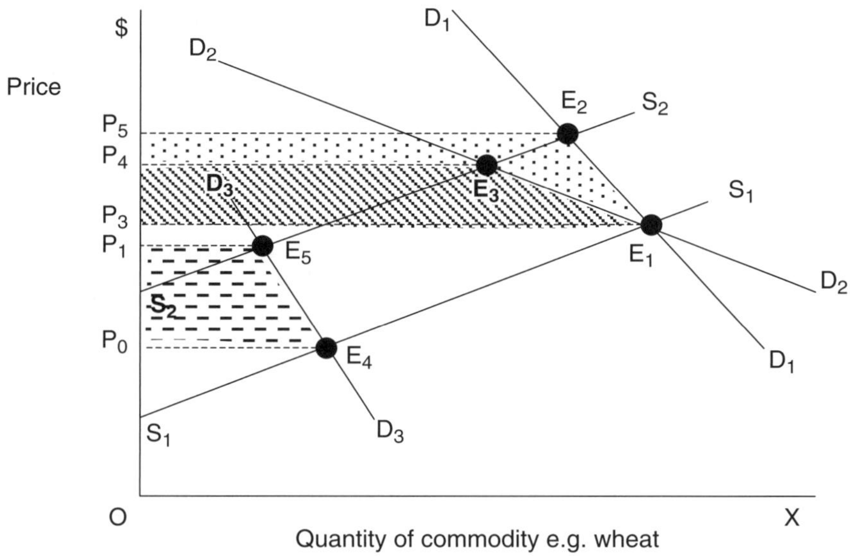

Figure 5.3 *An illustration of the proposition that consumers are likely (other things held constant) to benefit most from the conservation of the stock of germplasm of crops that produce food for which demand is high and relatively inelastic*

due to a reduction in the biological fitness of the varieties used to grow it and that this results in the industry supply curve shifting from the one marked S_1S_1 to S_2S_2. If the demand for X is as shown by the line marked D_1D_1, market equilibrium shifts from E_1 to E_2 and the market price of X rises from P_3 to P_5. The loss in consumer surplus is equivalent to the dotted plus the hatched area shown. If the demand curve is more elastic (for example, D_2D_2 rather than D_1D_1), the loss in consumer surplus is smaller, other things held constant. For example, when the demand curve is D_2D_2, it is equal to the hatched area. If the demand curve is very elastic, the loss in consumer surplus will be very small. In the perfectly elastic case, it is zero.

When the quantity demanded of the commodity is relatively small, the loss in consumer surplus as a consequence of the reduced productivity of the crop is also relatively small, assuming a declining demand curve. This is illustrated in Figure 5.3. The demand line D_3D_3 has the same slope as D_1D_1. In this case a shift upwards in the supply curve from S_1S_1 to S_2S_2 shifts the market equilibrium from E_4 to E_5 and consumer surplus declines by the flecked area. This involves a much smaller loss in consumer surplus than when the demand is at the higher level, D_1D_1. It follows, other things being held constant, that the more inelastic and the higher is the demand for the food produced by a crop, the greater is likely to be the benefit to consumers of conserving the stock of the crop's germplasm. Some staple crops may have these attributes. Nevertheless, additional factors to potential changes on consumer surplus need to be considered by governments in establishing economic priorities for conserving germplasm. For example, the extent to which a social group or the country is dependent on a particular crop for its income can be important.

Livelihood and Income Dependence

If there is a social group which depends heavily on a particular crop for its survival or livelihood (and which lacks alternative means of earning its sustenance), as is the case in some less developed countries, sustaining this group's source of livelihood may be an important goal. Furthermore, where a country depends heavily on supply of a particular crop to earn its export income (and if other alternative means of earning income would significantly reduce its income), sustaining the production of this crop can be an important goal. In both these cases, eco-geographic conditions may restrict the possibilities of crop substitution, for example in the Sahel. In the first case, another constraint can be restricted scope for individuals to earn income by moving to a new location in the economy. This is not unusual in many developing economies and can make it difficult to eradicate poverty in some regional areas.

In the above cases, maintaining genetic diversity in the crops grown may contribute to the maintenance of incomes. The problem of sustainability occurs in these cases because of lack of possibilities for substitution on the supply side of the economy.

'Predictable' Environmental or Ecological Changes, for example, Climate Change

Environmental and ecological conditions are subject to change. While these changes are often not perfectly predictable, general tendencies and probabilities can sometimes be determined given the available (scientific) evidence or knowledge.[8] The precision of these predictions varies according to the nature of the phenomena involved. However, even when they are not precise, they can still form a basis for rational action. While predictions about the impacts of global warming and climate change are not precise, the available evidence indicates that temperatures will rise in temperate zones and rainfall patterns will alter. Eco-geographic conditions will alter in several regions of the world as a result of global warming. This will affect the type and variety of crops that can be grown (or economically grown) in different regions of the world.

To cope with these changes, a suitable stock of crop germplasm needs to be conserved so that it can be transferred between regions as climate change occurs. For example, it is likely to be important to conserve the germplasm of heat-tolerant and drought-resistant cereals and to develop new varieties to cope with these conditions. However, other types of crop germplasm suitable for other types of environments may also prove to be valuable because some regions of the world will become wetter.

As a precaution against the economic impacts of climate change, it is rational to conserve a suitable stock of food crop germplasm to facilitate adaptation in cropping in different regions as environmental change occurs. The type of germplasm which should be conserved depends on the type, location and regional extent of the climate changes expected.

There are other 'predictable' ecological and environmental changes which can make it rational to conserve particular types of crop germplasm. For example, it may be predictable that the biological fitness of a particular new variety of a crop will decline with the passage of time. At some time in the future, yields from the crop will be higher if the new variety is able to be replaced by an existing variety which will disappear unless government takes action to conserve it.

Additional Comments

This section has identified economic factors of importance for setting priorities for the preservation of the stock of germplasm of food crops. However, it has not considered all possibilities. For example, economic conditions are subject to change with the passage of time. For instance, food which constitutes a large portion of diets now and which is in relatively inelastic demand may not be so in the future. The opposite possibility can also occur. Those living in some regions of the world may, in the future, become less dependent on a particular crop for their income. The theory needs to be adapted to take account of changing circumstances where these are relevant. In addition, account must be taken of the cost and the effectiveness of available methods for conserving crop germplasm. This does not appear to have been given much consideration in the economic literature. Let us consider some of the relevant issues.

5.6 METHODS FOR CONSERVING CROP GERMPLASM, FOR EXAMPLE, IN GENE BANKS

Given that it is likely to be beneficial to conserve stocks of crop germplasm, the costs and biological effectiveness of alternative measures to achieve this need to be assessed. It is a subject worthy of detailed study. However, because of lack of available data, it is only possible to cover some of the main issues here.

The following are some of the possible ways to conserve crop germplasm:

1. Particular crop varieties may be conserved by farmers cultivating them in areas where they were developed, or sometimes in other regions.
2. They may be cultivated or conserved on government land, for example on agricultural research stations and in botanic gardens.
3. Long-term storage of propagules (particularly, seeds of some food plants) is possible under controlled conditions, for example in cold storage in gene banks.[9]

If the cultivation of a particular crop variety is no longer profitable, persons will usually need to be subsidized to continue to cultivate it. It is also possible that as a result of climate change, cultivation of a crop in the area where it was developed may cease to be ecologically possible. It may be necessary to shift its cultivation to another area as climate change occurs in order to ensure its conservation.

Conserving crop varieties *ex situ* on state experimental farms does not guarantee that the germplasm targeted for conservation will be preserved. For cost reasons, only small plots of the crop varieties may be planted on state farms. This makes their populations vulnerable to environmental disasters. If they are out-crossing species, cross-fertilization may occur with other varieties. Furthermore, if the varieties to be conserved are planted on plots in a different environmental area from that of their origin, they may fail to survive.

The conservation of germplasm *in vitro* also poses biological challenges. Not all germplasm can be preserved in this way; for example, plants that reproduce vegetatively can be unsuitable for this type of conservation. According to Rao and Hodgkin (2002, p. 12):

> Even under optimum conditions, accessions held [under] *ex situ* storage [*in vitro*] will need to be regenerated after a number of years. The genetic diversity of conserved material must be preserved during germplasm regeneration and this is more complex and difficult in the case of out-crossing species than in-breeders (Porceddu and Jenkins, 1982).

Sustaining funding (from governments and aid donors) for *ex situ* crop conservation is a major problem. Sometimes this funding ceases or is slashed after gene banks are established and they have to be abandoned and their germplasm is lost. Furthermore, funding for regeneration of materials is often limited and their timely regeneration may fail to occur, as is evidenced by FAO (2010, pp. 76–7).

Accessions to the world's gene banks are dominated by cereals, followed by legumes. FAO (2010, p. 61) estimates that 45 per cent of global accessions to the world's gene banks consist of cereals; legumes account for 15 per cent, and vegetables, fruits and forage crops amount to 6.9 per cent each. Roots and tubers only make up 3 per cent of the accessions. This probably reflects the ease with which the germplasm of different types of crops can be conserved and the perceived relative economic loss from losing different types of germplasm.

It is estimated that there are now more than 1340 gene banks globally for conserving germplasm for food and agriculture (FAO, 2010, p. 55). FAO (ibid.), however, notes that accessions are becoming increasingly concentrated in fewer countries – nearly half of all accessions are now held in seven developed countries. Some countries from which some of this material has been accessed are concerned about whether they will have access to it in the future. This, and other property rights issues, will be discussed in Chapter 10.

A significant limitation of *The Second Report on the State of the World's Plant Genetic Resources for Food and Agriculture* (FAO, 2010) is that it lacks information about economic aspects of the conservation of plant germplasm. In fact, little economic literature appears to be available which directly concentrates on this subject. More information is needed about current expenditure on and investment in the conservation of crop germplasm, its distribution between different methods of its conservation, its geographical spread, allocation between different types of crops and so on. Additional studies of the cost-effectiveness of different methods of conserving germplasm, taking into account the probability of success, would be useful. Furthermore, the extent to which economies of scale can be achieved by conserving germplasm *in vitro* should be explored.

Nevertheless, the social cost–benefit analysis of conserving crop germplasm is likely to continue to be challenging. For example, a possible disadvantage of *ex situ* conservation is that it could stymie continuing co-evolution of germplasm.[10] The prime beneficiaries of some gene banks may, in the end, be large companies developing new germplasm *ex situ*. Furthermore, while cereal production is important for the continuing availability of food and dominates the stock of germplasm conserved in gene banks, meeting human nutritional requirements also calls for the continuing availability of other crops, such as vegetables, fruits and nuts. Consequently, even crops that contribute only a small amount to the human diet can play an important role in maintaining the health of humans and this can make it important to conserve their germplasm.

5.7 ASPECTS OF THE LOSS OF INDIGENOUS FOOD CROP VARIETIES IN THE PACIFIC ISLANDS

There has been a substantial decline in indigenous food crop varieties in the Pacific Islands since European contact and the integration of Pacific Islanders into the world community. The Pacific Islands, particularly New Guinea, are considered to be the original source of several cultivated food crops. These include bananas and plantains, breadfruit, coconut, sugar cane, taro and yams. New Guinea is believed to have been the epicentre for the origin of most of these food crops (see Fullagar et al., 2006; Tisdell, 2014a, Table 1). With the initial human settlement of and migration within the Pacific Islands, these food crops spread to other islands in the Pacific. Many different varieties were developed to suit the ecological conditions of the different islands. The sweet potato (*Ipomoea batatas*) was most likely introduced to the eastern and central Pacific islands some 3000 years ago from Central America and Peru (Library of the University of Hawaii at Manoa, 2012). Subsequently, many different varieties of sweet potato were developed in the Pacific Islands.

Following European contact with the Pacific Islands, the increased integration of Pacific Islanders into the world community, and important socio-economic changes, genetic erosion in the Pacific Islands of traditional food crops occurred. The following factors are identified in Tisdell (2014a) as significant contributions to this erosion.

- Marketing of farm produce has developed and market requirements increasingly determine what is grown. Varieties less suited to marketing are discarded in favour of those which are more suitable for this.
- Improved varieties have replaced some traditional ones.
- Greater decoupling of crop production from local ecological conditions (hence the increased availability of off-farm inputs) has resulted in some local varieties no longer being grown.
- New crops have been introduced to the Pacific and in some places have replaced traditional crops.
- Increased urbanization and centralization of populations have occurred in the Pacific Islands. This has resulted in reduced crop cultivation in remote areas and their local landraces have been lost.

- Food habits and tastes have changed. Alterations in the structure of society (including greater urbanization and changes in the nature of work) have resulted in lifestyle changes which affect the demand for different types of food. Demand for cereals and convenience food has (for example) increased.
- Subsistence food provision has been increasingly replaced by market supplies. The market system has become more pervasive and this facilitates the loss of crop biodiversity, for reasons given previously in this chapter.
- In most Pacific islands, there is increased availability of foreign currency. This enables food to be imported and to some extent it is substituted for local products. The sources of foreign currency vary, depending on the Pacific territory considered. Sources of foreign currency include remittances by family members living overseas, foreign aid and export income.[11] Brown et al. (2013) found, for example, that in Tonga, remittances are mainly used for consumption.
- Immigration has influenced and changed food consumption patterns in several Pacific islands. For example, the migration of Indians to the Fiji Islands (during the colonial period to produce mainly sugar cane) has changed food consumption there, as has the settlement of non-Polynesians in Hawaii.

As discussed in more detail in Tisdell (2014a), all of these factors have contributed in some measure to the loss of previously existing varieties of food crops in the Pacific Islands. However, similar socio-economic developments in other parts of the world have also occurred and have contributed to the global reduction in the stock of previously existing varieties of food crops. Socio-economic change has radically altered the global development path of the genetic evolution of crops. During a long interval of time (beginning early in the Holocene era about 9000 years ago), human efforts continually increased the general variety and diversity of food crops globally. However, in recent centuries (especially in the last 250 years or so) the size of this developed genetic stock has been greatly eroded as a result of socio-economic developments, many of which are still continuing. This has raised concerns about the future availability of global food supplies.

5.8 AUSTRALIA: THE LATE CULTIVATION OF INDIGENOUS FOOD PLANTS

On all other continents except Australia, cultivation of food crops began thousands of years ago. Agriculture only developed in Australia after it began to be settled by the British, commencing in 1788. Australian Aborigines (who are believed to have settled in Australia about 40112000 years ago) remained hunters and gatherers, despite agriculture having developed in nearby New Guinea about 9000 years ago. Why this is so is unclear. Contributing factors may have been the unsuitability of most indigenous food plants and animals for domestication and local climatic conditions.

The new settlers rapidly developed agriculture in Australia, relying on alien species of livestock and cultivated plants. As a result, Australia's landscape changed and natural ecosystems were transformed, for example tree cover was much reduced. Agricultural use of Australian native food plants has occurred only in recent decades and is still developing. Consequently, progress in the selection and development of varieties of Australian plants most suitable for cultivation is still in its early stages, compared, for example, to the process of selecting food crops, such as wheat in Eurasia.

Some Australian indigenous plants have been domesticated (in recent decades) for food production, but establishing new food industries is challenging both from a production and a marketing point of view. New cultivation methods have to be developed and the development and selection of improved plant varieties for cultivation takes time. Also, establishing and developing markets for new food products requires considerable investment. Those businesses pioneering the development of these industries face considerable economic risks.

In terms of the value of its production, the major food crop based on Australian wild germplasm is the macadamia nut. It was first cultivated in Hawaii. Its cultivation in Australia began slowly after World War II. In 1987, 4400 tonnes of macadamia nuts in the shell (with 10 per cent natural content) were produced in Australia. By 2012, this had increased to 40000 tonnes (Australian Macadamia Society, n.d., a) and is still increasing. The cultivation of macadamia nuts has spread to many countries with a suitable subtropical climate. The major global producers of macadamia kernels in 2012 and their percentage (in brackets) of world production of these kernels in 2012 were Australia (28 per cent), South Africa (25 per cent), USA, Hawaii (16 per cent), Kenya (14 per cent), Guatemala (5 per cent) and Malawi (3 per cent) (Australian Macadamia Society, n.d., b).

The Australian Macadamia Society (the industry's peak body) makes the following points:

- There is strong financial commitment to domestic and export market development and farm research funded by a compulsory grower levy on production. The industry spends about $2 million annually on research and development and marketing.

- Australia holds the only natural germplasm resources for macadamias, and has spent over $4 million over the last ten years on a comprehensive breeding program. Early indications are that yield increases of 30 per cent are possible from new varieties. (Australian Nut Industry Council, n.d.)

As far as Australian producers of macadamia nuts are concerned, the compulsory levy internalizes the externalities associated with the development of the industry. For a discussion of the externalities and thresholds involved in the development of markets for new products of this type, see Tisdell (2001). The economic relationship between producers in this type of new industry appears to be mutualistic in its early stages but as the industry matures, it is likely to become competitive (Tisdell, 2001; 2013, pp. 150–53). Currently, the macadamia nut industry still seems to be in the mutualistic stage of its development. The Australian industry is not yet concerned about the increasing production of macadamias overseas (Australian Macadamia Society, n.d., c) because in the early stages, expanding supply helps 'to grow' the market.

The comparatively late development of macadamias as a food crop implies that there has only been a short period for selection and development of varieties that are most valuable for cultivation. This is why significant increases in yields by human selection and development of varieties appear to be likely. At the same time, substantial losses in the stock of wild macadamia germplasm have occurred. This loss may have reduced the potential for increasing macadamia yields and sustaining commercial production. The Australian Macadamia Society (n.d., d) estimates 'that over 80 per cent of Australia's wild macadamias have been lost since European settlement. Four species of native macadamia can be found in small isolated remnants of rainforest between Lismore and Bundaberg' [in eastern Australia]. This society is implementing a plan to save the remaining wild stock. The large loss of wild macadamias resulted primarily from European settlers clearing rainforest to make way for agriculture.

Cultivation of several other native Australian food plants has commenced on a small scale. These include the production of bush tomatoes, finger limes, Davidson plums, Kakadu plums, lemon myrtle, mountain pepper and riberries. An unusual feature of the development of industries based on these food plants is the support given to it by Coles, one of Australia's two market-dominant supermarket retailers. It has established Coles Indigenous Food Fund (a charitable organization) to support indigenous business developments and promote Australian native bush foods (Coles, n.d.). Coles (n.d.) states that 'from Central Australia to the

Kimberleys and Cape York, it is helping these groups [Indigenous communities] develop commercial planting and harvesting of bush tomatoes, Kakadu plums and native herbs'. This case is unusual in that large supermarket chains usually prefer to stock standard food products in substantial demand rather than speciality products which are pioneering new markets. The policy adopted by Coles reduces barriers to market entry of suppliers of Australian native bush foods.

5.9 CONCLUDING COMMENTS

Over thousands of years (beginning early in the Holocene era), the number of varieties of cultivated food plants increased globally as a result of human selection and development of these. However, in the last few centuries, this heritage stock has fallen sharply for socio-economic reasons, which were identified in this chapter. This raises concerns about whether agricultural production can be sustained. Will this biodiversity loss result in agriculture being unable to increase its production to meet expected demands on it in the next few decades? Is it likely that some important food crops will fail to sustain their yields because of impoverishment of the stock of their germplasm?

Various efforts are being made to conserve biodiversity given that the demand for agricultural products is going to increase substantially in the next few decades. Land-saving (land-sparing) ecosystems involving intensification of agriculture have been proposed as a means of addressing the problem of biodiversity loss, but as discussed, agricultural intensification can (as well as agricultural extension) contribute to biodiversity loss. Trade-offs have to be evaluated.

Economic factors relevant to the cost–benefit analysis of conserving the germplasm of crops used for food were identified. This field of enquiry is relatively underdeveloped, especially the cost-effectiveness of alternative means of conserving crop germplasm. This discussion was followed by observations on the loss of varieties of traditional food crops in the Pacific Islands and Australia's late start in cultivating some of its native food plants. The Australian case indirectly illustrates the importance of path-dependence in the development of crop germplasm and associated agrosystems. The next chapter will investigate the loss of livestock breeds as a result of socio-economic change and complements the observations made in this chapter.

NOTES

1. Renfrew (2007) points out that agriculture began in the Near East (the Levant) about 9000 years ago and arose independently in the Holocene era in other parts of the world.
2. The UN estimated the global population in 2013 to be about 7.2 million. Therefore, its 2060 prediction indicates an increase in the global population of just over 30 per cent, that is, in a period of less than 50 years.
3. It is estimated that currently nearly 50 per cent of cereal production is fed to livestock (De Schutter, 2011, p.4) and that cereals are increasingly being diverted to feed livestock. This is an inefficient means of providing food for humans because of poor conversion ratios (Tscharntke et al., 2012, p. 55). Furthermore, biofuel support policies are also of significant concern (see, for example, Tisdell, 2011). By 2020, it is predicted that 15 per cent of global cereal and vegetable oil output will be used for biofuel as well as 30 per cent of sugar cane production (Tscharntke et al., 2012, p. 55).
4. This model supposes that the type of grain used for alternative purposes is the same. Perfect substitutability does not necessarily occur. For example, grain used for other than human consumption may be of poorer quality or even unsuited to human consumption. Nevertheless, provided there is some substitutability, the consequence suggested follows.
5. In some countries, marginal agricultural land has been abandoned as economic growth has occurred, for example, in Europe and Japan, but globally the amount of land used for agriculture has increased.
6. Taking into account growing income levels, increased urbanization, changing diets and demographic growth, demand for agricultural production is expected to rise by as much as 70 per cent by 2050 compared to its level in 2010 (Burney et al., 2010).
7. For some further discussion of the impact of the global economic system on the conservation of wild biodiversity, see Tisdell (1999, Ch. 6).
8. The concept of logical probability seems to be relevant here. See Keynes (1921), Carnap (1950) and Tisdell (1968, pp. 18–22).
9. The desirability of conserving the wild relatives of cultivated crops and the relevant methods for doing this should also be considered. Both *in situ* and *ex situ* conservation of these can occur.
10. For some discussion of co-evolution in agriculture, see, for example, Norgaard (1994) and Tisdell (2000).
11. The increased dependence of the Pacific Islands on food imports and their reduced food self-sufficiency has left them vulnerable to any significant decline in external revenue. Aspects of this problem are discussed in Tisdell and Fairbairn (1984) and in Tisdell (2014b).

REFERENCES

Anon (2014a), 'Millet', Wikipedia. Available at: http://en.wikipedia.org/wiki/Millet. (Accessed 9 July, 2014).

Anon (2014b), 'Sorghum', Wikipedia. Available at: http://en.wikipedia.org/wiki/sorghum. (Accessed 9 July, 2014).

Australian Macadamia Society (n.d., a), 'Domestic production and prices'. Available at: http://www.australian-macadamias.org/item/600-issue-five?lang=en&r=1&Itemid=97. (Accessed 17 July, 2014).

Australian Macadamia Society (n.d., b), 'Global production'. Available at: http://www.australian-macadamias.org/item/601-issue-five?lang=en&r=1&Itemid=97. (Accessed 17 July, 2014).

Australian Macadamia Society (n.d., c), 'The macadamia industry'. Available at: http://www.australian-macadamias.org/about-aussie-macadamias/the-macadamia-industry?lang=en&r=1&Itemid=96. (Accessed 17 July, 2014).

Australian Macadamia Society (n.d., d), 'Conservation'. Available at: http://www.australian-macadamias.org/about-aussie-macadamias/conservation-and-sustainability?lang=en&r=1&Itemid=99. (Accessed 17 July, 2014).

Australian Nut Industry Council (n.d.), 'Macadamias'. Available at: http://nutindustry.org.au/ANIC/Industry-Snapshots/Australian-Macadamias.asp. (Accessed 17 July, 2014).

Bentham, T.G., J.A. Vickery and J.D. Wilson (2003), 'Farmland biodiversity: Is habitat heterogeneity the key?', *Trends in Ecology and Evolution*, **18**, 182–8.

Brown, R.P.C., J. Connell and E.V. Jimenez-Soto (2013), 'Migrants' remittances, poverty and social protection in the South Pacific: Fiji and Tonga', *Population, Space and Place* [Online]. Available at: http://onlinelibrary.wiley.com/doi/10.1002/psp.1765/pdf. (Accessed 21 February, 2013).

Burney, J.A., S.J. Davis and D.B. Lobell (2010), 'Greenhouse gas mitigation by agricultural intensification', *Proceedings of the National Academy of Sciences*, **107**, 12052–7.

Carnap, R. (1950), *Logical Foundations of Probability*, Chicago: Chicago University Press.

Clark, C.W. (1957), *The Conditions of Economic Progress*, 3rd edn, London: Macmillan.

Coles (n.d.), 'Coles Indigenous Food Fund'. Available at: https://www.coles.com.au/helping-australia-grow/coles-in-the-community/national-charity-initiatives. (Accessed 18 February, 2014).

De Schutter, O. (2011), 'Report on the right to foods', *UN Human Rights Council* [Online]. Available at: http://www.srfood.org/images/stories/pdf/officialreports/20110308_a-hrc-16-49_agroecology_en.pdf. (Accessed 14 July, 2014).

Donald, P.F., F.J. Sanderson, I.J. Burfield and F.P.J. Van Bommel (2006), 'Further evidence of continent-wide impacts of agricultural intensification on European farmland birds, 1990–2000', *Agriculture Ecosystems and Environment*, **116**, 189–96.

FAO (1995), 'Regional trends in production and utilization of sorghum and millet', FAO. Available at: http://www.fao.org/docrep/t0818e/T0818E07.htm. (Accessed 9 July, 2014).

FAO (2010), *Second Report on the State of the World's Plant Genetic Resources for Food and Agriculture*, Rome: FAO.

Firbank, L.G., S. Petit, S. Smart, A. Blair and R.J. Fuller (2008), 'Assessing the impacts of agricultural intensification on biodiversity: A British perspective', *Philosophical Transactions of the Royal Society B: Biological Sciences*, **362**, 777–87.

Fullagar, R., J. Field, T.P. Denham and C. Lentfer (2006), 'Early and mid Holocene tool-use and processing of taro (*Colocasia esculenta*), yam (*Dioscorea* sp) and other plants at the Kuk Swamp in the highlands of Papua New Guinea', *Journal of Archaeological Science*, **33**, 595–614.

Goldman, A., S. Ramaswami and R.E. Kridor (2002), 'Barriers to the advancement of modern food retail formats: Theory and measurement', *Journal of Retailing*, **78**, 281–95.

Gustavsson, J., C. Cederberg, V. Sonosson, R. van Otterdijk and A. Meybeck

(2011), *Global Food Losses and Food Waste: Extent, Causes and Prevention*, Rome: FAO.
Keynes, J.M. (1921), *A Treatise on Probability*, London: Macmillan.
Library of the University of Hawaii at Manoa (2012), 'Sweet Potato (*Ipomoea batatas*)', Manoa: Traditional Pacific Island Crops. Available at: http://libweb.hawaii.edu/libdept/scitech/agnic/sweetpotato.html. (Accessed 17 February, 2014).
Norgaard, R.B. (1994), *Development Betrayed*, London: Routledge.
Phalan, B., A. Balmford, R.E. Green and J.P.W. Scharlemann (2011a), 'Minimizing harm to biodiversity of producing more food globally', *Food Policy*, **36**, 562–71.
Phalan, B., M. Orial, A. Balmford and R.E. Green (2011b), 'Reconciling food production and biodiversity conservation: Land sharing and land sparing compared', *Science*, **333**, 1289–91.
Porceddu, E. and G. Jenkins (1982), *Seed Regeneration in Cross Pollinated Species*, Rotterdam: A.A. Balkeme.
Rao, V.R. and T. Hodgkin (2002), 'Genetic diversity and conservation and utilization of plant genetic resources', *Plant Cell, Tissue and Organ Culture*, **68**, 1–19.
Reardon, T., C.P. Timmer, C.B. Barrett and J. Bordegué (2003), 'The rise of supermarkets in Africa, Asia and Latin America', *American Journal of Agricultural Economics*, **85**, 1140–46.
Renfrew, C. (2007), *Prehistory: The Making of the Human Mind*, London: Weidenfeld and Nicolson.
Ricardo, D. (1817), *The Principles of Political Economic and Taxation* (Reprinted 1955), London: Dent.
Tisdell, C.A. (1968), *The Theory of Price Uncertainty, Production and Profit*, Princeton, NJ: Princeton University Press.
Tisdell, C.A. (1999), *Biodiversity, Conservation and Sustainable Development*, Cheltenham, UK and Northampton, MA, USA: Edward Elgar Publishing.
Tisdell, C.A. (2000), 'Coevolution, agricultural practices and sustainability: Some major social and ecological issues', *International Journal of Agricultural Resources, Governance and Ecology*, **1** (1), 6–16. Reprinted in C.A. Tisdell (2003), *Ecological and Environmental Economics: Selected Issues and Policy Responses*, Cheltenham, UK and Northampton, MA, USA: Edward Elgar Publishing.
Tisdell, C.A. (2001), 'Externalities, thresholds and the marketing of new aquacultural products: Theory and examples', *Aquaculture Economics and Management*, (5), 289–302.
Tisdell, C.A. (2011), 'The production of biofuels: Welfare and environmental consequences for Asia', in M. Hossain and E. Selvanathan (eds), *Climate Change and Economic Growth in Asia*, Cheltenham, UK and Northampton, MA, USA: Edward Elgar Publishing, pp. 38–61.
Tisdell, C.A. (2013), *Competition, Diversity and Economic Performance*, Cheltenham, UK and Northampton, MA, USA: Edward Elgar Publishing.
Tisdell, C.A. (2014a), 'Genetic erosion in traditional food crops in the Pacific Islands: Background, socioeconomic causes and policy issues' – WP193 amended, 'Economics, Ecology and The Environment', Working Paper no. 196. Brisbane: School of Economics, The University of Queensland.
Tisdell, C.A. (2014b), 'The MIRAB model of small island economies in the Pacific and their security issues: Revised version', 'Social Economics, Policy and Development', Working Paper no. 58. Brisbane: School of Economics, The University of Queensland.

Tisdell, C.A. and T.I. Fairbairn (1984), 'Subsistence economies and unsustainable development and trade', *Journal of Development Studies*, **20**, 227–41.

Tscharntke, T., Y. Clough, T.C. Wanger, L. Jackson, I. Motzke, I. Perfecto, J. Vandermeer and A. Whitbread (2012), 'Global food security, biodiversity conservation and the failure of agricultural intensification', *Biological Conservation*, **151**, 53–9.

United Nations Department of Economic and Social Affairs, Population Division (2012), 'World Population Prospects: The 2012 Revision', Available at: http://esa.un.org/unpd/wpp/Documentation/pdf/WPP2012_Volume-I_Comprehensive-Tables.pdf. (Accessed 9 July, 2014).

6. Livestock and the socio-economics of biodiversity loss

6.1 INTRODUCTION

This chapter focuses on the socio-economic factors and processes that have contributed to the loss of genetic diversity of domesticated animals, particularly livestock. However, it also includes some discussion of the implications of livestock husbandry for biodiversity change generally, its consequences for ecosystems and environments and its impact on food security. These impacts vary with the type of animal husbandry practised and the species involved. Livestock are domesticated animals that have, or which previously had, considerable direct use value for humankind and this contrasts with much wildlife possessing mainly non-use values.[1] Nevertheless, some breeds of livestock, especially if rare or endangered, can have significant non-use values as well as use value. Furthermore, grazing by livestock can create favoured cultural landscapes (a type of favourable economic externality) and can help maintain global biodiversity.[2] This is well recognized in Europe. For example, the landscapes of the Yorkshire Moors in England and the Luneberg Heath in Germany owe their characteristics to sheep grazing. So in some European localities, agricultural practices creating cultural landscapes are subsidized, and this helps protect some rare breeds. Furthermore, the conservation of some breeds is directly subsidized. In some jurisdictions, rare breeds are also conserved in protected areas (World Conservation Monitoring Centre, 1992, p. 397).

Despite the fact that most domestic animals and their products are classified, in economic terms, as private goods because they are excludable and rivalrous, many breeds have been lost in the last 100 years or so (Alderson, 1994; Scherf, 2000). The World Conservation Monitoring Centre (1992, p. 397) reports:

> Pursuit of higher production targets, the commercial success of particular breed promoters, and, in developed countries, changes in consumer preferences have led to livestock development activities becoming concentrated in few breeds and breed groups. The corollary of this is that more breeds are declining in importance, many have been lost and the survival of many others is in considerable

> doubt. Concern for rare breeds has been most marked in northern temperate countries with a history of specialized livestock production, but it is becoming increasingly evident that declining breeds in less developed countries also represent genetic resources of great significance.

While there is considerable uncertainty about the magnitude of the loss in biodiversity of domestic animals, there is no doubt that loss is considerable. According to the website for the Civil Society Organizations and Participation Programme of the UNDP, 'Half of all Europe's domesticated animals have become extinct in [the 20th] century. A third of all remaining livestock species [breeds] in both Europe and North America are endangered' (UNDP/CSOPP, 2001, p.2). Shand (1997) reports that: 'Livestock breeds are disappearing at an annual rate of five per cent, or six breeds per month. In Europe, half of all breeds of domestic animals existing in 1900 are gone, with 43 per cent of those remaining endangered'.

However, these figures could exaggerate the size of the loss. Blench (2000, p.44) suggests that some lists of extinct or endangered livestock include non-indigenous species and breeds that have never left the research station, for example, the FAO Worldwatch List (Scherf, 1996). These are not species and breeds involved in co-evolution. Blench (2000, p.44) points out:

> The breeds most relevant to biodiversity concerns are those that have co-evolved with a particular environment and farming system and represent an accumulation of both genetic stock and management strategies in relation to a particular environment. These have usually taken a long time to evolve and have characteristics such as humidity resistance, that cannot be easily developed.

On the other hand, some breeds in developing countries which have not been identified may have already been lost.

Despite this, according to data collected by the World Conservation Monitoring Centre (1992) there were 3237 extant livestock breeds in 1992 and 617 breeds had become extinct since 1892. This suggests that almost one in six breeds became extinct in this time period. In addition, another 474 breeds were considered to be rare and endangered. This indicates that within a period of 100 years about 28 per cent of livestock breeds either became extinct or rare or endangered. Therefore, the magnitude of the loss is considerable, even on the basis of conservative estimates.

To a large extent this loss appears to have been accelerated by the extension of market systems and associated processes of globalization. These processes, together with the nature of technologies associated with the husbandry of particular breeds, have encouraged global concentration of economic activity on fewer breeds. Furthermore, the changing structures of societies, such as increased urbanization in developing countries, may favour breeds and associated technologies found initially to be of economic value by higher income countries. Development of the livestock industry in developing countries may be 'biased' in favour of breeds and technologies from higher income countries because of their 'prime-mover' advantage and the presence of a breed-technology 'lock-in' effect (Swanson, 1994; 1995a; 1995b).

In this chapter, the proximate causes of breed losses and the processes involved in this are outlined and the relevance of Swanson's theory (Swanson, 1994) of species extinction is considered. Then a discussion follows of how the extension of markets and economic globalization accelerates the loss of breeds and encourages the tendency to concentrate on a few breeds. Subsequently, the growing practice of decoupling the husbandry of animals from their natural environment is examined. This is partly a market-driven phenomenon which mostly involves industrial-type livestock production. It further adds to genetic erosion and raises serious sustainability issues. Loss of local breeds of pigs in Vietnam is introduced as an illustration of socio-economic changes influencing biodiversity loss in livestock. Next, consideration is given to the consequences of the development of livestock industries for the loss of other than livestock biodiversity, and the impacts of livestock husbandry on food security are discussed. It is emphasized that livestock production can, in specified circumstances, add to food supplies available to humans. The view that it reduces this supply needs to be qualified.

6.2 PROXIMATE CAUSES OF BREED LOSSES AND THE PROCESSES INVOLVED

Loss or substantial loss of extant breeds can occur as a result of their replacement by other existing ones considered to be superior from an economic point of view, by the formation of synthetic breeds that eventually replace existing breeds, and by stock obtained by stabilized crossbreeding (World Conservation Monitoring Centre, 1992, p. 395). The latter, however, requires pure breeds of parent stock to be maintained, and therefore, unless genetic introgression occurs in the breeding stock, is not a force for breed loss per se. However, it is possible that the crosses have superior quality and this results in some breeds being entirely replaced by the crosses.

Apart from breed substitution, economic change can result in the elimination of livestock in some regions in favour of other forms of agriculture such as the growing of crops. In such cases, breeds specific to a region that undergoes land-use changes may disappear.

Hammond and Leitch (1995) identify the factors listed in Table 6.1 as sources of the erosion of livestock biodiversity. Some of these sources have an economic basis, for example economic specialization, some are technologically based (but this change may be ultimately driven by economic considerations), and others depend on political and natural events.

Table 6.1 does not sufficiently specify the economic and market factors that accelerate erosion of biodiversity of breeds. The following economic factors can be important in biodiversity loss in livestock.

1. As discussed later, the extension of markets via economic globalization encourages regional economic specialization. This may result in particular types of livestock production becoming relatively uneconomic in a particular region and consequently in the loss of breeds peculiar

Table 6.1 Proximate causes of erosion of livestock biodiversity

Factor	Description[(a)]
Development interventions	Preference given to high-input, high-output breeds developed for benign environments. Commercial interest in donor countries promotes use of relatively temperate-adapted breeds and creates unrealistic expectations in developing countries.
Specialization	Emphasis on a single productive trait, for example, dairying, leading to exclusion of multi-purpose animals.
Genetic introgression	Cross-breeding and accidental introgression leading to loss of indigenous breeds.
Technology	Machinery replaces work animals.
Biotechnology	Cryopreservation equipment is as yet inadequate to store germplasm of threatened breeds.[(b)]
Political instability	Can eliminate local breeds owned by vulnerable populations.
Natural disaster	Floods, drought and epizootics preferentially affect remote or isolated human and livestock populations.

Notes:
(a) Adapted from Hammond and Leitch (1995) and Blench (2000).
(b) However, the adequacy of cryopreservation has improved but it is expensive and funding it is a problem (Pizzi et al., 2010).

to that region. However, according to the implications of neoclassical international economic theory, this adds to global economic efficiency, at least in the short term.

2. With economic globalization, it has become less costly to transfer breeds across international boundaries and this increases the possibilities for breed substitution.
3. Factor 2 enables the Swanson dominance-effect (discussed in the next section) to operate more easily. The Swanson dominance-effect suggests that breeds selected in more developed countries will tend to replace those in less developed countries.
4. The law of specialization according to comparative advantage suggests that specialized breeds will tend to replace multi-purpose breeds as markets expand and market transaction costs fall.
5. Changing tastes and demands can hasten breed erosion. Consumer preference for leaner meat is resulting in the demise of breeds of pigs that have meat with a high fat content.
6. Changes in the availability and price of imports, for example imported food for livestock, can change the economics of keeping different breeds.
7. The scope for altering environments in which livestock are held can change the economics of selecting different breeds. To a large extent, the keeping of livestock in developed countries has been divorced from dependence on surrounding natural environments. Much livestock in developed countries (and increasingly so in developing countries) is maintained in an artificial environmental capsule protected from the natural environment in intensive farming systems. Pigs and poultry are, for instance, usually kept in specially built housing. Local environments are either controlled or stabilized (for example, by use of fertilizer) and/or many inputs, including livestock food, may not be produced locally due to forces enabling market extension. This results in greater uniformity of livestock environments and therefore contributes to reduced diversity of livestock breeds.

Thus, it is clear from the above outline that the scope for economic forces to contribute to breed losses is considerable. These economic impacts are closely associated with the strengthening of the forces of globalization and market extension. Furthermore, the pattern of breed losses is influenced by the Swanson dominance-effect.

6.3 THE SWANSON DOMINANCE-EFFECT AND BREED LOSS

Swanson (1994) identified two important factors contributing to biodiversity loss generally. The first was the loss of natural habitat due its conversion to human use, mainly for agriculture. Tisdell (1999, Ch. 4) suggests that in addition to this, man-made activities have increased the uniformity of extant environments and that this has contributed, among other things, to reduced diversity of species. A further important influence on biodiversity conservation is the selective approach of humankind to conserving and husbanding species. This results in path dependence. As Swanson (1994, pp. 99–100) states:

> the entire roster of species is not being considered for use on any given parcel of land. It is more likely that the choice is only for a handful of 'commercialized' crops and livestock. The roster of species used to appropriate photosynthetic products for humans has converged to this very small select group of plants and animals.

Swanson (1994, pp. 101–106) argues that path dependence (a situation where initial and earlier conditions heavily influence the subsequent development path) is important in determining the survival of species. David (1985) observed a similar relationship in the development and survival of new technologies. Swanson argues that path dependence occurs in biological evolution because learning, investment and experience in developing the use of a species tends to be species-specific. This know-how and development cannot be easily shifted to other species but it may be shifted to other geographical regions. This theory is applicable to the evolution of breeds of livestock.

Swanson (1995a; 1995b) further elaborates on his hypothesis that choices of the species developed depend on the prime-moving regions and determine the subsequent choices made and paths taken later in other regions. Specifically, Swanson (1995a; 1995b) argues that:

> the degree of conversion witnessed in developing societies is predetermined by the conversion decision made by the firstdeveloping societies. These societies selected a set of locally available natural assets around which to develop, but many subsequent asset selections have taken their shape in response to those initial decisions. Now societies that are 'catching up' attempt to leapfrog intermediate stages of development made by previous developers in their own territories. In this way, development is biased toward the conversion of natural environments to the same set of assets across the globe. This is diversity decline as a result of the uniformity of the development process across heterogeneous states.

By analogy, the Swanson pathdependence hypothesis can be extended to the selection of different animal breeds. As economic development occurs, one might expect to witness the increasing global dominance of breeds selected in higher income countries and the displacement of breeds specific to less developed countries.

Swanson (1994) largely attributes these lock-in effects of choice of utilized species to non-rivalry in the use of knowledge and dynamic externalities of the type mentioned by Romer (1987; 1990a; 1990b). However, Romer's perspective was foreshadowed earlier by Myrdal (1956). In line with Romer's view, Swanson suggests that this leads to a non-convexity in development paths. Non-convexity arises, for example, if increasing returns (in contrast to decreasing returns) occur by specializing in the production and development of particular products. Alternatively, this phenomenon could also be envisaged as involving a form of hysteresis, that is, reduced plasticity or flexibility, and thus reduces adaptability in the relevant system. Furthermore, lock-in can conceivably arise in the absence of knowledge externalities. The latter could happen where, for instance, a monopolist obtains effective property-rights to new breeds or varieties of crops.

Because the initial selection of breeds or species for development tends to be partial or locally determined and is to a large extent uncoordinated, breeds or species may be selected for development that fail to maximize economic returns from a global or a long-term perspective. This is basically a result of decision-making being myopic and evolutionary. The array of breeds developed, although having some economic advantages, may not constitute the economically optimal choice. Because lock-in occurs, breeds can fail to survive that would have been superior from an economic viewpoint if they had been developed at an earlier time.

This can be illustrated by Figure 6.1. Two breeds, I and II, are assumed to be available initially and a 'decision' is to be made to develop one or the other. If no breed development takes place, the flow of the net economic value of Breed I might be as indicated by line CD and that of Breed II as indicated by line AB. If Breed II is developed rather than Breed I, the flow of net economic value from it might be as indicated by line EFG. On the other hand, if Breed I is developed, this flow might be as represented by line HJK.

It can be seen from Figure 6.1 that if Breed II is developed and Breed I is neglected, the economic benefits from Breed II eventually overtake those from Breed I. The opportunity cost of developing Breed I later increases, given the earlier development of Breed II and the sunk costs of investment in this development. Consequently, as time passes, it becomes increasingly clear that the development of Breed I is no longer economical. As time

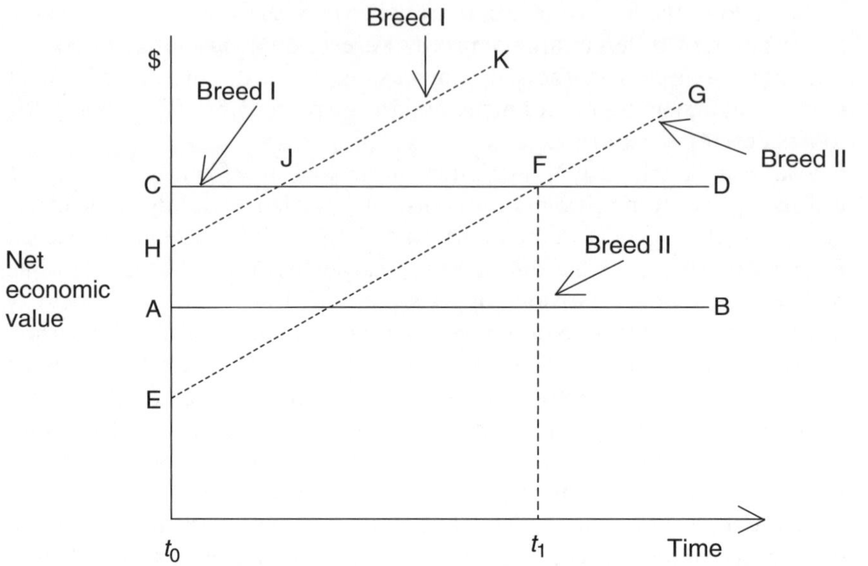

Figure 6.1 Illustration of the Swanson lock-in effect when applied to choice of breeds: it can result in the potentially most valuable breed not being selected

passes, its inherent superior genetic position is eroded by the early neglect of humans to develop its potential.

In the case illustrated in Figure 6.1, the initially inferior breed is shown to always remain inferior for the same level of investment in its development as the initially superior breed. In practice, this may not be so. The potential for what appears initially to be the inferior breed to respond to development may be greater than for the breed initially appearing superior in terms of its net economic value. Nevertheless, lock-in can occur, no matter which of the breeds is selected for development.[3]

Systems involving path-dependence can be extremely complex, particularly if coupled with the presence of radical uncertainty. They certainly add force to Clark's observation that 'predicting the future is a risky business at best, particularly where human activities are involved' (Clark, 1995, p. 143).

In his work, Swanson (1994; 1995a; 1995b) stresses that initial choices of species and associated technological development are the prime influences on biodiversity loss. While these processes and mechanisms are important, his approach does not put sufficient emphasis on the role of market extension and associated economic globalization as a contributor to the extent of biodiversity loss.

6.4 MARKET EXTENSION AND ECONOMIC GLOBALIZATION AS A SOURCE OF BIODIVERSITY LOSS

The widening of markets can help to magnify the types of initial persistent genetic biases identified by Swanson, and at the same time act as a prime-mover of such biases. Therefore, circular causation is present. Market extension often creates new avenues for extinction of breeds and species by eliminating protective economic niches (cf. Tisdell and Seidl, 2004; Tisdell, 2013, Ch. 8) and unleashing other competitive forces. While on balance the widening of markets can be expected to reduce the number of surviving breeds, occasionally some breeds are saved from extinction by this process because they possess special qualities that appeal to niche buyers. Market extension can increase the pool of niche buyers. The Schwabisch Hall (Swabian-Hall) pig has been suggested as such a case (Anon, 2014 a). In fact, the patterns of breed and species elimination arising from the extension of markets can be quite varied and complex. Here it is only possible to identify some of these patterns.

The spread of the market system favours selfish competitive forces and the survival of economic entities usually requires them to participate in economic rivalry. These forces also favour the adoption of least-cost technology (Svizzero and Tisdell, 2001). Thus when the market system is introduced to a region using a regional-specific breed inferior in productivity compared to an exotic breed, the exotic breed will replace the regional breed if introduced to the locality. Therefore, the regional breed will become extinct.

This is illustrated by the simple supply and demand curve analysis shown in Figure 6.2. Prior to market extension, the demand for the produce of a local breed might be as represented by line D_1D_1. Market equilibrium then occurs at E_1. Once the local region gets linked to wider markets, the demand for the same livestock produce from this region might be as represented by the line marked D_2D_2. The local breed of livestock has the supply curve for this produce of S_1S_1. But because of enhanced global links, the local region can obtain an exotic breed (new technology) for which the supply curve of the same livestock produced by the region is as shown by S_2S_2. The exotic breed can produce livestock output at lower cost. Consequently, competition will result in the exotic breed replacing the local breed. The local breed is driven to extinction by economic change. It matters not at all in this context whether the replacement breed has been made superior only because of a Swanson bias-type process. In the new market equilibrium at E_2, the economic surplus of livestock producers in the region increases.

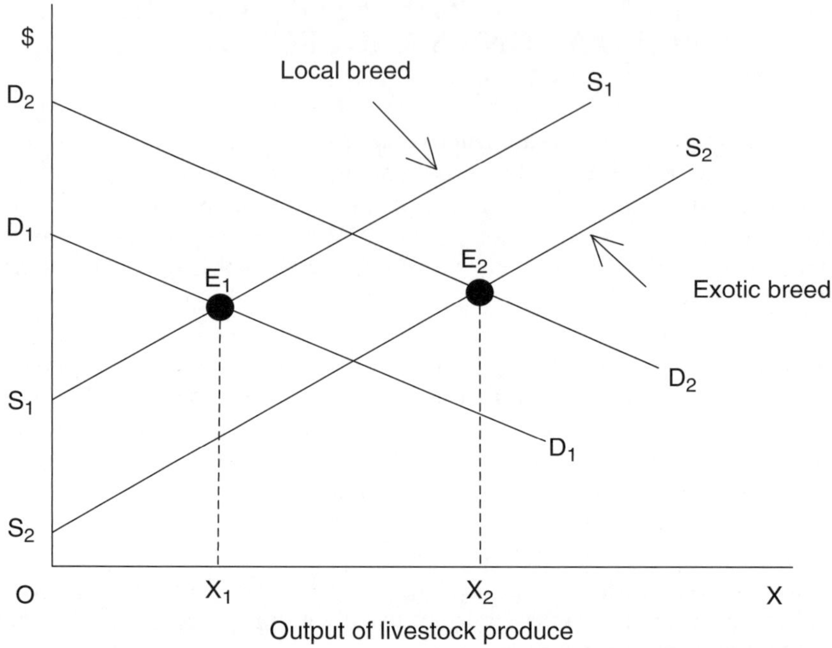

Figure 6.2 Market systems and global genetic opportunities result in this case in extinction of the local breed which is replaced by an exotic breed

Economic globalization, the process of extending markets, is extolled by Adam Smith (1776 [1910]) and is seen by David Ricardo (1817) as a powerful force for reducing economic scarcity by encouraging regional and international specialization in production according to comparative economic advantage. However, at the same time, it is a powerful force for loss of genetic diversity. Two different types of illustrations of this follow.

In the case shown in Figure 6.3, it becomes cheaper in a region to import livestock produce rather than supply it locally as a result of the extension of markets or growing economic globalization. The cost curve for supplying the produce locally using an indigenous breed might be as shown by S_1S_1, whereas the supply curve of the produce from outside the region is as indicated by S_2S_2. In a market system, the latter will usually be much more elastic than the former supply curve and could be perfectly elastic given that the global market is broad and a local market is narrow. It is shown as being perfectly elastic in the illustration, but this is just for convenience. Thus if DD represents the demand for livestock produce in the region, all livestock produce will be imported if the regional market is opened up.

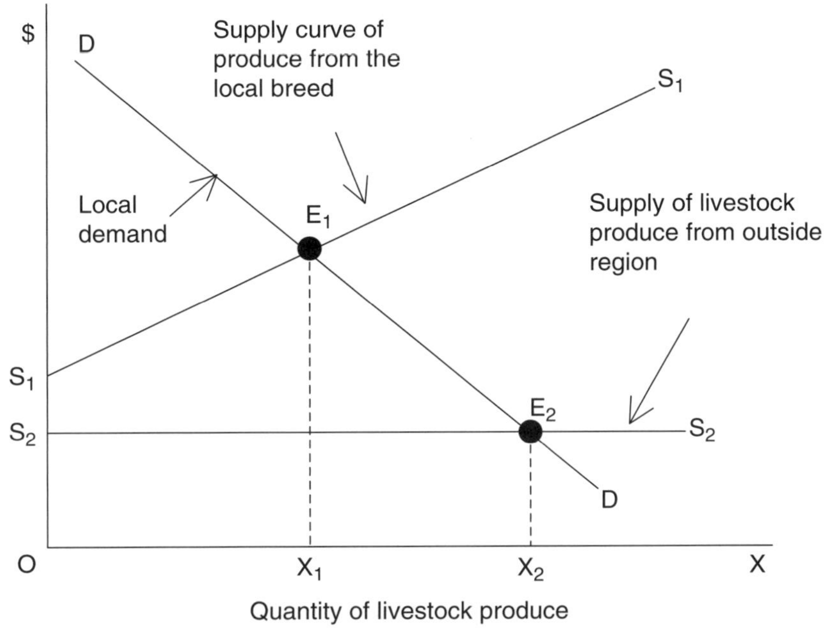

Figure 6.3 *A case in which local livestock breeds are rendered extinct by the import of livestock produce*

Endemic livestock breeds will disappear. Given a closed regional market, the initial market equilibrium will be at E_1 with all produce being supplied by local breeds. In the open market, the new market equilibrium will be at E_2.

Figure 6.4 illustrates the matter in a more holistic way. Assume, for simplicity, that individuals in a region are identical and have the same resources, preferences and production opportunities. Any one individual is representative of all. In Figure 6.4, the production frontier marked ABC represents the production possibilities available to an individual and the indifference curves marked I_1I_1 and I_2I_2 represent individual preferences. In the absence of trade, the mixed production system corresponding to the combination of crop and livestock production at B is ideal. But with the opening up of inter-regional trade, individuals in this region can engage in exchange, and exchange opportunities represented by the line CEF become available. This indicates that this region has a comparative advantage in crop production. By producing the level of crop production corresponding to point C and engaging in exchange, point E can be reached and regional economic welfare rises. The region specializes therefore in crop production

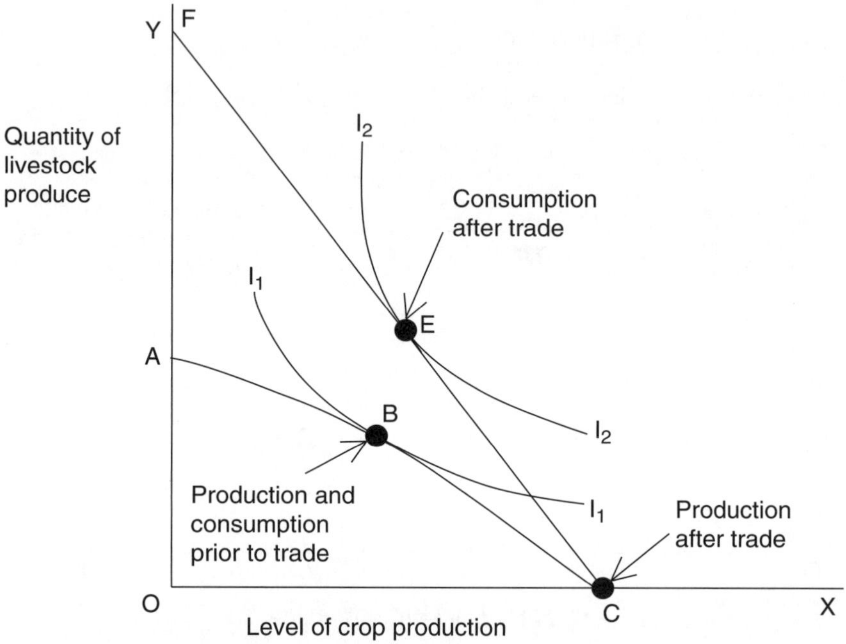

Figure 6.4 Another illustration of the potential elimination of local breeds of livestock as a result of market extension

and livestock production ceases. With market extension, individuals can move to equilibrium, E, and be 'better off'. However, any unique breeds of local livestock will become extinct.

Blench (2000, p. 45) provides a relevant example. He states: 'In many areas in Southern Nigeria, rising prices of tree-crops such as cocoa and palm-oil have caused the communities to dispense with their traditional dwarf cattle and goats to concentrate on these profitable crops'. These local breeds are in danger of disappearing. Blench in his FAO report continues with the following relevant statement: 'This is a perfectly rational medium-term strategy on their part. But it would be short-sighted of the national government to lose the genetic resource these livestock represent because of a temporary pattern in world trade'. This suggests that loss will reduce economic flexibility in an uncertain world and options should rationally be kept open at the national level by, at least, conserving a portion of this genetic resource.

6.5 LOSS OF MULTI-PURPOSE BREEDS OF LIVESTOCK

With the extension of markets and economic development, there is usually a switch from multi-purpose breeds of livestock to specialized breeds.[4] This switch may occur for at least three reasons. One may be the path dependence of development and differentiation of technology ancillary to the supply of different types of specialized produce of particular breeds. The technology and knowledge of husbandry needed for efficient milk production from cattle now differs to a considerable extent from that required for efficient beef production. Thus the development of different technologies for animal husbandry can contribute to the elimination of multi-purpose breeds or particular breeds. Soini and De Haas (2010, p. 29) give an interesting example. They state:

> Milking machines were introduced on a large scale in the 1960s and 1970s. They had been designed primarily for the udders of commercial dairy breeds (Holstein Fresians) . . . . The conformation of the udder of some local breeds (e.g. Groningen White Headed, Netherlands, Reggiana, Italy) was not satisfactory, and caused difficulties when attaching the milking machine equipment. For this reason, [some] local breeds were no longer favoured by many farmers.

A second reason has to do with market development. In a non-exchange subsistence economy, keeping multi-purpose breeds to meet human needs in a balanced way is likely to be an advantage. Market exchange may be absent in such economies because of the social system or because high market-transaction costs make markets uneconomical. But once markets become economic, pre-existing constraints to specialization are reduced or removed.

This case can be illustrated by Figure 6.5. Assume that three breeds of cattle, A, B and C, are available in a local region, and suppose that initially the local economy is a nonexchange one. For simplicity, assume that all in the region have the same resources and preferences and that their preferences are only for milk and beef. The indifference curves I_1I_1 and I_2I_2 in Figure 6.5 represent these individual preferences. The production possibility if Breed A only is used, is represented for each resource-holder by point A in Figure 6.5. Similarly, the production possibilities for herds consisting only of Breeds B or C are represented by points B and C respectively.

In the absence of exchange and assuming that mixed herds are not genetically or economically viable (that is, the existence of a divisibility problem), farmers will maximize their welfare by keeping only the multi-purpose breed, B. In the absence of exchange, choosing Breed A or C would place them on a lower indifference curve than I_1I_1. But if exchange

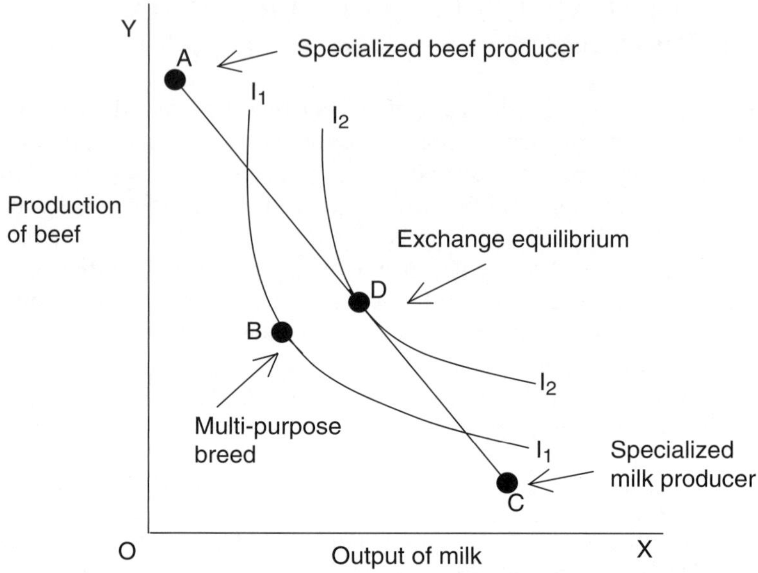

Figure 6.5 Illustration of a case in which the creation of markets eliminates multi-purpose breeds of livestock

became possible with zero (or minor) market transaction costs, farmers could gain by specializing in milk or meat production, that is, by having a herd either consisting entirely of Breed A or Breed C. For example, if the exchange line is ADC, they can reach point D on the indifference curve I_2I_2. Consequently, the multi-purpose breed, B, is eliminated. In fact, in many cases, specialization goes so far when market extension occurs that none of the by-products of one breed are marketed by a farmer. For example, beef producers in specialized (market) conditions do not also supply milk to markets. For example, the milk from rangeland beef cattle in Australia is not marketed: cows are left to feed their calves with it.

Technological developments not directly connected with the livestock industry are a third source of loss of multi-purpose breeds and some specialized breeds. Soini and De Haas (2010, p. 29) describe how the introduction of tractors resulted in the decline in the population of triple-purpose breeds of cattle in Spain. Presumably, also the decline in the demand for wool due to the increased availability of substitutes (for example synthetic fibres) has made it less economic to retain some breeds of sheep. Similarly, the conservation of horse breeds has been affected by technological change resulting in the advent of tractors, alternative methods of transport, and their replacement in war by machines.[5] The survival of several other species of farm animals is affected by the above-mentioned changes (see, for example, Anon, 2014b). Several NGOs (for instance the Rare Breeds Survival Trust in the UK and the American Livestock Breeds Conservancy) are active in promoting the conservation of rare breeds, and hobbyists and enthusiasts also play an important role in their conservation (Anon, 2014c).

In many developing countries, farming is still of a semi-subsistence type rather than pure subsistence or entirely non-exchange in nature. Nevertheless, in such cases, there are several ways by which local breeds disappear as market systems expand. For example, in Asia, breeds of livestock have traditionally been kept for multiple purposes. Cattle and buffalo, for example, provide fertilizer, draught power, milk and at the end of their working life may be sold for meat to obtain cash. In addition, they provide a store of value. But with the extension of market systems, the value of one or more of these functions may be reduced. For example, market extension makes chemical fertilizer available as a substitute for animal manure, the availability of motorized vehicles, stationary motors and electricity reduces the demand for animal draught power, and increased competition from other meat suppliers may reduce the 'retirement' price of an animal. All these circumstances reduce the utility of an animal as a store of value. Furthermore, the extension of the cash economy and banking provides an alternative and, in many respects, more convenient means to store value. Thus, because of changing economic circumstances fostered by market extension, it may no longer be economical for a farmer to keep a local multi-purpose breed.

In addition, the increasing possibility of off-farm work in some developing countries accelerates the loss of traditional local breeds used mainly for subsistence purposes or as a store of value. In such cases, purchased goods tend to substitute for home-produced ones by those having cash income, thereby reducing interest in keeping livestock for domestic use. Also, money increasingly becomes an alternative store of value for those earning cash income. As off-farm work expands, farm families have less time to care for livestock requiring considerable labour and attention, for example, herding of cattle and sheep and collection of fodder for livestock. Such economic changes can have important consequences for the survival of local breeds.

6.6 DECOUPLING OF BREEDS AND ANIMAL HUSBANDRY FROM LOCAL NATURAL ENVIRONMENTS

Modern agricultural technologies tend to decouple agriculture (divorce it) from the surrounding natural environment. This they do partly by the creation of man-made environments for domestic animals such as the provision of artificial housing, regulated water and food supplies for livestock managed under industrial-type farming. But even in the case of less-intensive modern agriculture, livestock is much protected as a rule from its surrounding natural environment, for example, via vaccinations and veterinary care, by improved pastures. Furthermore, none of the livestock food used to feed intensively managed livestock may come from the local environment. For instance, there is a heavy reliance on imported grains and food additives to feed livestock in some countries, for example in China and Vietnam. The environmental decoupling phenomenon is most pronounced for poultry and pigs kept in intensive conditions but can also be important for dairy cattle and feed-lot beef cattle.

Intensive animal husbandry seems to have been initially developed in higher income countries. It favours breeds that are highly productive under such conditions and it can be expected to result in strong lock-in (path-dependence of the type identified by Swanson). Therefore, a breed with very little environmental tolerance, say Breed I, is likely to be favoured in comparison to a breed with a high degree of environmental tolerance, say Breed II. In Figure 6.6, for example, curve ABC may represent production from Breed I in relation to a range of environmental conditions

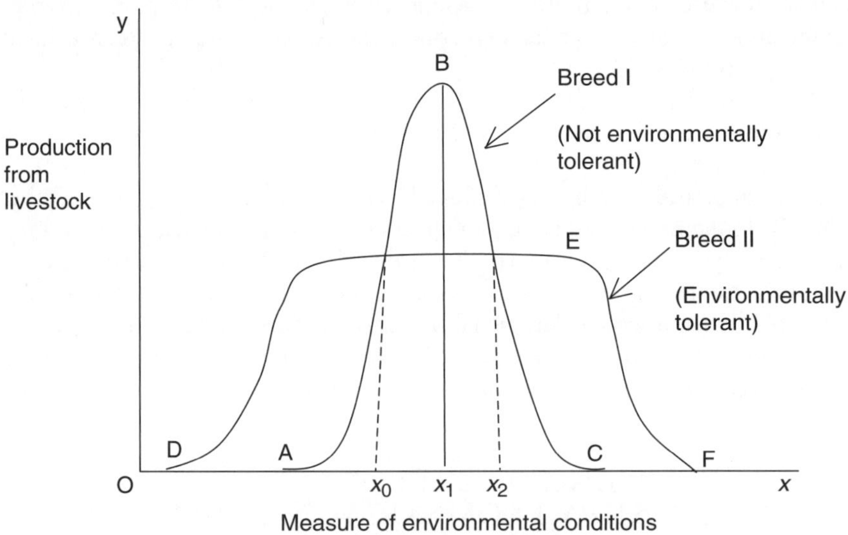

Figure 6.6 Modern livestock husbandry often favours breeds that are highly productive but show a low degree of environmental tolerance

and the corresponding curve for Breed II might be as indicated by curve DEF. Such curves reflect the biological law of tolerance (Tisdell, 1983). If it is economic by human manipulation to hold environmental conditions at or in the neighbourhood of x_1, Breed I will be favoured and Breed II may disappear. Thus, a high-yielding situation is chosen. Nevertheless, if for some reason, farmers cannot sustain ideal or near ideal environmental conditions for Breed I, production from it collapses. In contrast, Breed II is more tolerant and robust (cf. Tisdell, 1999, pp. 38, 46–7). For conditions in which $x < x_0$ or $x > x_2$, yields from Breed I are lower than from Breed II, and for some environmental conditions, they are much lower.

From a long-term point of view, the concentration of livestock production on high-yielding, environmentally-sensitive breeds may create a serious problem for the sustainability of livestock production. It is possible that farmers could lose their ability (or have reduced ability) at some time in the future to manipulate environmental conditions experienced by their livestock. If all environmentally-tolerant breeds are lost in the interim, the level of livestock production could fall sharply.

Reduced environmental control of conditions for livestock can occur for a variety of reasons. Given the close proximity in which animals are kept in large numbers when intensive animal husbandry is practised, there is a risk that new diseases will evolve to take advantage of these conditions, which constitute a new niche for microbes. Microbes have the ability to evolve quickly and they may make it less economic to engage in intensive animal husbandry, or the cost of natural inputs may increase to such an extent that intensive animal husbandry becomes less attractive from an economic point of view.

Observe that the problem illustrated by Figure 6.6 parallels that for the cultivation of high-yielding crops when compared to traditional crops. See Figure 4.7 and the surrounding discussion.

The decoupled environmental nature of modern animal husbandry has additional consequences for biodiversity conservation. It widens the geographical range of market competition. It increasingly enables much livestock production to become footloose. Such production is no longer necessarily tied to local environmental conditions and to local food supplies for animals. This footloose tendency is happening increasingly in the broiler industry but is not exclusive to it. To the extent that this raises market competition, it is liable to add to the demise of breeds not ideal for intensive husbandry. It accelerates genetic erosion.

Growing opposition in many higher income countries to industrial animal husbandry, and increasing demand for products from animals kept under more natural conditions, for example, increased demand for free-range eggs, could reverse past trends in livestock husbandry in more developed countries. Nevertheless, the tendency towards concentration of breeds may not be negated, especially given that modern supermarketing of food products puts a high premium on products conforming to a regular set of standards (cf. Young, 2001). Breeds able to deliver standardized products are favoured in such marketing systems. In addition, many of the breeds best suited for natural or near-natural conditions may already have been lost or could be in danger of being lost.

6.7 AN EXAMPLE: LOCAL PIG BREEDS DISAPPEARING IN VIETNAM

Vietnam began its transition from having a controlled communist economy to a market-led one in 1986 when it adopted *Doi Moi*, its economic renovation policy. Its economic reforms were carried out gradually but resolutely and as in China's case, were accompanied by the gradual opening up of its economy to the outside world, a high rate of growth and significant sectoral change (Tisdell, 2009). Demand for livestock products rose sharply, including the demand for pork, a major component of the Vietnamese diet.

After the commencement of Vietnam's market reforms and its opening up to the outside world, the genetic erosion of its local breeds of pigs appears to have accelerated. Several factors contributed to this erosion. Apart from market-related factors, Vietnamese government policy supported the import of foreign pig breeds to boost the productivity of pigs in Vietnam (Drucker et al., 2006). Imports of Australian Large Whites (Australian Yorkshires) proved to be very suitable for this purpose because of their resistance to heat stress, unlike some breeds from Europe (Tisdell and Wilson, 2003). In the period 1996–2006, Vietnam's pork production more than doubled due to a combination of higher yields of pork per pig and increased pig numbers (Tisdell, 2009).

Factors contributing to Vietnam's increased production of pork included its growing import of pig food and its infusion of imported genes into its pig stock. This raised the ability of its pigs to increase their yields under closely controlled husbandry dependent to a large extent on external (imported) food inputs.

The particular circumstances of the Vietnamese market resulted in the Swanson dominance-effect being marked. The Vietnamese strongly prefer fresh pork and are not inclined to buy chilled or frozen pork (Tisdell et al., 2010). As a result, the Vietnamese market for pork has a high degree of natural protection from pork imports, and these imports are low despite Vietnam not having an international comparative economic advantage in pork production (ibid.). Given that traditional pig breeds in Vietnam have lower yields under suitably controlled environmental conditions (see Figure 6.6), the policy of the Vietnamese government significantly increased the supply of pork in Vietnam. Its current policy of encouraging the development of larger enterprises in pig production will undoubtedly reinforce the replacement of local breeds by high-yielding breeds derived from imported germplasm. Nevertheless, the government's financial incentives favouring the development of larger pig enterprises may not be economically optimal in Vietnam's current stage of economic transition (Tisdell, 2010).

Note that genetic erosion of local pig breeds in Vietnam is not a result of pork imports. Nevertheless, it is partly a result of increased international trade, namely Vietnam's considerable import of pig food. In addition, Vietnam's growing range of international contacts has facilitated transfers of technologies and know-how of value for the development of its pig industry and this has favoured the use of imported pig breeds and has hastened the loss of its domestic breeds, that is, the occurrence of the Swanson effect. Although Vietnam's policies for developing its pig industry have been effective in substantially increasing Vietnam's domestic supply of pork, insufficient attention seems to have been paid to the possible long-term value of conserving at least minimum stocks of germplasm of its domestic breeds.

6.8 THE CONSEQUENCES OF LIVESTOCK HUSBANDRY FOR THE LOSS OF WILD BIODIVERSITY

In the previous chapter, it was pointed out how intensive animal husbandry can result in loss of human developed germplasm as well as wild biodiversity. Here consideration will be given to the consequences for wild biodiversity of more extensive livestock husbandry, for example, the use of rangeland for livestock production. Potentially, there are many variables that affect the consequences of pastoralism for wild biodiversity. These include stocking rates, the type of livestock kept, and the methods of managing the land used for rearing livestock.

Grazing by livestock usually alters the composition of vegetation on grazed land; some species of flora are favoured and others are disadvantaged. In addition, some species of wild fauna are likely to be favoured by environmental changes set in train by pastoralism, whereas others may disappear. The net effect of the commencement of livestock grazing seems, however, to be to reduce pre-existing wild biodiversity. If the grazing system is stable, eventually a stationary (or near stationary) state between the composition of nature and livestock grazing could be achieved in many cases. This depends on the nature of the natural environment. At the same time, any significant change in the grazing system will disrupt this equilibrium.

The development of livestock grazing in Australia illustrates these aspects. Prior to the commencement in 1788 of the British settlement in Australia, Australia had no livestock. After this settlement began, sheep and cattle grazing developed rapidly. This development favoured some wildlife species but adversely affected others. In general, species of large kangaroos (such as the Red Kangaroo, *Megaleia rufa*, Eastern Grey Kangaroo, *Macropus giganteus* and the Western Grey Kangaroo, *Macropus fuliginosus*) benefited, whereas several species of small macropods were disadvantaged and some were driven to extinction.

The large species of macropods (kangaroos) increased in number with the introduction of livestock because of the increased watering points (for example, the sinking of bores in remote areas to provide water for livestock) and pasture changes due to cattle grazing favoured them, whereas smaller species (such as hair-tailed wallabies and hare-wallabies) virtually disappeared (some became extinct) because livestock ate their shelter and increased their vulnerability to predators (Ealey and Suijendorp, 1959). In coastal areas, the clearing of scrub and rainforest for agriculture threatened other small macropods (Calaby, 1971). Similarly, Horsup (2004) also reports that cattle and sheep grazing have been major contributors to the Northern Hairy-nosed Wombat, *Lasiorhinus krefftii*, becoming critically endangered (Taggart et al., 2008). This is because this wombat species requires a permanent supply of native green grass close to its burrows, and this is usually eaten by livestock (Horsup, 1999). It also has been observed that in some areas grazing by sheep has led to the increasing presence of the shrub spinifex, and that this has contributed to an increase in the population of Euros, common wallaroos, because it is palatable to them in drought times (Newsome, 1971). Competitive and complementary relationships exist between different wild species and pastoralism in Australia (see Tisdell, 1973) and this is also the case elsewhere in the world. Not all wildlife species are adversely affected by pastoralism; some can benefit from it. Nevertheless, it is apparent that the net effect of pastoralism has been to reduce global wild biodiversity.

6.9 IMPACTS OF LIVESTOCK HUSBANDRY ON FOOD SECURITY

In the previous chapter, using Figure 5.2 for illustrative purposes, it was pointed out that the feeding of grain and other crops suitable for human consumption to livestock potentially reduces the calories of food available to humans because energy is lost by this indirect method of providing food for humans. This is sometimes used as one of the arguments in favour of vegetarianism. However, this 'competitiveness' does not occur for all methods of raising livestock, and in these cases, the keeping of livestock adds to human food security rather than reducing it.

For example, several livestock species are able to consume and digest food that is incapable of being digested by humans.[6] For example, ruminants (such as cattle, sheep and goats) are able to digest fibre including cellulose and hemicellulose, which is not able to be digested by humans (Anon, 2014d). Livestock enable areas to be used for food production which are unsuitable for growing food crops able to be consumed by humans. This is not to deny that the introduction of livestock to areas from which they were previously absent will not change the natural biodiversity of these areas. That they do change the biodiversity was illustrated in the previous section.

Furthermore, if seasonal movement of livestock is possible, this can minimize the adverse effects of livestock on natural environments. This is particularly important in arid and semi-arid regions subject to variable rainfall. Nomadism (transmigration) enables human communities to move their livestock to adjust to changing weather conditions and to continue to exist successfully in such environments, whereas they would be unable to survive if they relied on livestock confined to one locality or on cropping.

6.10 CONCLUDING REMARKS

The question of what breeds of domesticated animals should be saved from extinction is a large, complex and important topic in itself but beyond the bounds of this chapter. However, the World Conservation Monitoring Centre (1992, p. 404), states that 'a breed can be conserved (a stock maintained which continues to represent the foundation stock without too much genetic drift or inbreeding) for surprisingly small cost compared with possible economic benefits'. Furthermore, Smith (1984) demonstrates that the net economic benefits of conserving livestock breeds can be very great. Consequently, two conditions of Ciriacy-Wantrup (1968) in favour of maintaining 'safe' minimum populations of many livestock breeds appear to be satisfied, namely little cost and potentially high gains from such a strategy (see also Seidl and Tisdell, 2001 and see also Chapter 4 in this book). Nevertheless, it is unlikely to be economical to conserve all breeds even if Ciriacy-Wantrup's criterion is used (cf. Tisdell, 1990). Furthermore, given current social mechanisms, including market systems, human selection is unlikely to result in an optimal social choice of breeds to be conserved (cf. Perrings et al., 1995).

The extension of market systems is a powerful force for biodiversity loss, especially for the loss of breeds. This is not to deny that market systems may result in the development of new breeds better suited for marketing. However, this outcome may merely add to the erosion of existing breeds. The overall result of the extension of markets and economic globalization appears to be to reduce the number of extant breeds and to reduce biodiversity generally. Worldwide this loss is continuing as market systems penetrate developing countries more deeply. In this chapter, it is maintained that Swanson's lock-in or path-dependence effect identified by him in relation to species selection is also important in the selection of breeds of domesticated animals for conservation. In fact, it may be even more important at this level than at the species-selection level.

Swanson (1995a, p. 4), convincingly claims that choices of species for use in developing countries (and we can include here breeds of domesticated animals) are heavily influenced by these choices in more developed countries. However, although the Swanson effect contributes to the lock-in of choices of breeds, they are not pertinent to initial choices. Furthermore, while export of breeds from developed to developing countries appears to be the major route for transfer of breeds, there have also been some two-way flows, and consequently, exceptions to the Swanson effect. Consider, for example, the development of Brahman cattle, Brangus cattle and so on in the United States (Anon, 2014b) and Australia. This development relied on the use of imported breeds of Indian cattle.

It has also been noted that much development of animal husbandry in recent decades has resulted in its being decoupled from local natural environmental conditions. As a consequence, processes of co-evolution have largely been circumvented. This brings with it new environmental dangers and social problems (Tisdell, 2000). Apart from concerns for animal welfare (and in some cases human health) raised by industrialized animal husbandry systems, they may constitute a time-bomb for the collapse of livestock production. One cannot safely ignore the sustainability consequences of such methods of economic production. Whether or not consumer backlash against such methods will change such trends and result in more varied breeds being conserved is unclear.

Observe also that hobbyists, enthusiasts and NGOs in Western countries play a role in conserving rare and endangered breeds, but their role may be marginal. This is probably also true of the conservation of such breeds in protected areas. Nevertheless, these 'aberrations' in self-seeking economic market systems make some positive contribution to the conservation of breeds, and could, therefore, have social merit. Additionally, subsidies for preserving cultural landscapes by encouraging low intensity of land use, such as in some regions of the EU, also play a positive role as do direct subsidies for conserving specific breeds of endangered livestock. Nevertheless, the global trend of declining domestic animal diversity has not yet been halted.

The relationship between livestock husbandry and the conservation of wild biodiversity has been shown to be complex, and it has also been argued that, in identified circumstances, the keeping of livestock does not reduce human food security, but can add to it. Despite this, the actual way in which livestock production has developed appears to have reduced global wild biodiversity (including wild biodiversity in Australia), and depending on the nature of food fed to livestock, can reduce human food security with the risk of the poor being seriously affected. From this point of view, the increased consumption of grain by livestock is a legitimate concern.[7]

NOTES

1. For examples of wild species with high non-use value see Tisdell (2014).
2. Whether it adds to global biodiversity could depend, among other things, on whether it adds to the variety of niches available to wildlife.
3. Given Swanson's hypothesis, it is probably more likely that Breed I (or if a species, Species I) would be selected for development. However, despite its initial economic superiority, it could be that Breed II (Species II) could show greater responsiveness to human selection and development and could prove to be the superior economic choice. However, myopia may result in this selection not taking place.
4. Hiemstra (2010, p. 18), drawing on European experience, states: 'In the old days, domestic animals were multifunctional, they were used for draught work, clothes, manure, food and fuel. . . .After the industrial revolution, the traditional use of domestic animals for draught work, clothes and manure was slowly but steadily substituted by industrial products. With the increasing demand for protein of animal origin, breeds were intensively selected for food purposes and the development of specialized dairy and beef breeds began.' However, he does not mention the extension of markets as a factor influencing the development/selection of specialized breeds.
5. Note that in earlier times not all breeds of farm animals were multi-purpose ones. For example, horses were often bred for a particular purpose and some cattle breeds were used mainly for transport (for example, in Thailand). Nevertheless, many became victims of technological change.
6. This is one of the reasons why simple food conversion ratios for different species of livestock can be misleading. An article in the *National Geographic* (Bourne Jr, 2014, p. 101)

indicates that the food conversion ratio (body mass/meat production) of cattle is much lower than for pigs, which in turn is lower than for broiler chickens and fish, and states that 'by this measure, farming salmon is seven times more efficient than raising beef'. However, cattle and fish are able to consume different types of food. The supply of meat from one species is not necessarily in direct competition with its supply from others, but there can be overlap in consumption. For example, grain can be fed to cattle, pigs or poultry, and grain-based diets for fish are being developed.
7. This chapter contains extracts reprinted from one of my previously published articles, namely C.A. Tisdell (2003), 'Socioeconomic causes of loss of animal genetic diversity: analysis and assessment', *Ecological Economics*, **45** (3), 367–76, with permission from Elsevier.

REFERENCES

Alderson, L. (1994), *Rare Breeds*, London: Laurence King.

Anon (2014a), 'Swabian-Hall swine', Wikipedia. Available at: http://en.wikipedia.org/wiki/Swabian-Hall_swine. (Accessed 28 July, 2014).

Anon (2014b), 'Brahman (cattle)', Wikipedia. Available at: http://en.wikipedia.org/wiki/Brahman_%28cattle%29. (Accessed 28 July, 2014).

Anon (2014c), 'Rare Breeds Survival Trust', Wikipedia. Available at: http://en.wikipedia.org/wiki/Rare_Breeds_Survival_Trust. (Accessed 28 July, 2014).

Anon (2014d), 'Ruminant', Wikipedia. Available at: http://en.wikipedia.org/wiki/Ruminant. (Accessed 28 July, 2014).

Blench, R. (2000), *'You Can't Go Home'. Extensive Pastoral Livestock Systems: Issues and Options for the Future*, London: ODI/FAO.

Bourne Jr, J.K. (2014), 'How to farm a better fish', *National Geographic*, **225** (8), 92–111.

Calaby, J.H. (1971), 'The current status of the Australian Macropodidae', *Australian Zoologist*, **16** (Pt1), 17–29.

Ciriacy-Wantrup, S.V. (1968), *Resource Conservation: Economics and Policies*, 3rd edn, Berkeley, CA: Division of Agricultural Science, University of California.

Clark, C.W. (1995), 'Scale and feedback mechanism in market economies', in T.M. Swanson (ed.), *The Economics and Ecology of Biodiversity Decline: The Force Driving Global Change*, Cambridge: Cambridge University Press, pp. 143–8.

David, P.A. (1985), 'Cleo and the economics of QWERTY', *American Economic Review*, **75**, 332–7.

Drucker, A.G., E. Bergeron, U. Lemke, L.T. Thuy and A.V. Zarate (2006), 'Identification and quantification of subsidies relevant to the production of local and imported pig breeds in Vietnam', *Tropical Animal Health and Production*, **38**, 305–22.

Ealey, E.H.M. and H. Suijendorp (1959), 'Pasture management and the Euro problem in the North West of Western Australia', *The Journal of Agriculture of Western Australia*, **8** (3rd series), 273–86.

Hammond, K. and H.W. Leitch (1995), 'Towards better management of animal genetic resources', *World Animal Review*, **84/85**, 48–53.

Hiemstra, S.J. (2010), 'Introduction', in S.J. Hiemstra, Y. De Haas, A. Mäki-Tanila and G. Gandini (eds), *Local Cattle Breeds in Europe*, Wageningen: Wageningen Academic Publishers, pp. 16–21.

Horsup, A. (1999), *Northern Hairy-nosed Wombat* (Lasiorhinus krefftii) *Recovery Plan 1998–2002*, Canberra: Department of Environmental Heritage.
Horsup, A. (2004), *Recovery Plan for the Northern Hairy-nosed Wombat*, Lasiorhinus krefftii, *2004–2008*, Canberra: Department of Environmental Heritage.
Myrdal, G. (1956), *On International Economy: Problems and Prospects*, London: Routledge and Kegan Paul.
Newsome, A.E. (1971), 'Competition between wildlife and domestic livestock', *The Australian Veterinary Journal*, **47**, 577–86.
Perrings, C., K.-G. Mäler, C. Folke, C.S. Holling and B.-O. Jansson (1995), 'Introduction: Framing the problem of biodiversity loss', in C. Perrings, K.-G. Mäler, C. Folke, C.S. Holling and B.-O. Jansson (eds), *Biodiversity Loss: Economic and Ecological Issues*, Cambridge: Cambridge University Press, pp. 1–17.
Pizzi, F., D. Duales, H. Woelders and A. Mäki-Tanila (2010), 'The role and state of cryopreservation in local cattle breeds', in S.J. Hiemstra, Y. de Haas, A. Mäki-Tanila and G. Gandini (eds), *Local Cattle Breeds in Europe*, Wageningen: Wageningen Academic Publishers, pp. 78–97.
Ricardo, D. (1817), *The Principles of Political Economy and Taxation*, London: Dent. (Reprinted 1955).
Romer, P. (1987), 'Growth based on increasing returns due to specialization', *American Economic Review*, **77**, 56–62.
Romer, P. (1990a), 'Endogenous technological change', *Journal of Political Economy*, **98**, 245–75.
Romer, P. (1990b), 'Are nonconvexities important for understanding growth?', *American Economic Review*, **80**, 97–103.
Scherf, B. (1996), *World Watch List for Domestic Animal Diversity*, 2nd edn, Rome: FAO.
Scherf, B. (2000), *World Watch List for Domestic Animal Diversity*, 3rd edn, Rome: FAO.
Seidl, I. and C. Tisdell (2001), 'Neglected features of the safe minimum standard: Socio-economic and institutional dimensions', *Review of Social Economy*, **59** (4), 417–42.
Shand, H. (1997), 'Biodiversity threat: The steamroller of modern monoculture and factory farming in sweeping away our crop and domestic livestock diversity, putting future food production at risk', *New Internationalist Magazine* [Online]. Available at: http://newint.org/features/1997/03/05/bio/. (Accessed 1 May, 2014).
Smith, A. (1776 [1910]), *An Inquiry into the Nature and Causes of the Wealth of Nations*, London: Dent. (Reprinted 1910).
Smith, C. (1984), 'Economic benefits of conserving animal genetic resources', *Animal Genetic Resource Information*, **3**, 10–14.
Soini, K. and Y. De Haas (2010), 'Trends in cattle diversity and cattle production in Europe: From popular to niche', in S.J. Hiemstra, Y. De Haas, A. Mäki-Tanila and G. Gandini (eds), *Local Cattle Breeds in Europe*, Wageningen: Wageningen Academic Publishers, pp. 22–39.
Svizzero, S. and C. Tisdell (2001), 'Concepts of competition in theory and practice', *Revista Internazionale di Science Economiche e Commerciali*, **48** (2), 145–62.
Swanson, T.M. (1994), *The International Regulation of Extinction*, New York: New York University Press.
Swanson, T.M. (1995a), 'Why does biodiversity decline? The analysis of forces for global change', in T.M. Swanson (ed.), *The Economics and Ecology of*

Biodiversity Decline: The Forces Driving Global Change, Cambridge: Cambridge University Press.

Swanson, T.M. (1995b), 'Uniformity in development and the decline of biological diversity', in T.M. Swanson (ed.), *The Economics and Ecology of Biodiversity Decline: The Forces Drawing Global Change*, Cambridge: Cambridge University Press, pp. 41–54.

Taggart, D., R. Martin and A. Horsup (2008), '*Lasiorhinus krefftii*, The IUCN Red List of Threatened Species Version 2014.2'. Available at: http://www.iucnredlist.org/details/11343/0. (Accessed 29 July, 2014).

Tisdell, C.A. (1973), 'Kangaroos: The economic management of a common-property resource involving interdependence of production', *Economic Analysis and Policy*, **4** (2), 59–75. Reprinted in C.A. Tisdell (2002), *The Economics of Conserving Wildlife and Natural Areas*, Cheltenham, UK and Northampton, MA, USA: Edward Elgar Publishing.

Tisdell, C.A. (1983), 'The biological law of tolerance, average biomass and production in a variable uncontrolled environment', *International Journal of Ecology and Environmental Sciences*, **9**, 99–109.

Tisdell, C.A. (1990), 'Economics and the debate about preservation of species, crop varieties and genetic diversity', *Ecological Economics*, **2** (1), 77–90. Reprinted in C.A. Tisdell (2002), *The Economics of Conserving Wildlife and Natural Areas*, Cheltenham, UK and Northampton, MA, USA: Edward Elgar Publishing.

Tisdell, C.A. (1999), *Biodiversity, Conservation and Sustainable Development*, Cheltenham, UK and Northampton, MA, USA: Edward Elgar Publishing.

Tisdell, C.A. (2000), 'Coevolution, agricultural practices and sustainability: Some major social and ecological issues', *International Journal of Agricultural Resources, Governance and Ecology*, **1** (1), 6–16. Reprinted in C.A. Tisdell (2003), *Ecological and Environmental Economics: Selected Issues and Policy Responses*, Cheltenham, UK and Northampton, MA, USA: Edward Elgar Publishing.

Tisdell, C.A. (2009), 'Economic growth and transition in Viet Nam and China and its consequences for their agricultural sectors: Policy and agricultural adjustment issues', *Regional Development Studies*, **13**, 122–36.

Tisdell, C.A. (2010), 'The survival of small-scale agricultural producers in Asia, particularly Vietnam: General issues illustrated by Vietnam's agricultural sector, especially its pig production', in A. Salazaar and I. Riva (eds), *Sustainable Agriculture: Technology, Planning and Management*, New York: Nova Science Publishers, pp. 315–28.

Tisdell, C.A. (2013), *Competition, Diversity and Economic Performance*, Cheltenham, UK and Northampton, MA, USA: Edward Elgar Publishing.

Tisdell, C.A. (2014), *Human Values and Biodiversity Conservation: The Survival of Wild Species*, Cheltenham, UK and Northampton, MA, USA: Edward Elgar Publishing.

Tisdell, C.A. and I. Seidl (2004), 'Niches and economic competition: Implications for economic efficiency, growth and diversity', *Structural Change and Economic Dynamics*, **15** (2), 119–35.

Tisdell, C.A. and C. Wilson (2003), 'Genetic selection of livestock and economic development', in C.A. Tisdell (ed.), *Economics and Ecology in Agriculture and Marine Production*, Cheltenham, UK and Northampton, MA, USA: Edward Elgar Publishing, pp. 140–50.

Tisdell, C.A., M.L.A. Lapar, S. Staal and N.N. Que (2010), 'Natural protection from competition in the livestock industry: Analysis, examples and Vietnam's pork market as a case', in T.H. Lee (ed.), *Agricultural Economics: New Research*, New York: Nova Science, pp. 155–70.

UNDP/CSOPP (2001), '*Conserving Indigenous Knowledge: Integrating New Systems of Integration*', (no longer available). Available at: http://www.undp.org/csopp/CSO/NewFiles/dociknowledge2.html.

World Conservation Monitoring Centre (1992), *Global Biodiversity: Status of the Earth's Living Resources*, London: Chapman and Hall.

Young, J.A. (2001), 'Communicating with cod and others – some perspectives on promotion for expanding markets for fish', *Aquaculture Economics and Management*, **5**, 241–51.

7. Advances in genetic engineering and changes in biodiversity and ecosystems: economic and ecological considerations

7.1 INTRODUCTION

As a result of scientific advances in the last few decades, human beings are able to alter genomes directly, thereby creating a new type of organism. It is now possible (in many cases) to insert new DNA into a host genome, or to 'knock out' some genetic information contained in an existing genome. Consequently, the attributes of the targeted organism are altered. Genetic engineering or genetic modification has been defined as 'the process of inserting new genetic information into existing cells in order to modify a specific organism for the purpose of changing its characteristics' (Anon, 2014a, p.1). Types of genetic engineering include transgenic and cisgenic. Transgenic organisms have genes inserted into them which are derived from a different species, whereas in the case of cisgenic organisms, the transfers are from variants of the same species or from a closely related species (Anon, 2014a). In the case of transgenic organisms, the transfer of genes may even occur between kingdoms, for example, genes from bacteria have been successfully transferred to plants, and from animals to plants. A variety of methods exist for achieving such transfers (see, for example, Anon, 2014a), all of which are technically challenging. These processes (which no doubt will be refined and extended in their applications in the future) have significantly added to the ability of humankind to alter biodiversity.

Even prior to the introduction of genetic engineering, the actions of human beings in selective breeding and altering ecosystems had already greatly altered the evolution of species. With the emergence of genetic engineering, the influence of humankind on the evolution of species can be expected to increase (both directly and indirectly) relative to the occurrence of natural selection. A likely indirect effect of new genetically modified (GM) crops, for example, is that they may allow areas to be used for cropping that are now unsuitable for this purpose because they are drought-prone or subject to salting. This would enable agricultural extension to occur and result in the loss of some wild species.

In this chapter, the main focus will be on GM crops. These now account for more than 10 per cent of the total area of crop lands (Anon, 2014b). In particular, those crops that have been engineered for pest resistance or for herbicide resistance have been widely adopted by farmers.[1] Various crops have undergone genetic engineering to:

- improve their shelf life;
- improve their nutritional value, for example, by producing particular vitamins;
- resist stress, for example, as a result of drought or high soil salinity;
- resist herbicide use;
- increase their pathogen resistance;
- improve their qualities for biofuel production;
- produce useful pharmaceuticals and antigens as by-products; and
- create organisms which are efficient in removing some pollutants.

This, however, is only a sample (see Anon, 2014b, Section 7) of attributes which have been modified by genetic engineering.

The development of genetically modified organisms (GMOs) has generated both high expectations about their ability to increase economic production and help reduce economic scarcity as well as dire predictions about how they might reduce biodiversity, add to health risks and, in the long term, threaten economic sustainability. Furthermore, strong opposing views exist in the literature about their environmental benefits and costs. Proponents emphasize the benefits of reduced pesticide use whereas opponents argue that these benefits do not materialize in many cases (Zhao et al., 2011) and that when they do, they may not be sustained. For genetically engineered herbicideresistant crops, reduced tillage is also seen as an environmental advantage. However, assessment of the economic costs and benefits of the development and sustainable use of GMOs is complex, as is the evaluation of their ecological consequences (Xue and Tisdell, 2000; 2002).

A useful summary of the potential positive impacts of biotechnologies on economic development is given by Just et al. (2006a, pp. 710–12), who also express concern about the level of regulation costs borne by private firms in developing such technologies and in having them approved – a view that contrasts markedly with that of Batie and Ervin (2001). Furthermore, Qaim et al. (2008) provide a positive assessment of the economics of growing Bt (*Bacillus thuringiensis*) modified crops (including their use in China), but Zhao et al. (2011) suggest that these economic benefits are grossly overstated in China's case. These examples indicate that experts disagree in their assessments of the benefits of GMOs. Nevertheless, it is clear that GMOs vary in their ability to reduce economic scarcity and in their environmental and sustainability consequences. Individual cases need to be assessed. This view accords with that of Wolfenbarger and Phifer (2000, p. 2092), who state, 'Neither the risks nor the benefits of GEOs [genetically engineered organisms] are certain or universal. Both may vary spatially and temporarily on a case-by-case basis.' Furthermore, even in the case of those GMOs that reduce scarcity, their scarcity reducing benefits may diminish with the increased frequency of their use.

The use of GM crops results in ecological changes that usually have consequences for economic sustainability. It is argued in this chapter that the use of GM crops is likely to result, in the long run, in a decline in pre-existing biodiversity. Furthermore, some of the desired attributes of particular GM crops may be eroded in due course by the operation of ecological and evolutionary forces. Hence, their use can result in unsustainable economic development. The economics of developing and marketing GMOs (and this also has important implications for sustainable development planning) is also analysed in this chapter. It is claimed that when this is left to the private sector, the development of GMOs is heavily influenced by the nature of the economics of property rights in them. Economic factors favour the development and marketing of GMOs by large enterprises. Consequently, a few firms gain market power, which may also result in political power. Furthermore, economic considerations favour the development of GMOs to satisfy the requirements of large-scale commercial farms, and those farms are the prime (initial) markets for GMOs. Several other socio-economic factors are important when considering the development and use of GMOs. These include the building of patent walls, lack of co-evolution and related safety issues, provisions concerning legal liability and social conflicts over the use of GMOs. All these factors have sustainability implications.

The analysis in this chapter proceeds as follows. First, ecological sustainability issues raised by the introduction of agricultural GMOs are discussed. These include long-term biodiversity change caused by the use of GMOs, the ecological erosion of the fitness of GMOs with the passage of time, trade-offs of attributes of organisms as a result of genetic manipulation, and long-term changes in interspecies competition, for example, cases in which one pest is controlled as a result of genetic modification of a plant but this is countered by the increased occurrence of another pest. All these factors have consequences for the sustainability of agricultural production when a new agricultural GMO is adopted, and therefore are consequential for the sustainability of economic development.[2] Secondly, the implications for public policy and decision-making of environmental uncertainty and risks associated with the release of GMOs are also addressed. That discussion raises questions about how to make rational decisions about whether a new GMO should be allowed to be used, about how much caution should be exercised in this regard, the appropriate social criteria to adopt in such situations, and whether or not laws governing liability can adequately deal with possible damages caused by the use of GMOs.

The third relevant subject considered is the economics of developing and marketing GMOs. It is suggested that economic structures tend to evolve in industries supplying GMOs which lead to distortions in development and result in business strategies that can undermine the sustainability of development. Furthermore, it is claimed that there may be less sustainable use of GMOs in developing countries than in high income ones. Fourth, before concluding, some other socio-economic issues affecting the development and sustainable use of GMOs are considered. These include patent walls, lack of coevolution, social conflicts and knowledge imperfections among the users of GMOs. These all have consequences for the sustainability of development involving the use of GMOs. Economic, ecological and social considerations are all important in considering the scope for GMOs to contribute to sustainable development.

7.2 BIODIVERSITY LOSS AND THE USE OF GM CROPS: ECOLOGICAL PROCESSES

Initially, the introduction of a new GM crop adds to the extent of global biodiversity because some time is needed for economic and ecological change to occur. However, the initial situation is not an equilibrium one. The continuing use of the GMO is likely to change the ecological balance with the passage of time and may eventually reduce the extent of genetic diversity and/or change its composition. The change in biodiversity may come about as a result of a variety of processes. Note that because biological diversity is very heterogeneous, attempts to measure its extent by a single index are problematic, as has been pointed out by Scholes and Biggs (2005). Assigning economic or other values to its composition can be even more controversial. Nevertheless, as observed by Andow and Zwahlen (2006, p. 207), there is a risk of transgenic plants resulting in the 'loss of biological diversity, namely variability among living organisms including the ecological complexes of which they are part; this variability includes diversity within species, between species and of ecosystems'. It is worth considering some of the possible genetic changes that may come about as a result of the introduction of a GMO.

Processes Involving Long-term Biodiversity Change

Wolfenbarger and Phifer (2000) argue that the consequences of the introduction of a new GMO into an ecosystem can be likened to the introduction of a non-native species to a part of the world where it has not previously been present. It may or may not become invasive, and forecasts of its impact prior to its introduction can be very unreliable. They point out that 'not every risk associated with the release of new organisms, including transgenes, can be identified, much less considered' (ibid., p. 2090). Wolfenbarger and Phifer (2000) identify several of these ecological risks, as do Andow and Zwahlen (2006). These potential risks are complex and diverse, and only a partial indication of them is possible here.

Genetic flows involving GMOs can alter the set of extant biodiversity as the result of evolutionary processes that add new types of organisms to the set of living things and eliminate others. For example, cross-breeding of non-GMOs with a GMO may eventually eliminate some other varieties or strains of living organisms because the cross-breeds become dominant. Furthermore, some GMOs may escape into the wild and, in some instances, may outcompete and eliminate competitive organisms. However, this could be a rare event because many agricultural GMOs are only able to survive under artificially managed environmental conditions. Nevertheless, it is possible (Wolfenbarger and Phifer, 2000), and Andow and Zwahlen (2006, p. 208) point out that several new transgenic plant species (for instance, those with increased stress tolerance) could become feral and hybridize with wild relatives of existing agricultural transgenic plants, such as variants of maize. They state: 'As some of these newly transgenic species are already invasive in parts of their geographic range, there is a risk that these species could become more invasive, invade new habitats and cause a loss in biodiversity and ecosystem functions'.

An important indirect (economic) route by which GMOs can lead to a loss in genetic diversity is if they provide an economic incentive for the extension or intensification of agriculture, aquaculture or silviculture. Extensions of these land uses are likely to reduce the size and variety of habitats available to other species and reduce biodiversity in the wild. Intensification can also reduce the availability of diverse micro-habitats and reduce the presence of the varied species inhabiting these (see Chapters 5 and 8 in this book). The likelihood of these effects increases with a rise in the commercial economic success of a GMO and with the increased elasticity of demand for the produce of the GMO.

Nevertheless, the opposite view exists (expressed for example, by Just et al., 2006a, p. 711), namely that the development and widespread use of GMOs in agriculture will reduce the extension of agriculture globally and result in its intensification in a more environmentally friendly way than at present. However, this view seems too optimistic. Although the introduction of agricultural GMOs may raise yields on existing land used for agriculture, this does not ensure that the amount of land used globally for agriculture falls, other things being held constant. This outcome, predicted by Just et al. (2006a), depends on the demand for agricultural produce being relatively inelastic. However, an additional consideration (which is extremely relevant in a dynamic situation) is whether genetic engineering increases the extent of demand for a crop by increasing its uses. For example, corn (maize) has been transformed to facilitate its use for ethanol production, thereby adding to the demand for corn and to the type of potential problems previously raised using Figure 5.2 for illustrative purposes in this book. Also, the nature of biodiversity loss depends on the type of GMOs developed. Those GMOs that, for example, show greater stress tolerance can result in agricultural intensification on marginal agricultural land, for instance the replacement of grazing by cropping and consequently, the acceleration of wild biodiversity loss. It is claimed that the use of GMOs results in agricultural practices (such as reduced use of pesticides and minimum tillage) that are more environmentally friendly than existing ones. However, it is probably dangerous to generalize about this. For instance, in some cases, pesticide use has not fallen because secondary pests have increased. This has been reported for Bt cotton in China (Zhao et al., 2011).

Apart from possibly accelerating loss of wild biodiversity, a commercially successful GMO may displace similar traditional varieties of organisms in agriculture, aquaculture or silviculture due to economic or commercial selection. The mechanisms involved are similar to those described in Chapter 2 of this book. Consequently, a loss of previously human modified or developed genetic stock is likely to occur. Thus, in the long term, the sustained use of GMOs can reduce both pre-existing biodiversity in the wild as well as the previously available stock of human modified genetic material (see Figure 7.1).

The stock of genetic material can be divided into that existing in the wild and that contained in organisms cultured by humankind. These categories may partially overlap, but not all organisms cultured by humans are able to exist in the wild because processes of selection by humans result in many of these organisms being able to persist only in human managed environments. If human selected organisms are lost, they may be irreplaceable or they might only be replaced after incurring considerable cost and after a very long period of time. These might be called 'human heritage organisms'.

The suggested likely pattern of biodiversity change (illustrated in Figure 7.1) as the result of a new agricultural GMO can be summarized as follows:

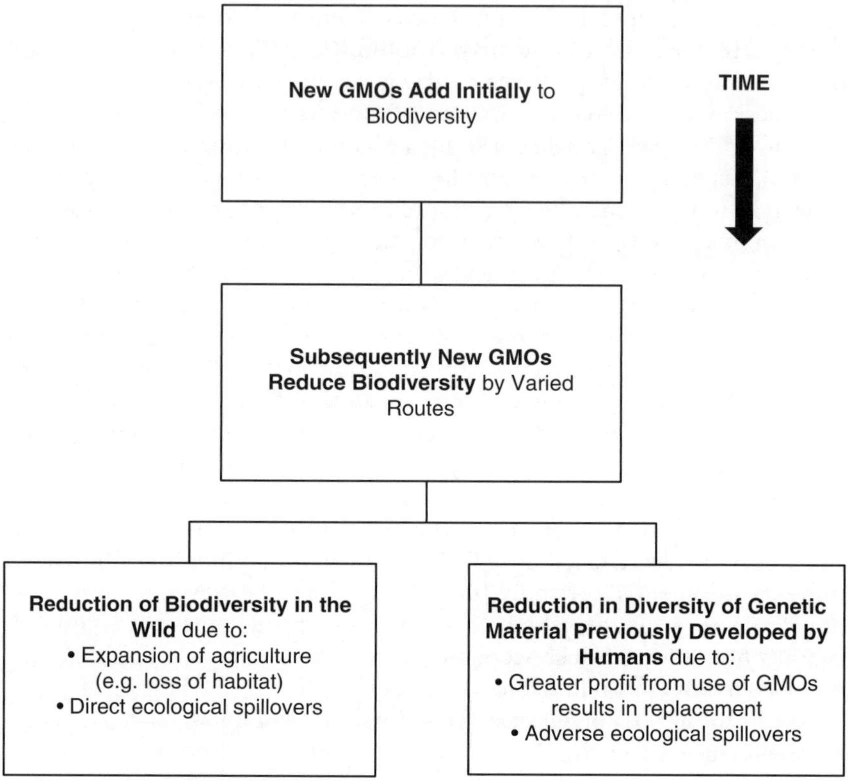

Figure 7.1 A hypothetical pattern of potential changes in the stock of genetic material once a GMO has been successfully adopted for a sufficient period of time: note that this pattern has similarities with biodiversity change likely to occur when a non-native species is introduced to a new region

- Initially (for a short period of time) the new GMO adds to biodiversity because of its limited time in the environment and its limited spatial presence.
- Subsequently (after the passage of sufficient time), the new GMO can be expected to reduce prior biodiversity.
- It may reduce the set of prior genetic material in the wild as well as that in human developed genetic material (heritage biodiversity) contained in organisms cultured by humans.
- Loss of pre-existing genetic material in the wild may occur (a) because agricultural land use is extended or intensified, resulting

in loss of habitat utilized by wild organisms or (b) as a result of gene flows from GMOs and/or competition from GMOs with wild organisms. Additions to wild genetic material could also occur in the long term.

- Human developed genetic material may be lost (a) because it is more profitable to culture GMOs rather than non-GMOs and (b) gene flows from GMOs or ecological competition from them may result in the disappearance of some human developed organisms.

There may also be other pathways by which GMOs alter the set of prior genetic material and reduce this set.

Some contrary views about the impact of GMOs on the stock of biodiversity are also found in the literature. For example, Just et al. (2006a, p. 711) state that new transgenic varieties increase yields and may reduce farmland use, thereby helping to preserve wild biodiversity. Qaim et al. (2005) argue that the use of genetically modified varieties of crops might increase crop diversity, although they do not completely dismiss the view that traditional local varieties could be replaced by a small number of genetically modified ones.

Although loss of some existing biodiversity (particularly some existing germplasm) is likely to be the consequence of the cultivation of a new GMO, it is possible also for no loss to occur. For example, this may occur if the use of the GMO is confined to a limited area, replaces a crop variety in that area which is also grown elsewhere, involves straight land-use substitution for this variety, and the GMO does not escape from where it is grown. One might, however, expect this case to be rare.

Selective Breeding, Genetic Engineering and Biodiversity Loss

It should also be noted that selective breeding may not be neutral in its effects on the stock of biodiversity. However, the process appears to be more gradual and is biologically more restricted than in the case of genetic engineering. Consequently, the ecological risks and uncertainties associated with the introduction of GMOs are normally considered to be much higher than is the use of selective breeding. For example, Wolfenbarger and Phifer (2000, p. 2088) state in their conclusions: 'Traditional breeding is limited by the available genetic variability in the target organism or its relatives. The great potential, as well as risk, of genetic engineering is that it removes those limits, providing a greater range of possibilities for transforming desired phenotypes into organisms.' It might be noted that the introduction of hybrids that are unable to reproduce themselves poses a smaller ecological risk than GMOs that can reproduce. However, the introduction of these hybrids can lead to the extension of agricultural land use and its intensification and, thereby, indirectly reduce biodiversity.

7.3 TRADE-OFF OF ATTRIBUTES AND EROSION OF FITNESS PROBLEMS INVOLVING GMOS: ECONOMIC ASPECTS

The quality of some desired attributes of GMOs may be diminished in the long run by evolutionary processes. For example, in the long run, a targeted insect pest (as a result of natural selection) could develop increased resistance to Bt modified crops (see Andow and Zwahlen, 2006, pp. 203–6). Another possibility is that pests not affected to any great extent by Bt (or another genetic modification) may eventually increase their populations and adversely affect the benefits otherwise obtained from these modifications of plants (Zhao et al., 2011). Ecological evolution and change may also occur in natural organisms to take advantage of new niches opened up by the presence of GMOs. In addition, adding a desired attribute to an organism by genetic modification can sometimes only be achieved by forgoing another valued property of the organism. Consequently, in this case (and the other circumstances mentioned above) economic considerations are relevant in determining the desirability of the modification. Let us consider these issues in more detail.

Genetic Modification Can Involve Trade-off Between Attributes

In genetically modifying an organism in order to increase its fitness for a particular purpose, there is a risk that other relevant traits will be weakened and become negative. The overall biological fitness of the organism may even decline because its fitness depends on the overall balance of its attributes (Dawkins, 1986). These trade-offs are not always anticipated by scientists. For example, Liu (2009) has reported that exudates from Bt cotton roots in China increase the susceptibility of some varieties of cotton to a fungal disease, cotton blast. The incidence of this disease has increased and is causing a substantial decline in cotton yields in China. This trait of Bt cotton was not predicted by scientists.

Lu (2009) has also reported that China has genetically modified some species of poplar trees to reduce their lignin content. Their lowered lignin content reduces the cost of processing these trees for paper production. But the lower lignin content makes the trees more prone to damage by wind and increases their susceptibility to some pests. Furthermore, it has been found that the development of Bt poplars to control attacks by one species of beetle has made them more vulnerable to attack by another species of beetle (Lu, 2009). Thus there is evidence that the development of organisms by genetic engineering to improve one or more attributes of an organism can occur at the expense of other desirable attributes.

Some of the trade-offs resulting from genetic modification appear to be predictable, but others are uncertain and can come as a surprise. When such a trade-off is unavoidable, there is an associated economic problem in the case of commercially utilized organisms, namely to ensure that the engineered genetic combination gives the greatest net economic benefit in relation to the attainable set of genetic combinations. The net economic benefit from using the GMO must be compared to that from using the alternative unmodified organism. However, making this optimal economic decision is often complicated by the presence of uncertainty. Consequently, there is difficulty in predicting accurately the possible trade-off of the relevant attributes. Furthermore, social decisionmaking under risk and uncertainty is complex because of heterogeneous attitudes to risk-taking in society. In addition, the sustainability of the attributes of the GMO needs to be considered because, as a result of evolutionary, ecological and other factors, the desired attributes of a GMO may be eroded with the passage of time. These dynamic considerations further complicate the economics of genetic manipulation and add to uncertainty. A simple economic conceptualization of the economic trade-off problem involved in genetic engineering and associated issues is given in Tisdell (2009b, pp. 344–8).

Changes in Interspecies Competition as a Possible Negative Consequence of Crops Engineered to Resist Particular Pests and Increased Resistance of Targeted Pests to their Modification

Several mechanisms have been mentioned that can result in the fitness of GMOs being undermined. These include the erosion of desirable attributes with the passage of time and sometimes the need to forgo some valuable traits in the process of adding other valued attributes. Another factor that can reduce the economic suitability of GMOs is that their presence may improve the competitiveness of competing species, some of which may be regarded as pests. The case of Bt poplars has already been mentioned. Another case involves herbicide-resistant GM rice developed experimentally in China. It has been found that it is prone to cross-fertilize with wild rice, which is regarded as a weed in rice crops (Lu and Fu, 2009). Because there is less rural labour available for weeding by hand in China due to the rural-to-urban migration drift, the increased resistance of GM wild rice to herbicide could become a serious economic problem if herbicide-resistant GM rice becomes available for commercial use.

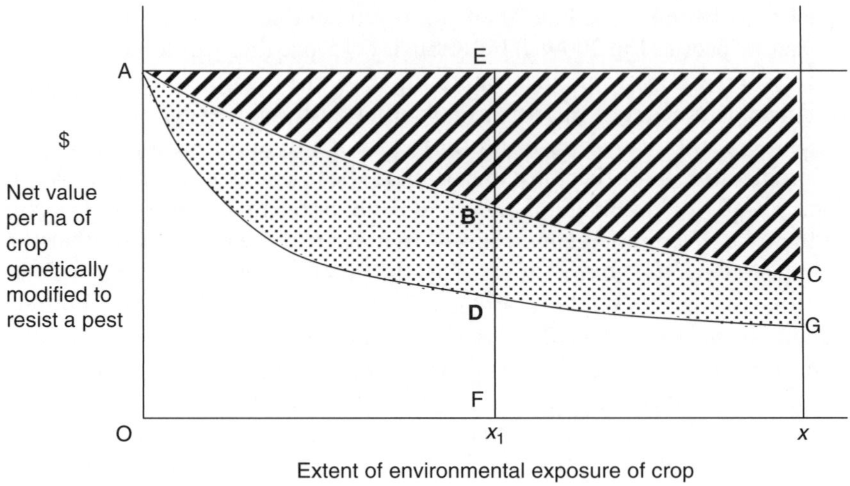

Figure 7.2 *An illustration that the net value per hectare of a crop genetically modified to resist pests or a particular herbicide is likely to fall as this crop receives greater environmental exposure*

Many crops (including cotton, soybeans, maize, alfalfa and canola) have been modified so that they are resistant to the herbicide glyphosphate. However, some weeds which were once susceptible to this herbicide have become increasingly resistant and weeds that are resistant but originally were not major pests have become more prevalent. Similar trends in reduced 'fitness' have been found in crops into which Bt protein crystals have been introduced. Therefore, as a result of increased exposure to crop modifications made to enhance herbicide or pest resistance, the *conceptual* pattern shown in Figure 7.2 might emerge for some crops. However, other patterns of the decreasing value of a crop genetically modified to resist particular pests are possible. The pattern shown indicates that once the primary pests become more resistant to the genetic modification, they eventually outcompete secondary pests. The net value per hectare of their use to resist targeted herbicides or pests has declined (as conceptually indicated by curve ABC) as their environmental exposure has increased. In addition, there has been a further reduction in their net value per hectare due to increased prevalence of non-targeted pests. As a result, the total net value per hectare of some GM crops has followed a path similar to ADG. For example, in the case illustrated when environmental exposure reaches level x_1, the total net value per hectare of the genetically modified crop is DF compared to its initial value of OA. The initial value of OA is reduced by BE on account of the increasing resistance of the targeted pest and by BD on account of the increasing prevalence of the non-targeted pests. Environmental exposure is likely to be a function of the amount of land cropped with the genetically modified crop, the efflux of time and any particular measures adopted to reduce the rate at which biological resistance develops. Therefore, x is used here as an index for the purposes of exposition.

Measures to Counteract Biological Erosion in the Fitness of a Genetically Modified Crop

It is sometimes possible to adopt measures to delay the growing resistance of pests to crops that have been modified to become insecticidal. Anon (2014b) points out:

> One method of reducing resistance is the creation of non-Bt crop refuges to allow some nonresistant insects to survive and maintain a susceptible population. To reduce the chance that an insect would become resistant to a Bt crop, the commercialization of transgenic cotton and maize in 1996 was accompanied with a management strategy to prevent insects from becoming resistant to Bt crops, and insect resistance management plans are mandatory for Bt crops planted in the USA and other countries. The aim is to encourage a large population of pests so that any resistance genes that are recessive are greatly diluted within the population.

However, all of these methods involve some costs and so the options must be weighed up from an economic point of view. These aspects will be considered in Chapter 9.

Another response is for companies specializing in the supply of GM crops to create even 'more powerful' engineered crops. For example, Anon (2014b) reports 'As weeds have grown resistant to glyphosphate and other herbicides used in concert with resistant GM crops, companies are developing crops engineered to become resistant to multiple herbicides to allow farmers to use a mixed group of two, three or four different chemicals'. However, this trend favouring 'cocktails' of herbicide use may worry some environmentalists.

7.4 RISKS, UNCERTAINTY AND SOCIAL DECISIONS ABOUT INTRODUCING AGRICULTURAL GMOS

Rational decisions about the development and use of GMOs are complicated by the fact that their biological and ecological consequences are usually subject to considerable uncertainty. In turn, this adds to the difficulty of estimating their social and economic impacts accurately – a difficulty compounded by the fact that their environmental impacts are not confined to the adopter of a GMO. Environmental spillovers or externalities from the use of GMOs create hazards (risk of losses) for non-users, as has been well documented (Batie, 2003; Batie and Ervin, 2001; Ervin et al., 2003). Consequently, all these factors have given rise to considerable discussion of optimal procedures for the introduction of GMOs and their regulation (see, for example, Just et al., 2006a). Questions debated include the following:

- To what extent and how should precaution be exercised before allowing a GMO to be used?
- In turn, this necessitates consideration of the appropriate social criterion to use in making this decision.
- To what extent should legal liability be used to improve the social desirability of choices about the development and use of GMOs?

Only brief comments on these aspects are required because of their extensive coverage in the relevant literature.

Exercising Caution

Views differ about how much precaution (care) should be exercised before a new agricultural GMO is released and about the nature of restrictions that should be put on its use. For example, Zilberman (2006) suggests that the nature and extent of the precautions legally required in the USA before a new GMO can be released are excessive from an economic point of view, whereas others, such as Batie and Ervin (2001), Batie (2003) and Ervin et al. (2003), argue that greater precaution should be exercised in the regulation of biotechnology in the USA. Ervin et al. (2003, p. 8) state: 'A rebalancing of US regulation to focus more on controlling type II errors appears prudent given the small amount of science to inform biosafety regulation'. According to these authors, biosafety regulation in the USA gives too much weight to type I errors – that is, the likelihood of rejecting the hypothesis that a new GMO has no serious adverse environmental consequences, when it in fact does have these consequences. A type II error involves the opposite mistake.

A useful way to think about the problem is to consider how much information should be collected on the possible environmental consequences of a GMO before it is released. The process of collecting existing information (by experimentation and so on) involves an extra cost and this has to be weighed against the extra benefits obtained. Baumol and Quandt (1964) contend that from an economic perspective the collection of information should only proceed to the point where the extra cost of its provision equals its extra expected benefit. However, a problem is that the extra expected benefit from extra search is often very uncertain, especially in the case of new GMOs. Thus it is not clear in advance how much search is a desirable precaution. Furthermore, not everyone will be interested only in expected benefits, that is, benefit on average. Many will be concerned about the range of possibilities, including the worst potential outcome from the adoption of a GMO. Attitudes to riskbearing will need to be taken into account and, within society, these are heterogeneous.

It has been suggested that the precautionary principle could be usefully applied in decision-making about the introduction of GMOs. However, there are many different interpretations of this principle, and some of these interpretations are problematic (Tisdell, 2010; see also Chapter 4 in this book). Moreover, they are mostly not very specific. Nevertheless, as a rule, other things being held constant, it is rational to take greater care (exercise more precaution) in making decisions about the release of GMOs when the changes set in train are irreversible or can only be reversed at a substantial cost (see, for example, Tisdell, 1970). However, it might sometimes happen that the introduction of a GMO creates no irreversible environmental change, for example if it is withdrawn sufficiently quickly after its initial introduction. Individual cases need to be assessed. However, sight should not be lost of the fact that unfavourable changes in biodiversity from the introduction of GMOs can have serious adverse economic consequences for future generations (Tisdell, 2011a).

The Appropriate Social Criterion

The adoption of new GMOs is likely to result in some members of society gaining and others losing, for example, as a result of adverse environmental spillovers. Where economic changes bring gains to some members of society and losses to others, some economists suggest that the potential Paretian improvement (Kaldor–Hicks) criterion should be adopted to decide if the change is socially advantageous. The change is judged to be socially advantageous if the gainers could compensate the losers and be better off than before the change. Posner (1971) has advocated use of this principle as a guide to just laws (Tisdell, 2011b). However, this criterion ignores the welfare effects of changes in income distribution. Furthermore, in the case of GMOs, it is not always clear in advance of their introduction and use (and it may not become apparent until long after their release) whether the gains of some are sufficient to more than compensate the losers.

Legal Liability

One way in which society may try to cope with the uncertainties associated with the development and use of GMOs is to make their developers and users legally liable for the adverse consequences of their introduction. The question then arises of whether that liability should be strict or should only apply if there is inadequate exercise of a duty of care (Tisdell, 1983). If strict liability applies, it will make for greater caution on the part of the developers and users of GMOs. But if the burden of proof is on the plaintiffs, it may be difficult and costly to succeed in bringing a case (even a class action) for damages. If liability is not strict but relies on the exercise of the duty of care, the extent to which a duty of care has been exercised is open to debate. Thus legal liability alone seems insufficient to protect society against all damages that might be inflicted by the use of a potentially damaging GMO. Furthermore, the maximum compensation that can be paid by those responsible for torts is limited by the amount of their available assets. These may fall well short of the amount of compensation required and, in some instances, monetary payment to victims may be deemed insufficient to compensate them for damages.

7.5 ECONOMICS OF DEVELOPING AND MARKETING GMOS

General Influences of Economic Factors on the Nature of Development and Marketing of GMOs

In addition to the ecological risks faced by nature and society at large from the use of some GMOs, high costs and considerable business risks are involved in developing GMOs. Therefore, when their development is left to private enterprise, only large firms usually engage in this activity. Furthermore, if businesses are to have an economic incentive to develop new GMOs, then they must be assured of intellectual property rights. Patents are used in many countries for this purpose.

However, it is important to note that patent laws and regulations covering the granting of patents in relation to GMOs are not the same in all countries. For example, in China patents are granted on the techniques for producing specific GMOs but not on the products produced as a result of the availability of those techniques. For example, a patent would be granted in China on techniques to produce a Bt modified plant but not in seed that transmits this genetic trait. However, in the USA both techniques for producing GMOs and the products produced as a result of applying these techniques are protected by patent law. The economic analysis of GMOs that follows is based on the assumption of strong and extensive patent laws of the type adopted in the USA.

Patents grant a monopoly to the patentee to use the patented invention for a specific number of years. During this period, the inventor has a monopoly in the use of the invention, in this case a new GMO. A business that has a patent for a GMO may, therefore, engage in monopoly-pricing of its use. While this is not socially ideal (Tisdell and Hartley, 2008, pp. 36–7), it is necessary for the inventor to be sufficiently rewarded for his/her inventiveness. Otherwise, there is no private economic incentive to engage in R&D effort. This is particularly important bearing in mind that not all research efforts are successful in developing an economically viable new product. Furthermore, the patent is only for a limited period of time and there is no guarantee that some other firm will not develop a commercially superior competitive GMO within this period. This will limit the profit possibilities available to the original patentee.

In addition, the level of the monopoly price of, say, GM seed will be influenced by the price of competitive traditional seed and by farmers' net profits when using this seed (Tisdell, 2005, p. 221). This constrains the price that the owner of GM seed will find it profitable to charge. In some instances, however, the GMO monopolist may deliberately charge a low price initially for the use of its GMO to encourage its adoption and eliminate the use of competing traditional organisms. This predatory behaviour can eliminate competing traditional varieties of a crop and can be a prelude to higher monopoly prices for the GMO (ibid., p. 222).

The legal costs of protecting patented GMOs against infringements of rights can be high. Once again, this tends to favour large firms as developers of GMOs, as does the high cost of obtaining approval for their release.

The establishment of legal property rights in new GMOs is only part of the economics of developing and marketing them. The transaction costs of marketing some GMOs (for example GM seed) are high because considerable effort is required to ensure that the buyers do not reuse or sell any GM seed they produce without permission. This adds to the cost of drawing up contracts with clients. Costs are incurred by sellers in monitoring the use of GMOs by buyers to ensure compliance with agreements, and there may be considerable legal costs if breach of contract occurs. Similar costs do not arise in the case of sterile hybrids.

These transaction costs favour the marketing of GMOs by large companies. They also favour their sale to larger-sized agricultural and related enterprises because market transaction costs per unit of the GMO sold tend to decline with the value of sales to an individual production unit.

Taking into account potential economic gains and transaction costs, there is an economic incentive to develop GMOs that are likely to be adopted by large-scale commercial farming enterprises, particularly in higher income countries. Furthermore, organisms (for example, crops) for which there is a large volume of demand are likely to be favoured, other things being equal. Private businesses have little incentive (at least initially) to develop GMOs to suit farming involving small-scale units or semi-subsistence agriculture. Therefore, crops that are important in less developed countries (but not in higher income countries) are unlikely to be initially targeted by private enterprises for the development of GMOs unless they are plantation crops, possibly involving foreign investment. Figure 7.3 provides an overview of predicted patterns of development of

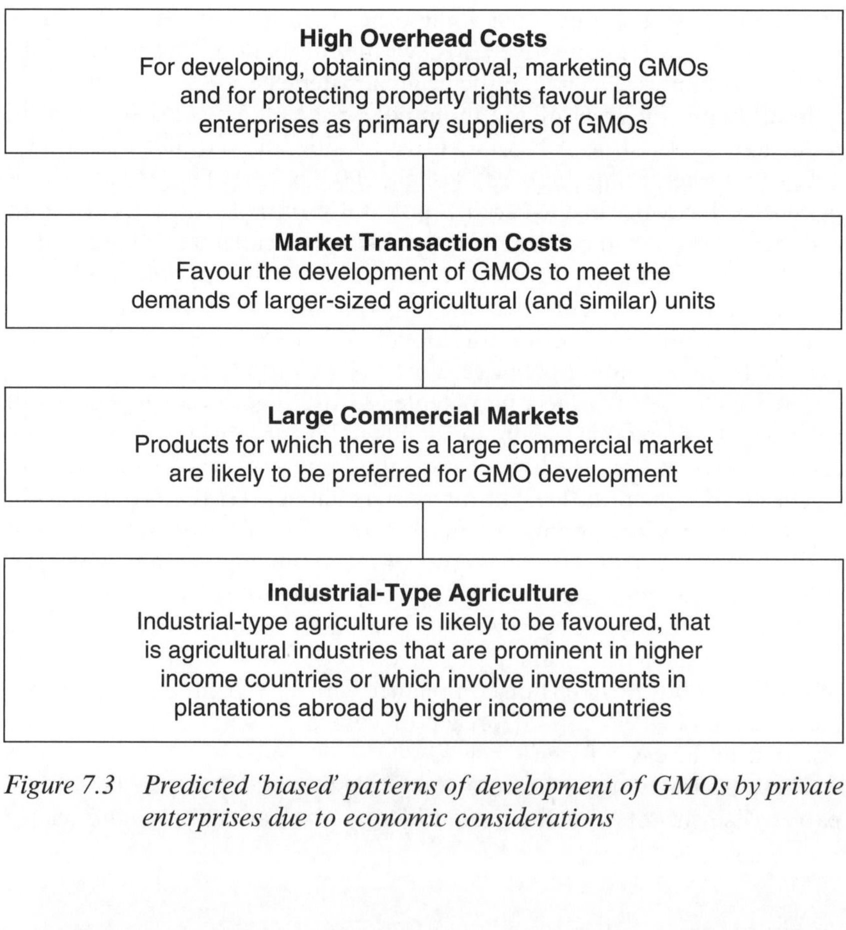

Figure 7.3 Predicted 'biased' patterns of development of GMOs by private enterprises due to economic considerations

GMOs by private enterprises when economic considerations are taken into account.

The view that the type of biases indicated in Figure 7.3 exist in the development and use of GMOs has been previously expressed by Batie and Ervin (2001, p.435). They state: 'Theory and empirical evidence suggest that there are significant incentives for private firms to discount and neglect certain environmental impacts and develop products to meet mainly the needs of those able and willing to pay.' They argue that the development of GMOs is largely 'technology push' driven and that in undertaking this development, private firms are unlikely to promote the social good because of missing markets (market failures). Although the developers of GMOs have to market and push their sales to would-be users who have no previous knowledge or experience with them, the economic incentives for their development are basically driven by (latent) private demand for their use and depend on prospective private economic returns, as is clear from the theory outlined below. Therefore, if Batie and Ervin's view is that advances in biotechnology are basically supply driven, it does not seem to be correct. The nature of demand seems to be influential.

Observe that while new GMOs are more likely to be developed in higher income countries (such as the USA) for the use of their farmers than in lower income countries, diffusion to less developed countries occurs. The economic theory of price discrimination suggests that a firm with a global monopoly in a GMO is likely to charge a lower price to users in markets where demand is lower and more elastic (developing countries) than in markets where demand is higher and more inelastic (higher-income countries). In addition, fewer precautionary measures may be required in the use of the GMO in a less developed country than in higher income countries. This adds to the commercial initial attractiveness of marketing GMOs in developing countries, but it is likely to accelerate the erosion of their economic value with the passage of time.

The possibility of biases in the development of biotechnology has already been noted in the relevant economic literature. For example, Just et al. (2006b, p.7) argue that large-crop, large-country bias exists in the development of GMOs. They state that 'the reason is that biotech firms require a large potential market to justify the large overhead costs of regulatory compliance in addition to research and development' (Alston, 2004; Bradford et al., 2006).

However, the strength of this bias is likely to weaken with the passage of time. This is because much of the knowledge obtained in initially creating genetically modified agricultural organisms (for example, crops) of a particular type can often be transferred to different organisms and markets. The extra cost of the R&D required to do this (for instance, insert the Bt gene construct into a new type of crop) is reduced as a result of learning-by-doing, which of course takes time.

Support from Economic Theory for the Above Propositions

Basic economic theory lends support to several of the propositions summarized in Figure 7.3 about likely patterns of development and use of GMOs. In order to show this, assume that the GM product is GM seed and consider the proposition that, other things being equal, the economic incentive to develop such seed is greater, the larger the expected size of the market for it. For simplicity, assume that the successful development of the GM seed will establish a monopoly for it.

In Figure 7.4, line BD_1 represents the demand for the GM seed if the market is small and the line D_2D_2 represents the demand for it if its market is larger. The average operating cost of supplying the GM seed once it has been developed is assumed to be a constant OA and, therefore, line AC represents the average variable cost of its supply. The line BMR is the marginal revenue line corresponding to the demand relationship BD_1. Hence, when demand is at level BD_1, the maximum operating profit of the producer of patented GM seed is equivalent to the area of rectangle $AFJP_1$. There is a mark-up

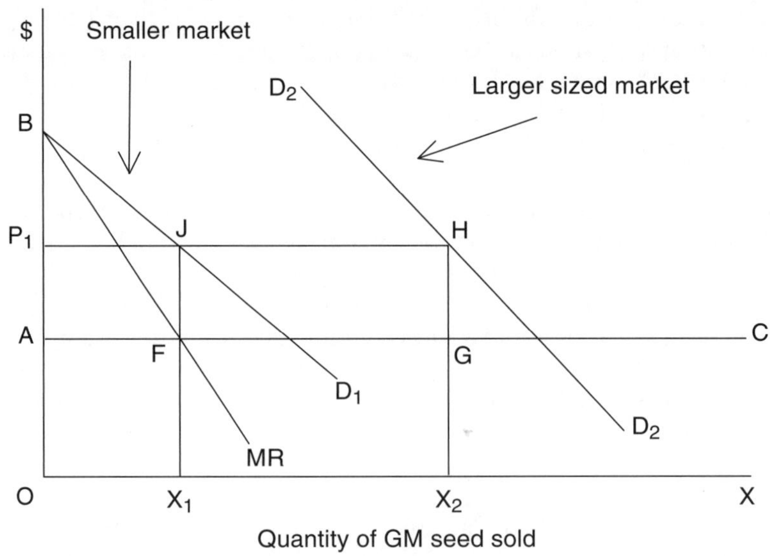

Figure 7.4 The proposition that the larger the expected market for a GMO, the greater is the profitability of developing it, other things held constant

on average operating costs for each unit of seed sold of AP_1. If this margin is retained when the demand for the seed is much higher at D_2D_2, operating profit increases substantially. It rises by an amount equal to the area of rectangle FGHJ, that is by an amount equivalent to $(X_2 - X_1)(P_1 - A)$. Therefore the economic advantages of a larger-sized market are apparent.

Observe also that if there are high overhead costs to be recovered (for example, the costs of the R&D invested in the creation of the GMO), this will also favour markets of larger size. When overhead costs are taken into account, the average total cost of supplying a new type of GM seed might be as indicated by the declining curve marked RST in Figure 7.5. Line BD_1 represents the demand for the GM seed if the market is relatively small and D_2D_2 represents demand if the market is larger. Because of the overhead costs involved, the development of GM seed for the smaller-sized market is unprofitable but it can be profitable for the larger-sized market, other things being equal.

The presence of high overhead costs and patents creates significant

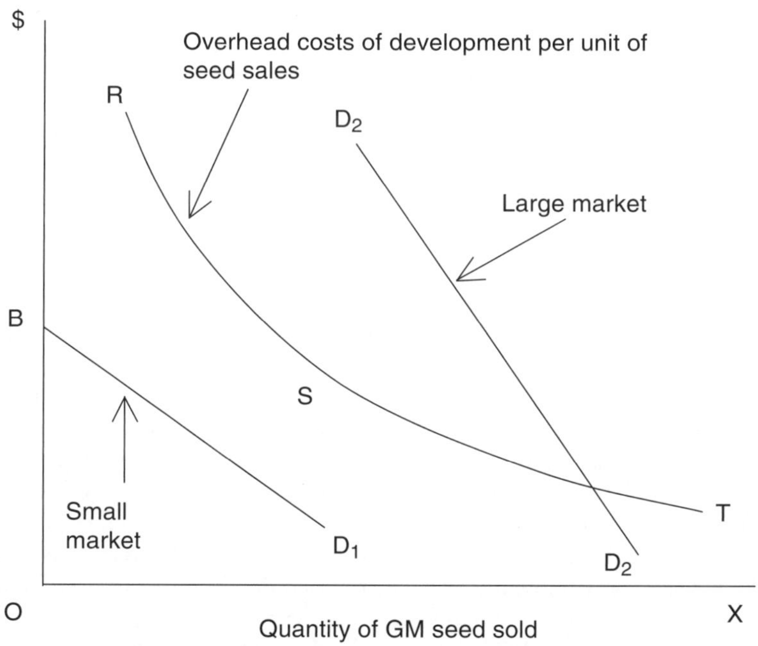

Figure 7.5 *The proposition that when overhead costs are large (as they are likely to be for the development and marketing of GM seed) the development of a GMO to satisfy smaller-sized markets is unlikely to be profitable*

barriers to entry into the agricultural biotechnology industry and fosters market concentration. High overhead costs include the costs of obtaining approval for the release and use of new GMOs. An additional 'overhead' cost in many countries (such as the USA) is the cost of political lobbying and engaging in publicity campaigns to counter opposition by some public interest groups (for example, environmentalists, some types of consumers) to the release of GMOs. Heisey and Schimmelpfennig (2006, p. 421) found that in the USA, 'for large agricultural biotechnology firms, there is a rough positive relationship between firms' R&D or net sales and the amounts they devote to the lobbying and campaign contributions'. These costs can reduce the private economic attractiveness of investment in the development of GMOs in high income countries.

When the demand for using an organism is relatively inelastic, this can increase the profitability of developing a GMO, other things held constant. The more inelastic is the demand for a resource, the higher is the monopoly price that can be secured by the monopolist from it and the greater are the profits obtainable from its supply, other things being equal. This is illustrated in Figure 7.6. If demand for GM seed corresponds in this diagram to BD_1, then making the same assumptions as were previously made for Figure 7.4, the patentee's monopoly price is P_1 and his/her profit is equal to the area of rectangle $AFJP_1$. Should, however, the demand curve for the GM seed be steeper, as for example indicated by the line D_2D_2, the profit-maximizing monopoly price will be higher. In the case illustrated it is P_2. Consequently, operating profit from selling the GMO rises by an amount equivalent to the area of the dotted rectangle P_1JKP_2. The steeper or more inelastic the demand curve, the greater is monopoly profit. Hence, if the size of the market for GM seed is large and the demand for it is relatively inelastic, this provides a strong economic incentive for the development of GM seed.

Note that, in Figure 7.6, the marginal revenue line (not shown) corresponding to D_2D_2 passes through point F, as does the marginal revenue line corresponding to BD_1. This is because, in the linear case, the decline in the marginal revenue line occurs at twice the rate of the decline of the average revenue line (equals the demand relationship), which implies that the distance of the average revenue line from the vertical axis is twice that of the marginal revenue line. Because the two demand lines are equal at L, the corresponding marginal revenue lines must also intersect at F, and the distance AF = FL. Observe also that, if the demand curves shown were for two separate markets, the monopolist would find it profitable to engage in price discrimination. It would be most profitable to charge a low price, P_1, in one market and a high price of P_2 in the other market where demand is more inelastic. This suggests that the price charged

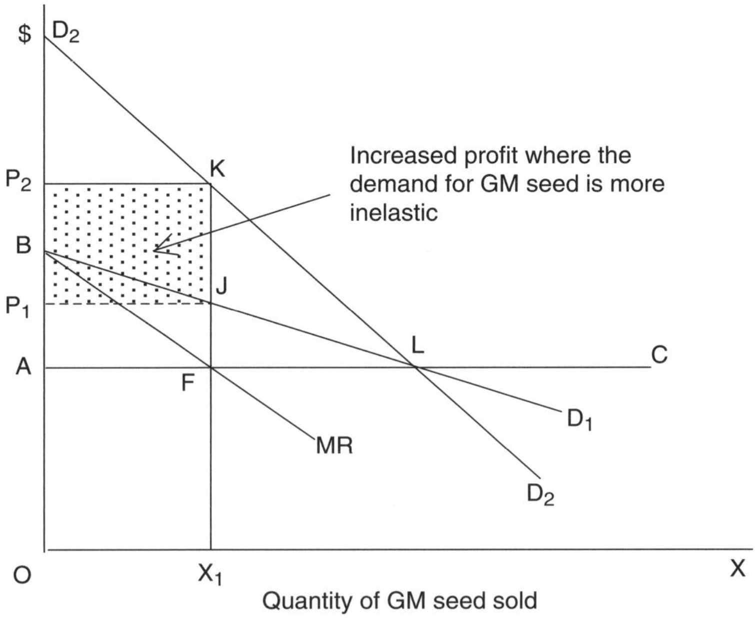

Figure 7.6 An illustration of the proposition that, other things held equal, the profitability of developing a new GMO for a monopolist is greater the more inelastic is the predicted demand for it

for GM seed is likely to be higher in a high income country than in one with lower income. Of course, the reason for doing this has little or nothing to do with the pursuit of social justice, but is motivated by profit maximization.

It was argued above that sellers of GMOs are likely to find it more profitable to sell to large-scale farms than smaller ones. This is based on the view that the average transaction costs involved in sales are likely to fall as the quantity sold to any production unit rises. This is illustrated by Figure 7.7, where the line AFC represents the average production cost of the supplier of GM seed and curve HJK represents the average total cost of supplying GM seed to a farming unit. The difference between line AFC and curve HJK represents the level of average marketing transaction costs involved in selling GM seed to a farm. Suppose that the seller of GM seed sells it at P_1 per unit. Then, it is unprofitable to sell GM seed to farms that wish to buy less than x_1 of it. For those farms wanting to buy more than x_1 of seed, the profit of the seller rises with the amount sold to a farm. Hence, the hypothesis stated above that developers of GMOs

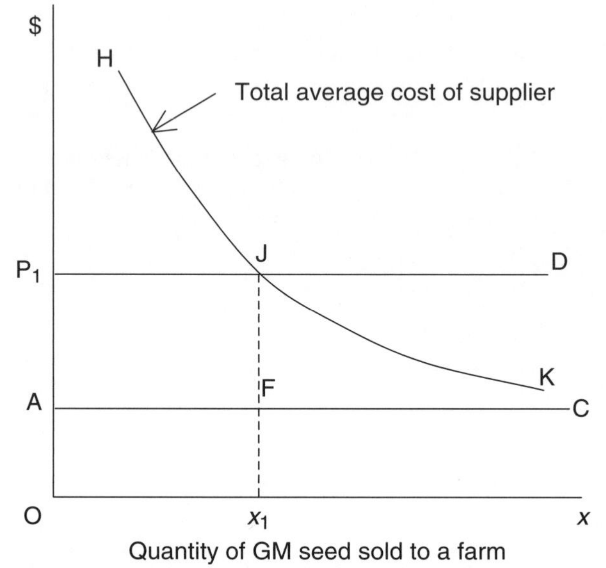

Figure 7.7 An illustration of the proposition that developers of GM seed are likely to find it more profitable to sell it to large commercial farms than to small farms

favour developing these for industries in which large commercial farms exist is supported.

If the marketing transaction costs of the supplier of GM seed include monitoring costs of their use incurred either voluntarily in pursuit of their own self-interest or as a result of legal obligations (checking on the presence of refuges, barriers to pollen drift), the level of these costs may vary by the types of economy in which GM seed is released. These costs might, for example, be lower in developing countries, in which case smaller agricultural producers might be accessed, particularly if the patent-holder's use of agents to distribute GM seed in developing countries involves reduced market transaction costs. In fact, large corporations involved in the sale and development of GM crops (such as Syngenta and Monsanto) prefer to license the production and commercialization of their technology in developing countries to local companies. Furthermore, less demanding biosafety rules (and or less ability to enforce rules) for farm use of GM seed in developing countries will tend to initially raise the demand by their farmers for GM seed, including those on smaller sized farms. However, as a result of accelerated environmental deterioration (for example, faster erosion of the value of the genetic composition of the GMO), this demand may not be sustained in the long run.

7.6 OTHER SOCIO-ECONOMIC FACTORS RELEVANT TO THE INTRODUCTION OF GMOS

Patent Walls and Property Rights

The patent system can be used in special cases by creators of new GMOs to erect patent walls. A range of unilateral patents makes it difficult for others to progress in developing new GMOs without infringing the rights of existing patents. This creates barriers to entry of potential competitors to existing suppliers of GMOs, reduces market competition and favours market concentration. In turn, this can slow technological progress (see, for example, Tisdell, 1999).

Furthermore, some representatives of LDCs believe that such a situation is unfair to LDCs since they are likely to be restricted in their future options for developing GMOs by the existence of patented GMOs, many of which are based on the modification of genetic material originally obtained from LDCs, for example, soya beans (Xue et al., 2004). A further bone of contention is that, in many cases, no compensation is paid to LDCs for the original genetic material, whose presence was essential for the creation of the GMO. The issue of whether compensation should be paid for the use of such material is, however, complex from an economic point of view. From a global development and economic welfare point of view it can be argued that payment for the use of extant genetic material of this type is only justified in special circumstances (Tisdell, 2009a). The matter is complicated, for example, by the fact that some LDCs make extensive use of genetic material derived from other LDCs (for instance, tomatoes and potatoes) and they also use genetic material from countries that are now higher income countries, for example some types of maize from the USA. This matter is given further consideration in Chapter 10 of this book.

Lack of Co-evolution

GMOs are usually developed in experimental conditions divorced from local conditions. Whereas most other methods of genetic change involve slow and gradual changes in genetic material, creation of GMOs can result in rapid and large shifts in the composition of genetic material. Therefore, the creation of GMOs is often considered to be relatively hazardous, especially if transgenic engineering is involved. Change may not be gradual, scope for learning by trial-and-error may be reduced and the potential for irreversible consequences are all factors seen as heightening the risk of releasing GMOs. Hence, there are considerable concerns about biosafety risks associated with the introduction of GMOs.

A further potential problem is the lack of social and ecological coevolution that can arise when new GMOs are released. Some writers, such as Norgaard (1994), regard lack of co-evolution as undesirable.[3] In the past, the social practices of many local communities evolved in step with changes in their agricultural and other productive practices, thus balance and social harmony existed between local communities and changes in production methods. In modern times, however, this harmony has been destroyed by rapid changes in production methods and in types of production (Tisdell, 2003, Ch. 19). The result is increasing stress for individuals and social dislocation. The introduction of GMOs potentially adds to this lack of coevolution, which seems to be symptomatic of modern economic systems.[4]

Social Conflicts about the Use of GMOs, and Knowledge Imperfections

The use of GMOs frequently creates social conflict. For example, some social groups are fearful of the possible biological consequences of their use, including their possible long-term impact on health. Some buyers wish to avoid products, particularly food products, that contain GMOs. This can result in some GM products selling at a reduced price compared to similar non-GMO products. This contributes in part to conflicts between growers of non-GMO crops and those cultivating GM crops.

Two problems can occur:

1. cross-pollination may occur between some GM and non-GM crops; and
2. if it is unclear to buyers which produce is based on GMOs and which is not, they may become suspicious of crops for which both GM and non-GM varieties exist and reduce their demand for these crops.

There is a problem of asymmetric information between sellers and buyers in this case (Akerlof, 1970; Varian, 2006, Ch. 37). In this case, sellers know the nature of the product but consumers are unable to determine this from mere inspection of the product. Consequently, there is scope for fraud by sellers. In extreme cases, this leads to the collapse of the whole market and in many cases results in reduced demand for some types of the marketed product. For example, in the case considered here, it could result in a decline in demand for produce claimed to be GM-free. Certification schemes can, however, help to reduce the problem and laws may be strengthened to require accurate disclosure of the nature of the goods being traded, whether GM-free or not.

A related problem is that many farmers have little or no knowledge about GMOs. Zhao et al. (2011, pp. 994–5) reported on the basis of their survey in China that Chinese 'farmers have virtually no knowledge about Bt cotton and genetic engineering. A low level of understanding implies that farmers have virtually no means to interpret possible agricultural production problems that they might encounter in the field.'

In order to reduce the problem of possible cross-pollination, GM crops and non-GM crops may be located in spatially well separated areas. For some crops that depend for pollination on drifts of pollen in the wind, border hedges may be required to lower the likelihood of pollen drift. However, this does not give complete protection. Furthermore, it is pertinent to note the following in relation to GMOs:

> Currently, insect resistance management advises a strategy that combines a high-dose exposure to toxin interspersed with planting refuges, areas without the transgenic crop, to minimize the spread of resistance of a population. Evidence indicates that a properly implemented refuge strategy can slow the rate of existence evolution but does not prevent it. (Wolfenbarger and Phifer, 2000, p. 2092).

Batie and Ervin (2001) argue that in the context of GMO introduction and use, missing information and missing markets are major reasons for social conflict and failures to make wise social choices. These factors can certainly add to social conflict. However, even with perfect information and perfect markets, social conflict is still likely to occur when new biotechnologies are introduced because as a rule there are gainers and losers in society (Just et al., 2006b, p. 4).

7.7 MORE ON LEGAL LIABILITY FOR THE USE OF GMOS

Legal liability for the use of GMOs is an issue. For example, to what extent should developers of GMOs be liable for any environmental damages caused by their use? To what extent should farmers who use GMOs be responsible for any environmental damages caused by their use? The socially appropriate extent of legal liability is difficult to determine. Strict liability for damages caused by GMOs will reduce economic incentives to develop and use them. The consequence could be loss of substantial economic benefits. Questions requiring consideration include how wide ranging should legal liability for damages be, should it be capped, should a developer be absolved from legal liability if the government approves a GMO for use and to what extent should liability be reduced if its developer acts with due care before releasing a new GMO (Tisdell, 1993, Ch.5). It seems that, as the level of income in countries rises, social preferences tend to shift from limited legal liability for possible damages caused by new products towards strict liability.

Establishing legal rights is one thing; enforcing them is another. Costs and other difficulties are involved in enforcing laws or taking advantage of the law. The costs of taking legal action sometimes exceed its expected benefits because legal solutions often entail high transaction costs and the outcomes are uncertain. The question of the adequacy of proof and that of burden of proof on parties affects the economics of pursuing legal action. If, for example, a non-GMO crop is 'contaminated' by a GM crop, it may be difficult to prove which farm(s) using GMs is (are) responsible. In Germany, this has resulted in laws being passed that make all growers of related GM crops in the locality legally responsible for contributing collectively to compensation. This means that the plaintiff does not have to identify the particular farm(s) that was (were) the source of the 'contamination'.

The stage at which it is socially optimal to release a GMO after its development and testing is difficult to determine. This is because, even after considerable testing, some environmental impacts may be uncertain. Further delay may reduce those uncertainties, but if the economic benefits from using the GMO seem to be substantial, delay involves considerable opportunity costs. It is usually too costly to wait for the release of a new economically promising GMO until virtually no doubts remain about its environmental impacts. Community standards and expectations have to be taken into account in deciding on the appropriate time at which to release a new GMO.

7.8 CONCLUDING COMMENTS

Planning for sustainable development is complex when this development depends on the introduction of new techniques with uncertain and potentially irreversible effects, as is the case for some new biotechnologies, particularly those relying on new GMOs. As shown in this chapter, ecological, economic and social factors associated with the adoption of GMOs all have to be taken into account in assessing the prospects for the sustainable use of GMOs and whether they are likely to permanently increase economic production and human welfare. Social agreement is lacking about what rules should be followed in allowing the adoption of GMOs, and particularly about how much and what type of precaution should be taken before permitting their use. Indeed, it may not be possible to achieve social unanimity about this. Nevertheless, consideration of the economic, ecological and social effects identified in this chapter as resulting from the introduction of GMOs should enable more informed decisions to be made about their impacts on sustainable development. This should assist policy-makers in determining the role that GMOs should play in their plans for sustainable development. Policy-makers need to be aware that the introduction of GMOs does not provide a miraculous means to achieve sustainable development and may even, in some areas, threaten its achievement. Some of the issues raised in this chapter will be given further consideration in the next chapter and in Chapter 9.[5]

NOTES

1. The two most widely adopted genetic modifications of crops are (1) their modification to express insecticidal crystal proteins contained in the bacterium, *Bacillus thuringiensis*, Bt, and (2) their modification to resist the use of the herbicides glyphosphate or glufosimate. Often both types of these modifications are made to a crop, as has been done, for example, for corn (maize).
2. Note, however, that not all GMOs are developed for agricultural use. Those GMOs developed for some uses (for example insulin production) may have no ecological spillovers since they are not released to an open environment.
3. While this may be so, it can also be argued that new GMOs are subject to greater scientific scrutiny prior to their release than are new cultivars obtained by selective breeding.
4. The development of GM crops seems to support the increasing 'commodification' of agriculture.
5. This chapter extends and revises one of my previous publications, namely C.A. Tisdell (2013), 'Economics, ecology and GMOs: sustainability, precautions and related issues', in M.A. Quaddus and M.A.B. Siddique (eds), *Handbook of Sustainable Development Planning: Studies in Modelling and Decision Support*, 2nd edn, Cheltenham, UK and Northampton, MA, USA: Edward Elgar Publishing, pp. 91–117.

REFERENCES

Akerlof, G. (1970), 'The market for lemons: Quality and the market mechanism', *Quarterly Journal of Economics*, **84**, 488–500.

Alston, J.M. (2004), 'Horticultural biotechnology faces significant economic and market barriers', *California Agriculture*, **58** (2), 80–88.

Andow, D.A. and C. Zwahlen (2006), 'Assessing environmental risks of transgenic plants', *Ecology Letters*, **9**, 196–214.

Anon (2014a), 'Genetic engineering', Wikipedia. Available at: http://en.wikipedia.org/wiki/Genetic_engineering. (Accessed 7 August, 2014).

Anon (2014b), 'Genetically modified crops', Wikipedia. Available at: http://en.wikipedia.org/wiki/Genetically_modified_crops. (Accessed 7 August, 2014).
Batie, S.S. (2003), 'The environmental impacts of genetically modified plants: Challenges to decision making', *American Journal of Agricultural Economics*, **85** (5), 1107–11.
Batie, S.S. and D.E. Ervin (2001), 'Transgenic crops and the environment: Missing markets and public roles', *Environment and Development Economics*, **6**, 435–57.
Baumol, W.J. and R.E. Quandt (1964), 'Rules of thumb and optimally imperfect decisions', *American Economic Review*, **54**, 23–46.
Bradford, K.J., J.M. Alston and N. Kaitzardonakes (2006), 'Regulation of biotechnology for speciality crops', in R.E. Just, J.M. Alston and D. Zilberman (eds), *Regulating Agricultural Biotechnology: Economics and Policy*, New York: Springer, pp. 683–97.
Dawkins, M.S. (1986), *Unravelling Animal Behaviour*, London: Longman.
Ervin, D.E., R. Welsh, S.S. Batie and C.L. Carpentier (2003), 'Towards an ecological systems approach in public regulation of transgenic crops', *Agriculture, Ecosystems and Environment*, **99**, 1–14.
Heisey, P. and D. Schimmelpfennig (2006), 'Regulation and the structure of biotechnology industries', in R.E. Just, J.M. Alston and D. Zilberman (eds), *Regulation of Agricultural Biotechnology: Economics and Policy*, New York: Springer, pp. 421–36.
Just, R.E., J.M. Alston and D. Zilberman (2006), *Regulating Agriculture Biotechnology: Economics and Policy*, New York: Springer.
Just, R.E., D. Zilberman and J.M. Alston (2006), 'Regulating agricultural biotechnology: Introduction and overview', in R.E. Just, J.M. Alston and D. Zilberman (eds), *Regulating Agricultural Biotechnology: Economics and Policy*, New York: Springer, pp. 3–35.
Liu, B. (2009), 'The effect of root exudates from two transgenic insect-resistant cotton lines and the growth of *Fusarium oxysporum*', in D. Xue (ed.), *Risk and Assessment and Safety Regulation of Genetically Modified Organisms*, Beijing: China Environmental Science Press.
Lu, B. and Q. Fu (2009), 'Commercialization of transgenic rice in China: Potential environmental biosafety issues', in D. Xue (ed.), *Risk Assessment and Safety Regulation of Genetically Modified Organisms*, Beijing: China Environmental Science Press, pp. 52–67.
Lu, M. (2009), 'Transgenic poplar-development and biosafety in China', in D. Xue (ed.), *Risk Assessment and Safety Regulation of Genetically Modified Organisms*, Beijing: China Environmental Science Press.
Norgaard, R.B. (1994), *Development Betrayed*, London: Routledge.
Posner, J.R. (1971), *The Theory of Justice*, Cambridge, MA: Harvard University Press.
Qaim, M., C.E. Pray and D. Zilberman (2008), 'Economic and social considerations in the adoption of Bt crops', in J. Romeis, A.M. Shelton and G.G. Kennedy (eds), *Integration of Insect-Resistant Genetically Modified Crops within IPM Programs*, Vol. 5, Dordrecht: Springer, pp. 329–56.
Quaim, M., C. Yarkin and D. Zilberman (2005), 'Impact of biotechnology on crop genetic diversity', in J. Cooper, L.M. Lipper and D. Zilberman (eds), *Agricultural Biodiversity and Biotechnology in Economic Development*, New York: Springer, pp. 283–307.

Scholes, R.J. and R. Biggs (2005), 'A biodiversity intactness index', *Nature*, **434**, 45–9.

Tisdell, C.A. (1970), 'Implications of learning for economic planning', *Economic Planning*, **10** (3), 177–92. Reprinted in C.A. Tisdell (1996), *Bounded Rationality and Economic Evolution*, Cheltenham, UK and Brookfield, USA: Edward Elgar Publishing.

Tisdell, C.A. (1983), 'Law, economics and risk-taking', *Kyklos*, **36** (1), 3–20.

Tisdell, C.A. (1993), *Environmental Economics: Policies for Environmental Management and Sustainable Development*, Aldershot, UK and Brookfield, VT, USA: Edward Elgar Publishing.

Tisdell, C.A. (1999), 'Diversity and economic evolution: Failures of competitive economic systems', *Contemporary Economic Policy*, **17** (2), 156–65.

Tisdell, C.A. (2003), *Ecological and Environmental Economics: Selected Issues and Policy Responses*, Cheltenham, UK and Northampton, MA, USA: Edward Elgar Publishing.

Tisdell, C.A. (2005), *Economics of Environmental Conservation*, 2nd edn, Cheltenham, UK and Northampton, MA, USA: Edward Elgar Publishing.

Tisdell, C.A. (2009a), 'Global property rights in genetic resources: Do they involve sound economics? Will they conserve nature and biodiversity?', in J.B. Aronoff (ed.), *Handbook of Nature Conservation – Global Environmental and Economic Issues*, New York: Nova Science Publishers, pp. 199–212.

Tisdell, C.A. (2009b), *Resource and Environmental Economics: Modern Issues and Applications*, Singapore, Hackensack (NJ), London and Beijing: World Scientific.

Tisdell, C.A. (2010), 'The precautionary principle revisited: Its interpretations and their conservation consequences', *Singapore Economic Review*, **55**, 335–52.

Tisdell, C.A. (2011a), 'Core issues in the economics of biodiversity conservation', in R. Costanza, Karin Limburger and Ida Kubiszewski (eds), *Annals of the New York Academy of Sciences*, **1219** (1), 99–112.

Tisdell, C.A. (2011b), 'Is Posner's principle of justice an adequate basis for environmental law?', in A.T. Riley (ed.), *Advances in Environmental Research*, 3, New York: Nova Science, pp. 347–55.

Tisdell, C.A. and K. Hartley (2008), *Microeconomic Policy: A New Perspective*, Cheltenham, UK and Northampton, MA, USA: Edward Elgar Publishing.

Varian, H. (2006), *Intermediate Microeconomics: A Modern Approach*, 7th edn, New York and London: Norton and Company.

Wolfenbarger, L.L. and P.R. Phifer (2000), 'The ecological risks and benefits of genetically engineered plants', *Science*, **290**, 2088–93.

Xue, D.Y. and C.A. Tisdell (2000), 'Safety and socio-economic issues raised by modern biotechnology', *International Journal of Social Economics*, **27** (7/8/9/10), 699–708.

Xue, D.Y. and C.A. Tisdell (2002), 'Global trade in GM food and the Cartagena Protocol on Biosafety: Consequences for China', *Journal of Agricultural and Environmental Ethics*, **15** (4), 337–56.

Xue, D.Y., Y. Lin and C.A. Tisdell (2004), 'Conflicts about global trade in GMOs: General issues and Chinese case studies', in C.A. Tisdell and R.K. Sen (eds), *Economic Globalisation: Social Conflicts, Labour and Environmental Issues*, Cheltenham, UK and Northampton, MA, USA: Edward Elgar Publishing, pp. 314–35.

Zhao, J.H., P. Ho and H. Azadi (2011), 'Benefits of Bt cotton counterbalanced by secondary pests? Perceptions of ecological change in China', *Environmental Monitoring and Assessment*, **173**, 985–94.

Zilberman, D. (2006), 'The economics of biotechnology regulation', in R.E. Just, J.M. Alston and D. Zilberman (eds), *Regulating Agricultural Biotechnology: Economics and Policy*, New York: Springer, pp. 243–61.

8. Human developed and modified ecosystems: their economic value

8.1 INTRODUCTION

In recent times, considerable attention has been given (both by economists and natural scientists) to the economic value of natural ecosystems and in particular to their importance as suppliers of natural capital and providers of valuable goods and services. Much less consideration has been given to the economic value of human developed and modified (natural) ecosystems, such as those present in agriculture, aquaculture and silviculture. In fact, many discussions of the economic value of ecosystems (such as that of Brown et al., 2007) focus entirely on the value of natural ecosystems. Even when the economic value of human developed or modified ecosystems is recognized, they usually receive minimal coverage, as in the TEEB project examining the economics of ecosystems and biodiversity (Kumar, 2010), an aspect discussed in Tisdell (2014a). Why does this apparent bias exist?

One of the reasons might be a belief that human developed or modified ecosystems do not satisfy the basic requirements for the existence of an ecosystem. This possibility is discussed below. Another reason could be that individuals have typically failed to appreciate the total economic value of natural ecosystems but are more aware of the economic value of those developed by human beings. Therefore, this bias could be an attempt to redress this imbalance. A third possibility is that this emphasis is designed to offset a perceived political imbalance. Those seeking to convert natural ecosystems to private use by altering them usually proclaim the economic value of this transformation. By contrast, those who favour the retention of natural ecosystems face organizational difficulties in publicizing the economic value of these systems and in politically opposing their conversion. This is because a large proportion of the economic benefits from such ecosystems stem from their components which constitute collective or public goods. This can result in both political and market failure. For example, political free-riding is likely to occur in such cases. Although the strong focus of scholars on the value of natural ecosystems may be well intended, balanced analysis also requires concerted consideration of human developed and modified ecosystems, particularly their economic value compared to one another and to natural ecosystems. Therefore, the emphasis in this chapter is mainly (but not entirely) on human developed and modified ecosystems and additional consideration is given to the modification of natural ecosystems in Chapters 16 and 17.

This chapter is developed as follows. First, the concept of an ecosystem is discussed, paying particular attention to its relevance to human developed and modified ecosystems. This extends the discussion in Chapter 2. General features of the dynamics (the functioning) of ecosystems (such as their robustness or resistance, resilience, sustainability and equilibration) are subsequently examined, and this is followed by a discussion of the economic valuation of human developed or modified ecosystems and the ways in which cultivation alters the functioning of ecosystems. Attention is then given to the relevant properties and functioning of some specific agroecosystems (for example, non-organic and organic farming) and to systems for aquaculture and for silviculture, paying particular attention to the intensification of such systems. The subsequent discussion provides a general theoretical perspective on the sustainable intensification debate and its consequences for the state of the environment.

8.2 THE CONCEPT OF AN ECOSYSTEM AND ITS RELEVANCE TO HUMAN DEVELOPED ECOSYSTEMS

An ecosystem consists of a community of organisms, the nature of which depends on their abiotic environment and their relationships with one another. Both the biotic and the abiotic components of an ecosystem are linked by network relationships and cycles, such as nutrient cycles and energy flows, and form a holistic system.

The concept was first introduced into the literature by Tansley (1935). Although Clements (1916) had emphasized the importance of considering the holistic nature of communities of organisms and the succession of these communities after disturbance, Tansley was of the opinion that both biotic and abiotic factors need to be taken into account in considering the nature of these communities and their dynamics. Willis (1997, p. 270) (following his discussion of the history of the concept) suggests that an ecosystem can be defined 'as a unit comprising a community (or communities) of organisms and their physical and chemical environment, at any scale, desirably specified, in which there are continuous fluxes of matter and energy in an interactive open system'. According to Willis (1997, p. 270), concerns about sustainable development have highlighted the need for a science of ecosystems 'to safeguard ecosystems at risk'.

In response to these concerns, several governments have sponsored studies by ecologists and by economists to determine the value of ecosystems, the threats to these, and to determine policies that may be used to counter these threats. These studies include the Millennium Ecosystem Assessment (2003; 2005) by natural scientists and the study of the Economics of Ecosystems and Biodiversity, mainly by economists (see, for example, Kumar, 2010). While these studies have highlighted the contributions of natural ecosystems to human well-being, they have failed to give enough attention to the comparative economic value of human engineered changes to natural ecosystems and to human dominated ecosystems. In general, the economic value of human developed ecosystems and natural systems altered by human actions has received inadequate attention given their economic and environmental importance (see Tisdell, 2014a). Also the scale at which the functioning of ecosystems is assessed is too wide in many cases and sometimes is too narrow for policy purposes.

Are these shortcomings due to the inadequacy of the concept of an ecosystem and the inadequate ways in which the nature of ecosystems has been analysed? Sagoff (2003) argues that the ecosystem concept has several shortcomings from the point of view of scientific methodology and that too often ecosystem analysis has been based on theories divorced from observations. While all the limitations of ecosystem analysis mentioned by Sagoff have substance, the important one (in the context of this chapter) is his assertion that ecologists (and one can add economists examining the economics of ecosystems) have failed to show how ecosystem analysis 'can help solve environmental problems [and associated economic problems] both in pristine and in human-dominated systems' (Sagoff, 2003, p. 529).

Lack of attention to human dominated or modified ecosystems probably reflects the view of many ecologists that ecosystem analysis is really only applicable to natural systems. For example, Willis (1997, p. 270) states that 'the ecosystem concept has provided a productive basis for the understanding of natural systems with a high degree of [self] organization'. This seems to imply that ecosystem analysis has little relevance to other than natural ecosystems, especially if they are not self-organizing. Although most (probably all) human developed or modified ecosystems are not self-organizing (because they rely on human effort or input to retain their organization), nevertheless their performance depends on their biotic and abiotic components in interrelated ways. Therefore, ecosystem analysis seems to be relevant. The nature of this interdependence influences the sustainability of such systems. For example, some agroecosystems accelerate soil loss and nutrient loss from areas where crops are grown and result in a deterioration of soil structure. In the absence of remedial action, this eventually reduces crop yields. While the application of chemical fertilizers can compensate for nutrient loss, changes in soil structure and loss of soil may continue to create problems for the sustainability of crop yields. The nutrient run-off from agriculture may also result in adverse environmental externalities, for example, unwanted acceleration of weed growth in water bodies.

Schröter et al. (2014) also acknowledges that the concept of what constitutes an ecosystem service is controversial and discuss contending views about this. They express the view that only the contributions of natural processes, such as fertile soil and available water, should be seen as providing ecosystem services. They correctly point out that some forms of intensive agriculture and aquaculture have a very low degree of dependence on natural processes. They argue that the concept of ecosystem services 'should be limited to the contribution of natural processes to the production of these "man-made" goods [that is, the products of agriculture and aquaculture] and not consider these goods themselves as ES [ecosystem services]' (Schröter et al., 2014, p. 8). This differs from the approach in the Millennium Ecosystem Assessment (2003; 2005), which considers such products to be ecosystem services. It seems that Schröter et al. (2014) want to limit ecosystem analysis as far as possible to natural processes. It is true that the products of most bio-industries are not produced entirely by natural processes but nevertheless ecological systems (managed by humans to varying degrees) are involved in their production, and therefore ecosystem analysis is relevant to their assessment. Whether or not the 'man-made' products from such systems should be described as being ecosystem services seems to be of secondary importance. They are not the outcome of natural processes but their production is subject to natural laws and is not entirely independent of the natural world, particularly if humans are regarded as part of that world. In this book, the view is taken that ecosystem analysis is relevant both to natural systems and human dominated ones. It makes no difference to the assessments made in the book whether the 'man-made' products of bio-industries are regarded as being ecosystem services or not.

8.3 THE DYNAMIC PROPERTIES OF ECOSYSTEMS: RESILIENCE AND RESISTANCE

The dynamic properties of natural or near-natural ecosystems have received much attention in the literature. In particular, a central concern has been the resilience of such ecosystems, that is their ability (or otherwise) to re-establish their prior equilibrium following an exogenous stock. Ecosystems displaying greater ability to re-establish their equilibria following such shocks are considered to be more sustainable than those that lack this resilience. It is frequently argued that ecosystems containing greater biodiversity are more sustainable than those that are not biologically diverse because they show greater resilience. However, this view needs to be qualified, as I have argued elsewhere (Tisdell, 1999).[1]

The concept of resilience is closely related to that of the stability of the equilibrium of a system. A system that is resilient tends to return to its equilibrium once disturbed (Chapin III et al., 2002, pp. 282–3). However, the speed and the path by which it returns to that equilibrium can be quite varied. The extent of this variation is not captured by the simple concept of resilience. Furthermore, some systems have multiple equilibria. After some types of disturbance, an ecosystem may not return to its original state but will gravitate to a new equilibrium state. For example, after a natural ecosystem has been transformed by agriculture and agriculture ceases, the ecosystem may gravitate towards a new equilibrium state with a different set of biological communities to the original one.

Apart from the concept of resilience, another concept used in considering the dynamics of ecosystems is their resistance to shocks. For instance, for a shock of equal magnitude, some ecosystems are more prone to be disturbed than others. I have described ecosystems that are more resistant to shocks as being more robust (Tisdell, 1999, Ch. 4). However, in addition, I also classified those systems that are able to continue to fulfil specified functions when parts of the system are removed as being robust. For example, some ecosystems may continue to supply their wanted services if some species are removed. However, the removal of parts of a system could be regarded as a type of shock to the system. Therefore, the conventional specification of resistance will be used here. Nonetheless, the concept of how well a system functions as parts are removed is worth further exploration.

Given the conventional perspective, four types of ecosystems can be identified. These are:

1. resilient and resistant;
2. resilient but not resistant;
3. not resilient but resistant; and
4. not resilient and not resistant.

However, these indicators are qualitative rather than quantitative, and this is a limitation.

Moreover, the concept of a shock is not straightforward. The size, duration and attributes of the shock need to be specified. How sensitive, for example, is the functioning of an ecosystem to loss of different parts? If the shock is in the form of an environmental event, its consequences are likely to depend on its magnitude and its duration.

The question also needs to be considered of whether the equilibrium of a natural ecosystem[2] which is resilient and resistant is desirable. The fact that a natural ecosystem is relatively sustainable is not enough to make it desirable. For example, Chapin III et al. (2002, p. 267) indicate that the tsetse fly is a keystone species in Africa and its removal would significantly change the nature and functioning of African ecosystems. Nevertheless, its removal is likely to be wanted by Africans. Consider another example. The addition of species (such as feral animals in Australia) alters ecosystems and may eventually result in new equilibria. However, these new equilibria may be less desirable than the previous 'natural' ones.

Most human developed or modified ecosystems are not self-regulating. They need sustained management by humans in order to keep them in a desired state. For example, the cultivation of soils usually leads to nutrient loss from the cultivated area by leaching, soil erosion and by crops (containing nutrients and organic matter) being taken off the land for use elsewhere. This loss may be compensated for, to some extent, by the addition of artificial fertilizers or organic fertilizers brought to the cultivated area. Crops also have to be tended by humans to prevent weeds and other competing vegetation from growing, that is, to prevent the succession of new communities of organisms which would compete with crops on the cultivated land. However, the loss of trees on cultivated land interferes with ecosystem cycles. For example, it interferes with nutrient recycling. Despite this, a reduction in natural nutrient recycling may not be considered by farmers to be an economic loss. This will be so if these lost nutrients can only be recycled (retained) in a relatively natural system, the products and services of which are little valued by agriculturalists. In such circumstances, it can be economic to forgo some natural nutrient-recycling on land by cultivating it in order to obtain products and services of greater value. From an economic perspective, natural nutrient recycling is not an absolute good.

8.4 THE ECONOMIC VALUATION OF HUMAN DEVELOPED AND MODIFIED ECOSYSTEMS

Determining the economic value of human dominated ecosystems and comparing this value with ecosystems not managed by human beings is extremely challenging, particularly when environmental spillovers occur and interdependence exists between different ecosystems and patterns of land use (biosphere use).[3] These factors should be taken into account in assessing the economic value of different patterns of land use. In addition, patterns of land use frequently give rise to a multiplicity of different economic values, some of which may be positive and others negative. These values can be classified in different ways. One classification (currently widely adopted) is to divide the total economic value of environmental resources into their use value and their non-use value. These categories can then be further subdivided depending on the valuation problem being considered (see, for example, Pascual et al., 2010). Furthermore, it is widely recognized that natural ecosystems which have been altered by human use (and consequently become human dominated ecosystems) can have multiple values consisting of both use and non-use values (see, for example, Durand and Van Huylenbroeck, 2003). This underlines the view that agriculture can possess multi-functionality. Both natural as well as human altered ecosystems can have multiple economic values. For example, ecosystems and landscapes in the Yorkshire moors and dales have changed as a result of sheep grazing and associated agricultural practices. These landscapes (ecosystems) possess direct use value (supply of mutton and wool) and their scenery is valued by visitors and locals, and presumably (as a non-use value) by some who do not visit the area. That is not to suggest that there is an unlimited demand for such landscapes and ecosystems but they are, nonetheless, valued resources (Vanslembrouck and Van Huylenbroeck, 2005). Beyond a point, the spillovers from the supply of such ecosystems are liable to become Paretian irrelevant (Tisdell, 2009a).

Estimating the comparative total economic value of replacing a natural ecosystem by a human developed one is a formidable task, given the likely inadequacy of knowledge and the complexity of the calculations required to produce accurate estimates. Furthermore, the flow of economic values as a function of time varies and therefore the question of whether and how to discount future economic values must be considered. Various approaches to discounting are discussed in Gowdy et al. (2010) and different economic approaches to valuing ecosystem services and biodiversity are outlined by Pascual et al. (2010).

Figure 8.1 can be used to highlight some of the discounting problems and valuation issues given the flows of total economic value involved in the replacement of natural ecosystems or land use by human developed systems. It is assumed (for simplicity) that if a natural ecosystem is not altered by human beings, it will generate a flow of total economic benefits indicated by the broken line ABF. If the system is allowed to return to an unmanaged state after having been altered by human actions, it is assumed that the flow of total benefits will be lower than in its original natural state.

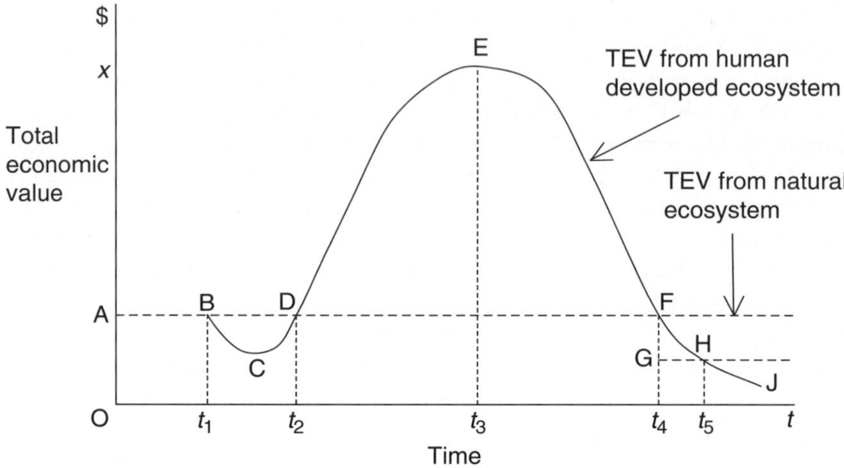

Figure 8.1 A theoretical comparison of flows of total economic value from a natural ecosystem with those flows following its conversion (alteration) by human beings

This is indicated by the broken line GH. If natural ecosystem conversion begins at time t_1, the flow of total economic benefits from the human dominated ecosystem may follow the path BCDEFJ. This path eventually fails to sustain the total economic value of land use.

Given the possibilities shown in Figure 8.1, it is 'optimal' to return to an unmanaged ecosystem at t_5. Note that the switch back to an unmanaged ecosystem is delayed because the total value of the unmanaged ecosystem is lower at t_4 than was that of the original natural system.[4] Despite the total economic value of the human developed ecosystem not being sustainable, it may still be chosen in preference to sustaining the original ecosystem because it provides a higher value of total benefit for a considerable period of time.

It should, nevertheless, be pointed out that the possibilities shown in Figure 8.1 involve considerable abstraction. For example, they do not consider the scale and alternative patterns (mosaics) of conversion of ecosystems. The economic value of different patterns of conversion of natural ecosystems will be explored in Chapter 17.

Valuing Alternative Ecosystems Solely on the Basis of Single Monetary Values for Each is Problematic

Using monetary values to compare the desirability of different economic alternatives is an appealing prospect and has been widely used in economics following the contribution of Pigou (1932). While this approach to welfare economics has been useful, it also has significant limitations. As Pigou (ibid.) pointed out, such estimates should be only part of the input to making decisions about public policy. Attempts to compare the total economic value of alternative ecosystems or land use on the basis of single monetary estimates of their value face difficulties. For example, as was mentioned in relation to Figure 8.1, some type of discounting (possibly even a zero discount rate) or allowance for time is required to specify a single monetary value for each alternative set of flows of economic benefits from biosphere use. Furthermore, considerable uncertainty about the patterns of these flows can occur for a variety of reasons. These reasons include uncertainty about biophysical relationships, for example how the non-biotic component of an ecosystem will change with time. In addition, there may be uncertainty about how social valuation of different ecosystems will alter with the passage of time. This is likely to be a particular problem if the valuation period is very long and encompasses many generations of human beings. Social values do change over long periods of time (Passmore, 1974; Tisdell, 2014b). To complicate matters further, individual preferences and social values may not coincide, as discussions of communitarianism highlight (Etzioni, 1988). Moreover, if the span of analysis covers several generations, intergenerational equity issues arise; for example, the weight to be given to the welfare of different generations must be considered in trying to reduce a flow of economic benefits to a single monetary value.

Value judgements underline all methods of estimating economic welfare. These need to be made explicit and kept in mind in assessing the economic value of alternative ecosystems. Most approaches to welfare economics are based on the view that individual valuations should count. Two alternative measures of these values are commonly used, namely willingness to pay (WTP) and willingness to accept compensation (WAC). It is well known that economic values obtained by using these two alternative approaches can differ substantially (Kahneman et al., 1991; Tisdell, 2009a, pp. 189–90; Knetsch and Sinden, 1984). If the aggregate economic value of alternative possibilities is based on the summation of the WTP of individuals, it enables those with greater ability to pay to exert a much greater comparative influence on this valuation than if the valuation is based on WAC. Furthermore, those belonging to as yet unborn generations are unable to express their preferences so this curtails the use of this approach when valuation stretches over a very long period. Again, not all those who are asked to express their preference for an environmental alternative may be well informed and consequently, in some cases, it may be unwise to respect their preferences.

An additional problem is that the values and preferences of individuals about resource use frequently conflict. The question then arises of how this conflict is to be taken into account in calculating aggregate economic values of the resource alternatives being considered. The approach commonly used by economists to resolve this problem is to adopt the Kaldor–Hicks criterion, also known as the potential Paretian improvement criterion. However, the application of this criterion involves equity issues because its application can result in some individuals losing and others gaining. While some scholars, for example Posner (1981), strongly support the moral worth of the type of wealth maximization likely to result from the application of the Kaldor–Hicks principle, others are more circumspect about its application (Tisdell, 2009b, Ch. 4).

It would greatly facilitate social decision-making if the value of each alternative system of resource use could be expressed as a single monetary value acceptable to all. However, economic methods have not yet achieved this result and probably never will. Although economic methods of valuation provide some worthwhile insights into the values of alternative ecosystems, this valuation needs to be supplemented by additional considerations when making social decisions about resource use.

8.5 GENERAL OBSERVATIONS ON THE FUNCTIONING OF ECOSYSTEMS ALTERED BY HUMANS FOR CULTIVATION PURPOSES

Human modification of ecosystems involves alterations to their biotic and abiotic components, usually both. In areas where humans change the biotic component of an ecosystem, they usually reduce genetic diversity. Probably, the main reason for this is to protect the cultured species from biological competitors. The extent to which humans alter a natural or pre-existing ecosystem and manage it depends on the economic value of doing this, which, in turn, varies with the state of technology, knowledge and other factors.

A common contemporary pattern of modifying a terrestrial ecosystem for cultivation purposes is illustrated in Figure 8.2. There the area within the quadrilateral ABCD represents a part of a pre-existing (natural) ecosystem which has been altered to grow a crop. This system requires human effort for its creation and continuing human effort to manage it. Whether or not human beings should be regarded as endogenous to this system or exogenous to it has been debated. However, this is not as important as being aware of the fact that the sustainability, or more generally, the trajectory of the system depends on human efforts. Figure 8.2 highlights some

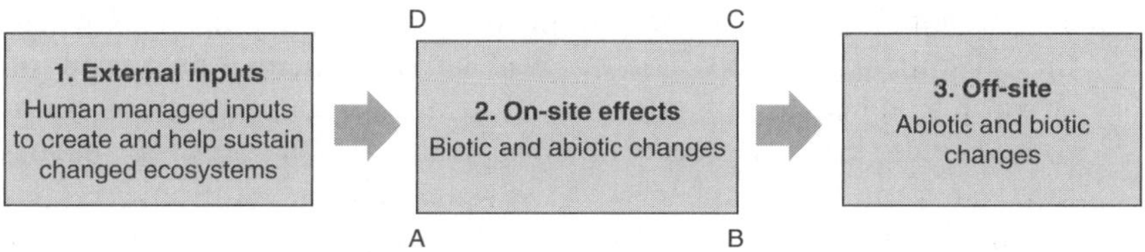

Figure 8.2 *When an ecosystem is altered to cultivate crops by human effort using external inputs, biotic and abiotic changes occur in the cultivated area and there are also biotic and abiotic spillovers to other areas*

of the common effects of such ecosystems, including their environmental consequences.

Three components that affect the functioning of human altered ecosystems are highlighted in Figure 8.2. Their relative importance will vary with the particular type of cultivation method adopted. Despite this, it is worthwhile considering particular aspects of each of these components.

1. Human effort is required to prepare sites for cultivation and today this is mostly facilitated by the use of machines. External inputs to cultivated areas also include in particular cases, fertilizers, pesticides, irrigation water and selected propagules. Fertilizer may be applied to compensate for nutrient loss from the cultivation of the soil or to increase its nutrients, or both. Nevertheless, increased application of fertilizer may increase nutrient run-off from cultivated soils. The backward environmental effects of the use of external inputs should also be considered. For example, offtake of water for irrigating crops can be expected to reduce the availability of water in rivers and streams with adverse consequences for populations of some wild species (Tisdell, 2009b, Ch. 9). Dams built to supply water for cultivated areas can flood important wildlife habitats. The production of chemical fertilizers may also have some adverse environmental consequences.
2. On-site preparation of land for cultivation usually involves the removal of pre-existing vegetation, and methods are adopted to reduce biological competition with crops such as mechanical weeding or herbicide use. The soil is prepared for planting sometimes by ploughing, although minimum tillage has become more common. This tends to make the site prone to soil erosion. Cultivation usually alters ecosystem functions. The nature and extent of the changes depend on the location of the site and its pre-existing biophysical properties. For example, hydrological functioning in the cultivated area may alter.

The watertable can rise, bringing salt to the surface in areas where the subsoil contains salt. The watertable may rise either as a result of irrigation or because of the removal in non-irrigated areas of deep-rooted trees (such as species of eucalyptus), which transpire subsurface water and reduce the level of the watertable. In addition, less CO_2 is likely to be sequestrated in organic matter on most cultivated land than in natural areas, and this contributes to rising CO_2 levels in the atmosphere.

3. Off-site biotic and abiotic changes also occur as a result of cultivation. Cultivation usually increases the rate at which nutrients seep from an area. This accelerates eutrophication of water bodies and the growth of aquatic weeds: it alters the composition (diversity) of species in aquatic areas. Water quality can be adversely affected, resulting in an extra economic burden for society. Wind and water may dump soil off-site at a greater rate than in the absence of cultivation. This can have several adverse economic consequences. These include the accelerated silting of dams and ports. Water run-off is likely to become more irregular in its flow as a result of cultivation.

Nutrients are also lost from cultivated sites by their produce being exported for consumption off-site, for example in cities and towns. The organic and related wastes produced from the use of this produce are rarely returned to cultivated areas but often end up in landfill and water bodies, for instance as a result of the disposal of sewage into water bodies.

Traditionally, economists have modelled production relationships involving the cultivation of land by means of production functions. Frequently, the output of a particular private good on cultivated land is depicted as being a function of a number of controlled inputs and several uncontrolled inputs. The latter mostly consist of environmental factors such as the quantity of rainfall and levels of temperature. The following is a simple example of such a production function:

$$y = \mathrm{f}(\boldsymbol{x}, \boldsymbol{z}) \tag{8.1}$$

where y represents the output of product (for example, wheat) and $\boldsymbol{x}$ and $\boldsymbol{z}$ are respectively vectors of controlled and uncontrolled inputs. More sophisticated models involving time lags and dynamics have been developed. These production models can also be extended to take account of some of the external effects of the cultivation of crops. While this mathematical approach can help in understanding the economic and environmental consequences of changes in ecosystem functioning resulting from cultivation, it needs also to be supplemented by other insights given the complexity of the issues raised when ecosystems are altered.

Given this background, let us now consider selected attributes of alternative agroecosystems. The attributes considered are those that are likely to influence the comparative valuation of alternative ecosystems.

8.6 THE ATTRIBUTES AND VALUE OF ALTERNATIVE AGROECOSYSTEMS

Agriculture is believed to have first developed about 9000 years ago in South-West Asia (Renfrew, 2007) and it also independently developed at several other centres on earth during the Holocene era. It eventually revolutionized the way in which humankind obtained the necessities for its existence. Prior to the advent of agriculture, humankind depended on hunting and gathering to supply these necessities. The adoption of agriculture resulted in the Neolithic Revolution, which enabled the global human population to increase substantially (Childe, 1965[1936]; Ashraf and Galor, 2011). Without this revolution, it seems unlikely that the Industrial Revolution would have happened. In the beginning, the environmental impact of agriculture was small because its extent was limited, but eventually it became a force which greatly altered the earth's ecosystems, and which continues to do so.

Since agriculture's inception, many new methods or systems of agricultural production have evolved. Here it is only possible to consider a selection of these systems. Inevitably, each of those considered is a stereotype. Most attention will be given to high external input agroecosystems and to organic agriculture.

High External Resource-using Agroecosystems – Conway Revisited

High external resource-using agroecosystems appear to be the most common ones in use today. Some are present on very small-sized farms, as in most parts of Asia, whereas others involve broadacre farming as, for example, in the Americas and Australia. In Asia, these systems tend to be relatively labour-intensive whereas in most higher income countries they are capital-intensive. They usually rely on improved varieties of cultivated organisms which have a low degree of environmental tolerance. Conway (1985; 1987) has argued that these types of modern agroecosystems provide higher yields than traditional systems but these yields are less sustainable and more variable than those based on traditional systems. Therefore, with the passage of time, the yields from the adoption of modern agroecosystems could become less than those from the traditional system they replaced. In addition, Conway suggests that the adoption of modern agroecosystems increases inequality in farm incomes. Aspects of his theory were discussed in section 4.5 of this book. Although Conway's model provides a useful starting point for assessing alternative agroecosystems, it is also subject to several limitations, only a few of which were mentioned in section 4.5. Additional qualifications will now be considered.

The first problem to note is that 'traditional' agroecosystems differ and not all have provided sustainable yields in the past. For example, slash-and-burn agriculture (which was an early traditional form of agriculture) fails to sustain yields as the cycle of slashing and burning shortens. If practised in hilly terrain, it can result in considerable soil erosion. Moreover, cultivation on some soils using traditional methods may result in net nutrient loss from soils and declining yields. Furthermore, the irrigation of crops in the Levant in early times is believed to have increased soil salinity and reduced crop yields. Probably many early agricultural innovations were prompted by the need to offset declining agricultural yields associated with the use of 'traditional' methods of agricultural production. In other words, the sustainability of agricultural yields was an important issue prior to the widespread use of artificial fertilizers and pesticides, that is, the adoption of modern agroecosystems.

Conway (1985; 1987) envisaged agricultural yields from modern agriculture declining each time an episodic event occurred, but declines may also happen slowly and continuously. Furthermore, it is unclear what Conway assumes in his model about human behaviour. One possibility is that the adopted method of production and amount of human effort is constant. This being so, yields may follow a declining path, such as ABC in Figure 8.3. However, the rate of decline of this path may be reduced if new agricultural methods are introduced. For example, if a new agricultural method is introduced at t_1, path BEF may be followed rather than path BCD. It is even possible that by a series of innovations, yields do not decline but rise. If the decline in yields is due to nutrient loss in cultivated areas, it may be possible to compensate for this by increasing fertilizer applications, but this would need to be economic, and the external input would need to be available.

Human beings can be considered (for some types of modelling) to be a part of an agroecosystem. They are often in a position to offset (by their behaviour) adverse environmental effects on agriculture production which would otherwise occur, as was illustrated in Figure 8.3.

Conway also contends that yields are likely to be more variable when modern rather than traditional agroecosystems are adopted. However, this depends on how much control human beings are able to exert over

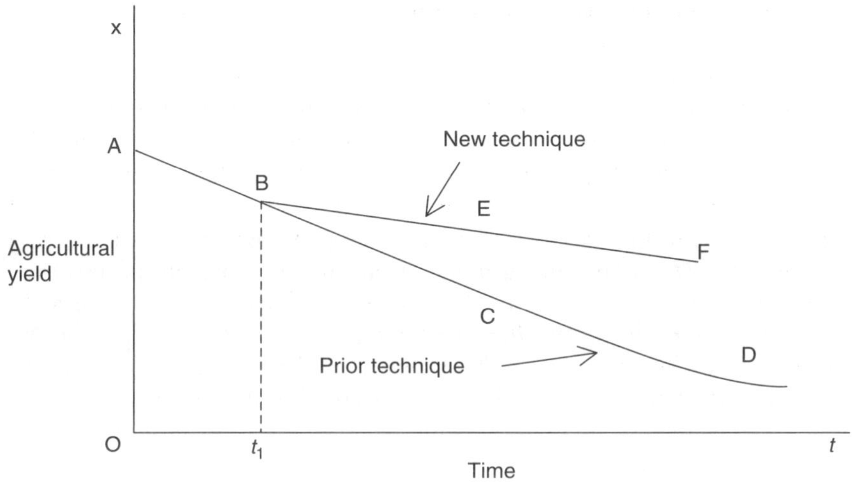

Figure 8.3 Changes to cultivation methods (techniques) may offset declining yields from the use of modern agroecosystems

the environmental conditions influencing agricultural yields. For example, although the tolerance of modern varieties of a crop to lack of water may be less than for a traditional variety, human beings, by carefully regulating the supply of irrigated water to the modern variety, may avoid significant declines in its yields in dry periods. Consequently, the actual variations in its yield may be less than the variability of yields of traditional varieties grown in situations in which human beings are unable to regulate the availability of water. The ability of human beings to control environmental conditions must, therefore, be taken into account when choosing an agroecosystem because this ability influences the performance of the system.

An important consideration in choosing between agroecosystems is their external environmental consequences. Modern agroecosystems have major environmental consequences because of the extensive area in which agriculture is practised globally and the type of methods used. For example, high rates of artificial fertilizer use and high densities of livestock result in high nutrient loads being released into water bodies. These can have adverse ecological consequences. High density of crops and livestock in modern agriculture (based usually on monoculture) creates ideal conditions for epidemics and the evolution of new diseases, which in turn, can threaten the sustainability of agricultural production. Also, as previously stressed in this book, loss of genetic diversity (partly as a consequence of modern agroecosystems) is another serious threat to the sustainability of agricultural yields.

Sustainable Agricultural Intensification

Modern agricultural systems are usually seen as being intensive systems involving high levels of external inputs and high yields. At the same time, they often generate significant negative environmental spillovers.

As discussed in Chapters 5 and 6 of this book, a substantial increase in demand for agricultural products is expected to occur in this century as a result of global population increases, rising per capita incomes and as a consequence of the increasing range of end-uses of agricultural products for other than food production. This has led to consideration of whether it would be better to meet this increased demand by intensifying agricultural production on land already used for agriculture rather than extending the area used for this purpose. Advocates of agricultural intensification argue that it can result in less environmental damage than its extension.

Despite the fact that the economic valuation of the alternatives remains speculative, many policy advisors find the intensification strategy to be attractive, given the available data. For example, Ringler et al. (2014, p. 43) assert the following in the *2013 Global Food Policy Report of IFPRI*:

> By sustainably increasing agricultural production, we can enhance food availability, which is a necessary condition to combat hunger. Experts agree that greater production must be achieved by increasing yields while using fewer resources and minimizing or reversing negative environmental impacts. This sustainable agricultural intensification approach is fundamentally about making the current agricultural system more efficient through the use of new technologies or by improving current production systems.

On the basis of simulation modelling, these authors provide projections indicating that substantial increases in yields of the cereals (maize, wheat and rice) can be obtained if the various 'alternative' agricultural technologies listed in Table 8.1 are adopted. For example, it is suggested that the widespread adoption of no-till technology could raise maize and wheat yields by around 16 per cent and that more efficient nitrogen-use could increase rice yields by about 20 per cent. The technologies listed in Table 8.1 are arranged in declining order in accordance approximately with their estimated ability to increase yields of the cereals mentioned. However, the on-farm economics of making each of these choices is not specified. Furthermore, while the expected environmental changes from adoption of these technologies are sometimes indicated, they are not quantified. Consequently, the overall results are impressionistic. It is even possible that some of the suggested technologies, such as increased water harvesting by constructing more earthen dams, could generate negative environmental externalities. For instance, this practice could reduce the availability of

Table 8.1 The set of agricultural technologies listed by Ringler et al. (2014) which, if adopted, are projected by their chosen modelling to increase significantly maize, rice and wheat yields by 2050[a]

1.	Increased nitrogen-efficiency
2.	No-till[b]
3.	Heat-tolerant crop varieties
4.	Precision agriculture
5.	Integrated soil fertility management
6.	Improved protection against weeds
7.	Improved crop protection against insects
8.	Drought tolerance
9.	Drip irrigation
10.	Water harvesting
11.	Sprinkler irrigation (rather than flood irrigation)

Notes:
(a) Lower numbers correspond to higher percentage effects or yields.
(b) Does not apply to rice.

water downstream, and so deprive downstream users of water and alter the ecological balance, causing wild biodiversity to change.

On the other hand, it is true that many of the techniques mentioned have the ability to reduce adverse environmental spillovers from cropping without decreasing yields and can, in some cases, increase them. Furthermore, adoption of some techniques (such as no-till) have proven to be profitable for some farmers (for example, some wheat growers) and the adoption of this practice has increased.[5] Also there has been increasing adoption of agricultural techniques involving greater precision in delivery of fertilizers, water and pesticides. The extent to which these techniques are likely to be adopted by farmers in the absence of government incentives for their adoption (or penalties for not doing so) requires investigation.

The economics of adopting different types of techniques which could contribute to sustainable agricultural intensification varies geographically and depends on local environmental and economic conditions. For example, economically optimal techniques for agricultural intensification can be expected to be different in developing countries where labour is relatively abundant and capital scarce (and farms mostly small) compared to high income countries in which labour tends to be expensive and capital is relatively abundant. For example, labour-intensive techniques for intensification may be economic in the former countries but not in the latter ones, and vice versa. Therefore, it is stressed that appropriate techniques for sustainably intensifying agriculture (taking into account their social acceptability and their economic and environmental consequences) are geographically sensitive.

There has been considerable debate among biologists and ecologists about whether the strategy of meeting the increased demand for agricultural products by means of sustainable agricultural intensification is likely to be more favourable to the conservation of wild biodiversity than agricultural extension. In this respect, most contemporary support among ecologists seems to be for the former strategy (see, for example, Balmford et al., 2012). In practice, however, both agricultural intensification and extension tend to reduce wild biodiversity. However, the set of wild species lost is likely to differ in each of these cases. For example, if the sets of lost species are represented respectively by **A** and **B**, they may be disjoint sets. If a combination of agricultural intensification and extension is adopted to increase agricultural production, a different set of species, **C**, may be lost but it would probably include some species from both **A** and **B**. To make a rational choice between these alternatives, it is necessary to value the species affected in some way. However, so far this does not seem to have been explicitly accounted for in the relevant literature. In fact, very little attention has been given to the explicit valuation of the environmental consequences of sustainable intensification of agriculture (and its alternatives) for increasing agricultural production.[6]

Garnett et al. (2013) point out that the major aim of promoting sustainable agriculture is to ensure that techniques are adopted that increase yields and also significantly lower adverse environmental impacts. They point out (ibid., p. 33):

> In some areas, increases in yield will be compatible with environmental improvements. In others, yield reductions or land reallocation will be needed to ensure sustainability and deliver benefits such as wildlife conservation, carbon storage, flood protection, and recreation. An overall increase in production does not mean yields should increase everywhere or at any cost. The challenge is context- and location-specific.

Organic Agriculture

The question has been raised of whether organic agriculture would satisfy the criteria for the sustainable intensification of agriculture. In considering this matter, it needs to be recognized that there are different forms of organic agriculture, some of which appear to be more sustainable than others. For example, slash-and-burn agriculture is a possible form of organic agriculture but, as pointed out above, the yield from it is unlikely to be sustained if the frequency of cycles of cultivation is reduced in an effort to maintain total production. Furthermore, it can have severe adverse environmental effects. For instance, particularly in hilly areas, it usually accelerates soil erosion and, of course, alters natural ecosystems.

As for current organic systems (not based on shifting agriculture), Kirchmann et al. (2008, p. 39) report that 'in a global perspective, the scientific literature shows that organic yields are between 25 and 50 per cent lower than conventional yields, depending on whether the organic system has access to animal manure'. These systems may also generate nutrient losses especially if a significant amount of animal manure is used. Compared to conventional agriculture, crop yields are usually smaller for organic agriculture because of lower nutrient availability. Organic farming often requires the wider spacing of vegetable plants than conventional farming and this is associated with lower yields.

Despite this, some consumers prefer organic to non-organic products and are prepared to pay a price premium for them. The principle of consumer sovereignty suggests that they should have the right to buy this produce. However, a large reduction in agricultural supplies would occur if all land now used for agriculture was to depend only on organic agricultural production. Additional land would be required for agriculture in this case to maintain existing production. Furthermore, adverse consequences for the functioning of ecosystems would not disappear. Negative environmental externalities from agricultural land would still be present and there could be increasing loss of natural systems as a result of the extension of agriculture. Furthermore, in order to maintain agricultural production, more marginal and less fertile land would need to be used for organic agriculture. This land would have lower yields than that on existing agricultural land. Furthermore, agricultural yields from less fertile land are, according to Kirchmann et al. (2008), less responsive to organic than conventional agricultural practices. This would magnify the amount of agricultural extension needed to maintain production levels if organic agriculture was widely adopted.[7]

Nevertheless, an increased amount of recycling of organic material in agriculture would improve the functioning of current agroecosystems. Despite this, economic considerations and the changed structure of societies (their increased urbanization) currently restrict the economics of recycling of organic matter on farms. In addition, composting has become uneconomic on many farms compared with the cost of using artificial fertilizers.

Although yields from organic agriculture are usually lower than from the adoption of conventional agriculture,[8] in some cases switching from conventional to organic farming has increased yields and farm incomes.

Bennett and Franzel (2013) give examples of this in less developed countries. However, they point out that the switch appeared in most cases to be associated with better agricultural management and marketing after organic production was adopted. Therefore, the results were not based on 'controlled experiments'. It is also true that farmers who rely on few external inputs or on organic agriculture can sometimes increase their yields by adopting improved agroecological practices.

Agroecology

Agroecology involves the application of ecological principles to improve agricultural management and production. By applying these principles, it is sometimes possible to increase agricultural production in an economic manner without using extra agricultural inputs. Some innovations for sustainable intensification of agriculture are based on agroecology. For example, Juma et al. (2013, p. 7) report that the growing of maize under a canopy of *Faidherbia albida* in parts of Africa has more than doubled maize yields and has had several environmental benefits, such as reduced soil erosion. This practice has also helped to maintain soil fertility.

It is important to take ecological principles into account in trying to achieve sustainable agricultural intensification. However, the economics of applying those principles must also be taken into account, bearing in mind that this is likely to vary geographically. For example, while some forms of mixed cropping can raise the overall quantity of agricultural production, this practice is likely to be uneconomic in some regions when the extra cost of cultivating and harvesting the mixed crops is taken into account. This was discussed in Chapter 5 of this book. Although mixed cropping can sometimes increase agricultural yields, it is rarely economic in higher income countries. If it were economic, the profit motive would result in its greater adoption in higher income countries than at present. Mixed cropping may be more economic in countries where labour is more abundant.

From a social economics perspective, greater application of agroecological principles than at present may be desirable when the external environmental impacts of agriculture are taken into account. However, their increased application in this case is likely to depend on greater government intervention in agriculture.

Some Other Agroecosystems

Several other types of agroecosystems exist but currently they appear to generate a small proportion of global agricultural production. One such system is that involving permaculture. It involves the planting of diverse plants and is based on patterns (designs) that imitate nature. Its essence has been described as follows:

> Permaculture design emphasizes patterns of landscape, function and species assemblies. It determines where these elements should be placed so they can provide maximum benefit to the local environment. The central concept of permaculture is maximizing useful connections between components and synergy of the final design. The focus of permaculture, therefore, is not on each separate element, but rather on the relationships created among elements by the way they are placed together, the whole becoming greater than the sum of its parts. (Anon., 2014a)

As yet, however, permaculture does not appear to be economically viable on a significant commercial scale, particularly in higher income countries.

A very different type of crop production is by means of hydroponics. In this case, plants are grown in a mineral solution in water and without soil (Anon, 2014b). They may (in addition) be grown in greenhouses where atmospheric conditions are controlled. Hydroponics is a commercially economic method of growing some crops, such as lettuce, but it is not economic for many crops, such as cereal crops. The production of crops using hydroponics requires intensive management and depends almost entirely on inputs external to cultivated areas. Consequently, there is a high degree of decoupling of crop production from natural environmental conditions.

8.7 COMMENTS ON AQUACULTURE AND SILVICULTURE

In addition to agriculture, several other bio-industries contribute to economic production and raise similar sustainability issues. The most important of these are those involving aquaculture and silviculture (managed forestry). Both intensive and extensive forms of these industries exist and as in the case of agriculture, a variety of production methods are available with different consequences for yields, their sustainability, and for the state of the environment. For example, forestry may be practised by managing natural stands of timber or by means of plantations in which a single tree species is grown. Prawns (shrimp) may be grown in large ponds relying on naturally occurring food and controlled flows of water from a larger natural water body, or be intensively reared and artificially fed in pools supplied with pumped water.

Just as there is debate in agriculture about whether it is socially more desirable from a sustainability point of view to foster intensive or extensive agricultural production, a parallel debate occurs in relation to aquaculture and silviculture. In this regard, biologists and ecologists have focused on whether a land (biosphere) sharing strategy is socially more desirable than a land-saving (sparing) strategy. The former is associated in this debate with greater extension of bio-industries, whereas the latter is associated with greater intensification of these industries. Basically the argument is about whether forms of mixed land-use (allowing for nature conservation and environmental effects) are socially more desirable (or less desirable) than compartmentalized land (biosphere) use that also allows for the conservation of nature and environmental considerations. A popular current view is that by 'sustainability' intensifying bio-industries, a larger area of the biosphere will be made available for the conservation of nature (and the natural environment) and, at the same time, the increasing demands of humankind for the produce of bio-industries will be able to be met. However, in practice, there is no guarantee that this strategy will stop the extension of bio-industries, given expected increases in demands for their produce due to population increases, demands for higher levels of per capita consumption and so on.

Sustainability issues involving different types of aquatic systems and forms of silviculture will only be considered here briefly because most of these issues are similar to those already raised in relation to alternative types of agroecosystems. I have also given attention elsewhere to these matters and to changes in the genetic stock of fish (Tisdell, 2009c; 2013; 2014c; 2003). Furthermore, I intend to provide a general perspective on the intensification debate (which is applicable to most bio-industries) in the discussion section of this chapter.

Aquaculture

Although aquaculture has been practised for many centuries, it developed later than agriculture. Nevertheless, aquaculture production has increased enough in recent decades to displace wild caught fish as the major source of fish supplies. However, as discussed in Tisdell et al. (2012), there are increasing obstacles to sustaining the growth of production from aquaculture, for example, lack of availability of suitable supplies of water and shortages of fish meal to feed carnivorous species such as Atlantic salmon.

A variety of organisms are grown by means of aquaculture. They include plants (such as seaweed), fish that mostly consume plants (for example, grass carp) and carnivorous animal species such as barramundi, trout and salmon. Aquaculture is practised in fresh water, brackish water and marine areas. Many different aquatic habitats are impacted by it. Its environmental effects depend on types of aquaculture practised and its location. Table 8.2 lists and comments on some of the major environmental

Table 8.2 Some possible adverse environmental spillovers from aquaculture

Type of adverse spillover	Comments
1. General soil subsidence due to use of water from underground aquifers	Buildings may subside and land may be flooded, for example, by high tides. This is a problem in Taiwan
2. Withdrawal of underground water may reduce the quality of the remaining water	For example, intrusion of salt water may occur
3. Brackish water used for aquaculture, for example, of shrimp, may seep into underground water and cause it to become saline	As a result, the water may become unfit for human consumption and for agriculture or reduce agricultural yields
4. Nitrates and chemicals used in aquaculture may contaminate water bodies	This may make the water unfit for aquaculture, agriculture and human consumption
5. Clearing of trees and vegetation for aquaculture, for example, by removal of mangroves for shrimp farms	Results in coastal erosion in some cases, loss of cleansing functions of vegetation, loss of habitat of some commercial species and their recruiting grounds
6. Spread of diseases and pests	Aquaculture may accelerate spread of disease and pests injurious to aquaculture and to aquatic wild species
7. Competition indirectly with the capture fisheries because food of wild fish is harvested to supply aquaculture	'Trash' fish caught to produce pelleted food or meal for cultured fish such as eel or Atlantic salmon
8. Sometimes juveniles or fry of wild stock are captured for culture	Reduces wild stocks, for example, collection of prawn juveniles in Bangladesh
9. Conversion to aquaculture farms of habitats used by wild stocks	Reduction in wild stocks
10. Loss of recreational space to aquaculture	Can affect swimming, boating, recreational fishing and other pursuits
11. Intrusion of salt into agricultural land from brackish water aquaculture	Caused by seepage from brackish water ponds. May adversely affect the production of rice, coconut, etc.
12. Frequent and widespread use of pharmaceuticals, especially antibiotics, may accelerate disease resistance	Reduced effectiveness of veterinary treatments in aquaculture, possible reduced effectiveness of drugs for human treatment

Table 8.2 (continued)

Type of adverse spillover	Comments
13. Crowding of filter feeders such as molluscs, crowding of organisms extracting nutrients from shared water columns, for example, seaweed	Reduced yields from aquaculture or smaller specimens of reduced value. Increased risk of disease transmission within aquaculture
14. Some types of aquaculture produce nutrient-rich wastewater, for example, water high in nitrates and phosphorus	May result in aquatic weed growth, eutrophication, and stimulate red tides

Source: Tisdell (2013, Table 12.1, pp. 283–4) slightly modified.

consequences associated with the development of aquaculture. All of these alter the functioning of ecosystems.

Techniques are being developed and applied in aquaculture that are designed to decrease the cost of production and that incidentally can also reduce the adverse environmental consequences from aquaculture. For example, more efficient methods for feeding salmon have been developed that reduce food waste. Consequently, pollution of marine water from food wastes declines. Scope exists for further developing techniques for the sustainable intensification of aquaculture. However, there is no guarantee that these technological developments will be sufficient to prevent significant extension of agriculture, especially in marine areas.

Silviculture (Forestry)

Systems for managing forestry and using forests are quite varied and their environmental impacts are also diverse. They include the following:

- Clear felling of natural forests to produce products such as wood chips. This has a high negative environmental impact.
- Thinning of natural stands of forest to remove competing vegetation of little commercial value or vegetation of negative economic value. This also may be done to improve the spacing of stands of valued trees.
- Selective logging intended to conserve vegetation of value to particular types of wildlife or to help maintain other particular types of ecosystem services. This is a form of reduced impact forestry.
- The establishment of forest plantations and their ongoing management. This often involves the planting of a single tree-species, such

> as exotic *Pinus radiata* in parts of Australia. The planting of exotic species of trees usually has negative effects on wild biodiversity and can have similar environmental consequences to those arising from agriculture. For example, some plantation stands are chemically fertilized and this increases run-off of nitrates.

Sometimes the planting of trees, such as eucalypt species, reduces water tables and consequently has a negative effect on shallow rooted plants; as a result, the actual environmental effect can be different to the intended one. For example, it has been claimed that the planting of trees in some arid and semi-arid areas of China (which is intended to reduce the advance of desertification) has been counterproductive because it has reduced ground cover, thereby making the soil more prone to wind erosion (Cao, 2008; Cao et al., 2010; Tisdell and Xue, 2013).

Some scientists favour increased intensification of forest plantations to increase yields of timber and are of the view that this will help conserve a greater natural area than otherwise (see, for example, the discussion in Tisdell, 2005, Ch. 8). Although the sustainable intensification of some forest plantations seems to be possible, the extent to which that will result in greater preservation of natural forests is unclear.

The land sparing versus land sharing debate extends to the assessment of the different spatial ways of supplying forest products. Economic aspects of this issue are discussed in Tisdell (2014b) and are given further consideration in Chapter 17 of this book.

8.8 DISCUSSION: EXTENSION OF HUMAN DEVELOPED ECOSYSTEMS VERSUS THEIR INTENSIFICATION

Extension versus intensification of bio-industries has become an important environmental sustainability issue. While many of the issues involved cannot be resolved a priori, simple diagrams can help to illuminate some of the decision-making problems needing consideration. In Figure 8.4, the curve ABCEF indicates the relationship between the output, Y, of a bio-industry and the supply of other ecosystem services, X, when using a given technique of producing Y. Increased output of Y is obtained at the expense of X, other ecosystem services, as the production of Y is increased from zero to Y_1. Attempts to raise the production of Y above Y_1, using the existing technique, not only reduce the availability of other ecological services but also decrease the output of Y as well.

In Figure 8.4, the curves marked W_1W_1 and W_2W_2 are social welfare

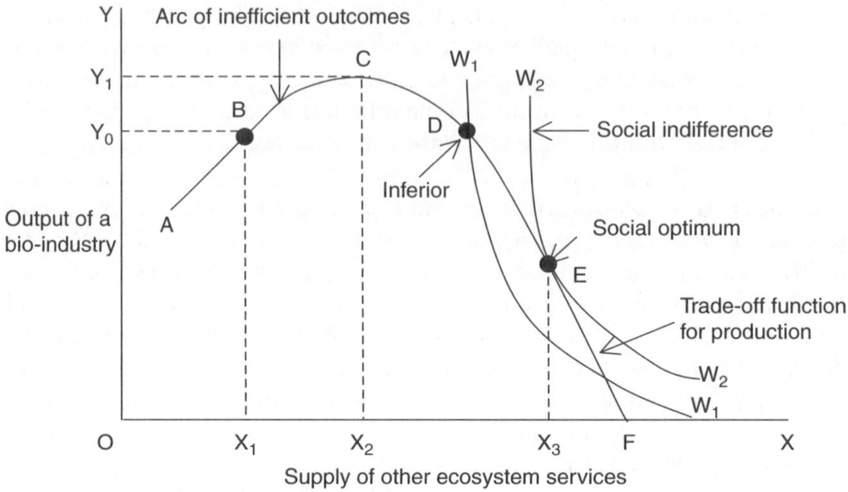

Figure 8.4 Illustration of a trade-off problem involving the supply of a commodity, Y, by a bio-industry and the supply of other ecosystem services, X

indifference curves, that is Bergson-type curves (Bergson, 1938), and are assumed (in this application) to have similar properties to consumers' indifference curves, for example more of both commodities are preferred. Given this approach, the combination of X and Y at point E is socially ideal. However, it is unlikely to be achieved in practice due to market and political failures. For example, the combination at point D could eventuate in practice, or even worse, point B on the arc of inefficient outcomes, ABC, could occur.

Improved social choices, however, might occur if a new technique or method of producing Y is developed and applied that also helps to reduce the loss of other ecosystem services, X. As a result of the application of a new production technique, the production possibility function may shift to the right, as shown in Figure 8.5. In this case, the new production technique shifts the production possibility frontier from the relationship indicated by curve ACF to that identified by GJF. The social optimum subsequently alters from the combination at point E to that at point M. Before the adoption of the new technique, the actual combination in the economy may have corresponded to point D, because of market and political failures. Afterwards, it is unclear to which point movement will occur. However, it is unlikely that the system will gravitate to point M, the new social optimum. It may move to point L, in which case supply of X

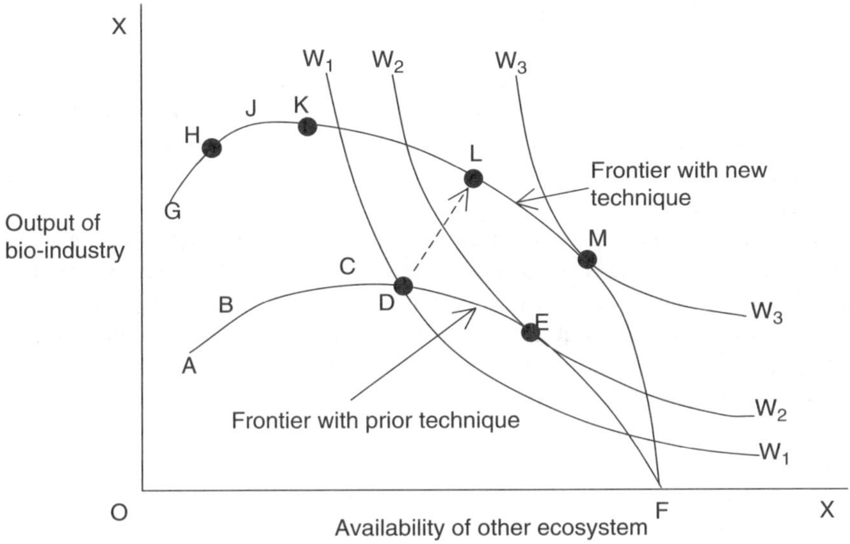

Figure 8.5 *An illustration of the possible consequences for social welfare of the introduction by a bio-industry of a new technique of production which is productivity enhancing and which potentially increases the availability of other ecosystem services*

and all of the ecosystem services increases. Social welfare is increased compared to that corresponding to the original point D. However, there is no guarantee of this type of outcome. It is possible (depending on the nature of political and market failures) for the system to gravitate to a point like K (or even H). In that case, social welfare is reduced despite the increased efficiency (in supplying both X and Y) that can be obtained by using the new technique of production.

Clearly, if it is intended to increase social welfare, it would be a folly to rely only on the introduction of new techniques that increase their productivity of bio-industries, and that simultaneously increase the potential availability of other ecosystem services. The introduction of such techniques does not necessarily ensure the increased availability of ecosystem services. The opposite is possible, as should be clear from the above exposition. Technological progress alone will not solve environmental problems.

8.9 CONCLUDING COMMENTS

The view has been stressed recently, particularly by ecologists, that in valuing the total economic value of changes in economic production account should be taken of all associated changes in the supply of ecosystem services. However, in assessing the economic value of alterations to ecosystems, excessive attention has been given in most studies to the loss of natural ecosystem services caused by such changes and there has been inadequate attention to the benefits obtained. Possible reasons for this bias were suggested in this chapter. Furthermore, many ecologists and some economists focus almost exclusively on the functioning and valuing of natural ecosystems and do not consider these aspects of human developed and modified ecosystems. Some authors, such as Sagoff (2003), have highlighted significant methodological limitations of contemporary ecosystem analysis, and Sagoff (2003) argues that insufficient attention is paid to human dominated ecosystems. This accords with the view expressed in this chapter. Despite this relative neglect, human dominated and modified ecosystems can be analysed using systems of ecological interdependence. However, such systems are not self-regulating but depend on continuing human inputs (management) for the characteristics of their functioning. Human management influences the resilience of such systems and may also affect their resistance. The human element of ecosystem functioning was not accounted for in Conway's (1985; 1987) agroecosystems analysis, as was illustrated. Some limitations of the contemporary analysis of the dynamics of ecosystem functioning were also noted.

Ways in which cultivation changes ecosystem functioning were identified, several of which also occur when other human alterations to the biosphere are made. This was followed by a consideration of the consequences for ecosystem functioning of different types of agroecosystems, and some different systems of aquaculture and silviculture or forestry.

Particular attention was paid to the concept of the sustainable intensification of agriculture (as well as the intensification of aquaculture and silviculture) as a means to increase the level of production of bio-industries and to maintain or to increase the supply of valued (natural) ecosystem services. Many scientists are optimistic about this possibility being achieved as a result of technological progress and the adoption of more environmentally friendly technologies and methods. However, as was illustrated in this chapter, even if such techniques become available and are adopted, the desired results may not be achieved. Garnett et al. (2013, p. 33) state that sustainable intensification 'denotes a goal but does not specify *a priori* how it should be obtained or which agricultural techniques to deploy'. One can add that present discussions of sustainable intensification have not determined whether sustainable intensification will be attained as a result of sufficient technological progress and whether adequate institutional changes (improvement in governance) will be adopted to bring it about. The possibility cannot be dismissed that sustainable intensification will not be achieved because the required amount and type of technological progress will not be forthcoming or because continuing institutional failures hamper its achievement. Currently, the possibility of achieving sustainable intensification is primarily based on faith, even though it is a worthy goal.

NOTES

1. For example, if the species in the ecosystem show a wide range of environmental tolerance, the system is likely to be more resilient (more sustainable) than a species-rich system that is comprised mainly of species that exhibit little environmental tolerance (Tisdell, 1999, Ch. 4). In the former case, the species in the system can survive a major environmental shock but not in the latter case. Furthermore, it is important to state the attributes of a system that are being assessed for their stability when an ecosystem is subjected to particular shocks. Usually, ecologists assess this in relation to its impacts on ecosystem processes such as energy flows, primary production, decomposition of organic matter and nutrient recycling, or sometimes in relation to the ability of the system to maintain its biodiversity (see, for example, Chapin III et al., 2002, Ch. 12). When some sets of biodiversity are lost in an ecosystem, the system may be relatively stable in sustaining some of its ecological processes but not others. This depends on the composition of the sets of biodiversity lost. There seems to be insufficient evidence to support the categorical assertion that increased species diversity always increases the sustainability of an ecosystem or all its processes, although many ecologists believe that this proposition is theoretically correct (see Chapin III et al., 2002, pp. 276–7).
2. This is despite the fact that it is often believed that self-regulating natural systems are best. For example, the competitive market system often gains moral support on the basis that it is self-regulating in a natural way. Adam Smith (1776[1910]) even described equilibrium prices in such markets as natural prices.
3. Land use should be interpreted to mean any types of uses of the biosphere, including uses of aquatic areas as well as terrestrial areas.
4. It is only for simplicity that the total value of unmanaged ecosystems has been assumed to be constant. Both the total value of natural ecosystems and ecosystems that revert to being unmanaged after having been previously altered by humans can change with the passage of time. Natural systems can change over time independently of human actions. For example, the Sahara Desert was once a savannah but became a desert as a result of climate change not brought about by humans.
5. However, note that increased use of no-till technologies usually results in increased herbicide use, and this can have negative environmental effects.
6. It has been predicted that a 70 per cent increase in agricultural production is required (compared to that early in this century) to meet increased demand around the middle of this century (Burney et al., 2010). There is probably little prospect of an increase of this magnitude in intensification of agriculture. Ringler et al. (2014) indicate that an increase of about 20 per cent in maize, rice and wheat yields is possible by sustainable intensification. Even if the figure is 30 per cent, this is well short of the 70 per cent requirement, and there is no guarantee that a similar increase is possible for other crops. In addition, the view that the intensification of agriculture will result in the minimum loss in

environmental value and its extension will result in its maximum loss may be flawed. It is possible that a judicious combination of intensification and extension of agriculture minimizes environmental costs for the targeted increase in agricultural production (see, for example, Tisdell, 2015).

7. Actually, for economic and environmental reasons, if all agriculture was forced to depend only on organic methods of production, the area of the earth used for crop production is likely to decline because crop production on less fertile land would probably be uneconomic. Crop yields on land subject to heavy leaching of nutrients and rapid loss of organic matter as a result of cultivation (mostly in tropical and sub-tropical areas) would be unsustainable. It is possible that global agricultural production could fall by more than 50 per cent (given the estimates of Kirchmann et al., 2008) if all agriculture was required to depend on organic methods. Agriculture would most likely not be able to support the current level of world population, let alone its predicted increase in this century, and living standards would decline sharply. A major economic crisis would occur.
8. On the other hand, advocates of organic culture might contend that it is more sustainable in its yields than conventional agriculture. Much, however, depends on the actual specific systems that are being compared.

REFERENCES

Anon (2014a), 'Permaculture', Wikipedia. Available at: http://en.wikipedia.org/wiki/Permaculture. (Accessed 19 August, 2014).

Anon (2014b), 'Hydroponics', Wikipedia. Available at: http://en.wikipedia.org/wiki/Hydroponics. (Accessed 19 August, 2014).

Ashraf, Q. and O. Galor (2011), 'Dynamics and stagnation in the Malthusian epoch', *American Economic Review*, **101**, 2003–2041.

Balmford, A., R.E. Green and B. Phalan (2012), 'What conservationists need to know about farming', *Proceedings of the Royal Society B*, **297**, 2714–24.

Bennett, M. and S. Franzel (2013), 'Can organic and resource-conserving agriculture improve livelihoods? A synthesis', *International Journal of Agricultural Sustainability*, **11**, 193–215.

Bergson, A. (1938), 'A reformulation of certain aspects of welfare economics', *Quarterly Journal of Economics*, **52**, 310–34.

Brown, T.C., J.C. Bergstrom and J.B. Loomis (2007), 'Defining, valuing and providing ecosystem goods and services', *Natural Resources Journal*, **47**, 329–76.

Burney, J.A., S.J. Davis and D.B. Lobell (2010), 'Greenhouse gas mitigation by agricultural intensification', *Proceedings of the National Academy of Sciences*, **107** (26), 12052–7.

Cao, S. (2008), 'Why large-scale afforestation efforts in China have failed to solve the desertification problem', *Environmental Science and Technology*, **42**, 1826–31.

Cao, S., G. Wang and L. Chen (2010), 'Questionable value of planting thirsty trees in dry regions', *Nature*, **465**, 31.

Chapin III, F.S., P.A. Matson and H.A. Mooney (2002), *Principles of Terrestrial Ecosystem Ecology*, New York: Springer.

Childe, V.G. (1965[1936]), *Man Makes Himself*, London: Watts.

Clements, F.E. (1916), *Plant Succession: An Analysis of the Development of Vegetation*, Washington, DC: Carnegie Institution of Washington.

Conway, G.R. (1985), 'Agroecosystem analysis', *Agricultural Administration*, **20**, 31–55.

Conway, G.R. (1987), 'The properties of agroecosystems', *Agricultural Systems*, **24**, 95–117.
Durand, G. and G. Van Huylenbroeck (2003), 'Multifunctionality and rural development: A general framework', in G. Van Huylenbroeck and G. Durand (eds), *Multifunctional Agriculture: A New Paradigm for European Agriculture and Rural Development*, Aldershot, UK and Burlington, VT, USA: Ashgate, pp. 1–16.
Etzioni, A. (1988), *The Moral Dimension: Towards a New Economics*, New York: The Free Press.
Garnett, T., M.C. Appleby, A. Balmford, I.J. Bateman, T.G. Benton, P. Bloomer, B. Burlingame, M. Dawkins, L. Dolan, D. Fraser, M. Herrero, I. Hoffman, P. Smith, P.K. Thorton, C. Toulmin, S.J. Vermuelen and H.C.J. Godray (2013), 'Sustainable intensification in agriculture: Premises and policies', *Science*, **341** (July), 33–4.
Gowdy, J., R. Howarth and C. Tisdell (2010), 'Discounting, ethics and options for maintaining biodiversity and ecosystem integrity', in P. Kumar (ed.), *The Economics of Ecosystems and Biodiversity: Ecological and Economic Foundations*, London and Washington, DC: Earthscan, pp. 257–38.
Juma, C., R. Tabo, K. Wilson and G. Conway (2013), *Innovation for Sustainable Intensification in Africa*, London: The Montpellier Panel, Agriculture for Impact, Imperial College.
Kahneman, D., J.L. Knetsch and R.H. Thaler (1991), 'The endowment effect, loss aversion and *status quo* bias', *Journal of Economic Perspectives*, **5**, 193–206.
Kirchmann, H., L. Bergström, T. Kätterer, O. Andrén and R. Andersson (2008), 'Can organic crop production feed the world?', in H. Kirchmann and L. Bergström (eds), *Organic Crop Production – Ambitions and Limitations*, Amsterdam: Springer, pp. 39–72.
Knetsch, J.L. and J.A. Sinden (1984), 'Willingness to pay and compensation demanded: Empirical evidence of an unexpected disparity in measures of value', *Quarterly Journal of Economics*, **99**, 507–21.
Kumar, P. (ed.) (2010), *The Economics of Ecosystems and Biodiversity: Ecological and Economic Foundations*, London and Washington, DC: Earthscan.
Millennium Ecosystem Assessment (2003), *Ecosystems and Human Well-being: A Framework for Assessment*, Washington, DC: Island Press.
Millennium Ecosystem Assessment (2005), *Ecosystems and Human Well-being: A Synthesis*, Washington, DC: World Resources Institute.
Pascual, U., R. Muradian, L. Brander, E. Gómez-Baggethun, B. Martin-López and M. Verma (2010), 'The economics of valuing ecosystems services and biodiversity', in P. Kumar (ed.), *The Economics of Ecosystems and Biodiversity: Ecological and Economic Foundations*, London and Washington, DC: Earthscan, pp. 184–256.
Passmore, J.A. (1974), *Man's Responsibility for Nature: Ecological Problems and Western Traditions*, London: Duckworth.
Pigou, A. (1932), *The Economics of Welfare*, 4th edn, London: Macmillan.
Posner, R. (1981), *The Economics of Justice*, Cambridge, MA and London, UK: Harvard University Press.
Renfrew, C. (2007), *Prehistory: The Making of the Human Mind*, London: Weidenfeld and Nicolson.
Ringler, C., N. Cenachhi, J. Koo, R. Robertson, M. Fisher, C. Cox, N. Perez, K. Garret and M. Rosegrant (2014), 'Sustainable agricultural intensification:

The promise of innovative farming practices', *2013 Global Food Policy Report*, Washington DC: International Food Policy Research Institute (IFPRI).
Sagoff, M. (2003), 'The plaza and the pendulum: Two concepts of ecological science', *Biology and Philosophy*, **18**, 529–52.
Schröter, M., E.H. van der Zanden, A.P.E. van Oudenhoven, R.P. Remme, H.M. Serna-Chavez, R.S. de Groot and P. Opdam (2014), 'Ecosystem services as a contested concept: A synthesis of critique and counter-arguments', *Conservation Letters*, **7** (6), 514–23.
Smith, A. (1776[1910]), *An Inquiry into the Nature and Causes of the Wealth of Nations*, London: Dent. (Reprinted 1910).
Tansley, A.G. (1935), 'The use and abuse of vegetational terms and concepts', *Ecology*, **16** (3), 268–71.
Tisdell, C.A. (1999), *Biodiversity, Conservation and Sustainable Development*, Cheltenham, UK and Northampton, MA, USA: Edward Elgar Publishing.
Tisdell, C.A. (2003), 'Aquaculture's potential impacts on conservation of wild stocks and biodiversity', *Aquaculture Economics and Management*, **7** (1–2), 155–65.
Tisdell, C.A. (2005), *Economics of Environmental Conservation*, 2nd edn, Cheltenham, UK and Northampton, MA, USA: Edward Elgar Publishing.
Tisdell, C.A. (2009a), 'Complex policy choices regarding agricultural externalities: Efficiency, equity and acceptability', in V. Beckmann and M. Padmanabhan (eds), *Institutions and Sustainability. Political Economy of Agriculture and the Environment – Essays in Honour of Konrad Hagedorn*, Heidelberg: Springer, pp. 83–106.
Tisdell, C.A. (2009b), *Resource and Environmental Economics: Modern Issues and Applications*, Singapore, Hackensack (NJ), London and Beijing: World Scientific.
Tisdell, C.A. (2009c), 'The economics of fish biodiversity: Linkages between aquaculture and fisheries – some perspectives', in K. Ninan (ed.), *Conserving and Valuing Ecosystem Services and Biodiversity: Economic Institutional and Social Challenges*, London, UK and Sterling, VA, USA: Earthscan, pp. 47–57.
Tisdell, C.A. (2013), 'Aquaculture and sustainable development: Allowing for environmental externalities and common-pool resources', in M.A. Quaddus and M.A.B. Siddique (eds), *Handbook of Sustainable Development Planning*, 2nd edn, Cheltenham, UK and Northampton, MA, USA: Edward Elgar Publishing.
Tisdell, C.A. (2014a), 'Ecosystem functions and genetic diversity: TEEB raises challenges for the economics discipline', *Economic Analysis and Policy*, **44** (1), 14–20.
Tisdell, C.A. (2014b), *Human Values and Biodiversity Conservation: The Survival of Wild Species*, Cheltenham, UK and Northampton, MA, USA: Edward Elgar Publishing.
Tisdell, C.A. (2014c), 'Conserving forest wildlife and other ecosystem services: Opportunity costs and the valuation of alternative logging regimes', in K.N. Ninan (ed.), *Valuing Ecosystem Services: Methodological Issues and Case Studies*, Cheltenham, UK and Northampton, MA, USA: Edward Elgar Publishing, pp. 227–44.
Tisdell, C.A. (2015), 'Agricultural development and sustainability: a review of recent and earlier perspectives', *Economics, Ecology and the Environment* Working Paper no. 198, Brisbane: School of Economics, The University of Queensland.
Tisdell, C.A. and D. Xue (2013), 'Managing ecosystem services for human

benefit: Economic and environmental policy challenges', in E. Creighton and P. Davovich (eds), *Environmental Policy: Management, Legal Issues and Health Aspects*, New York: Nova Science Publishers, pp. 87–106.

Tisdell, C.A., N. Hishamunda, R. Van Anrooy, T. Pongthanapanich and M. Arjuna Upare (2012), 'Investment, insurance and risk management for aquaculture development', in R.P. Subasinghe, J.R. Arthur, D.M. Bartley, S.S. De Silva, M. Halwart, N. Hishamunda, C.V. Mohan and P. Sorgeloos (eds), *Proceedings of Global Conference on Aquaculture 2010, Farming the Waters for People and Food*, Phuket, Thailand, 22–25 September 2010, Rome: FAO and Bangkok: NACA, pp. 303–35.

Vanslembrouck, I. and G. Van Huylenbroeck (2005), *Landscape Ameneties: Economic Assessment of Agricultural Landscapes*, Dordrecht: Springer.

Willis, A.J. (1997), 'The ecosystem: An evolving concept viewed historically', *Functional Ecology*, **11**, 268–71.

9. Declining effectiveness of techniques designed to control pests and diseases: economics, evolution and human choices

9.1 INTRODUCTION

Human beings have been very successful in developing techniques to control pests and diseases. However, the effectiveness of these techniques often declines with their increased use and the passage of time. Frequently, this is due to the selective evolution of populations of organisms which it is intended to control. This is because, as a result of using the control technique, members of a population which are resistant to this control multiply and increase as a proportion of their population. Also in the case of some higher order organisms (such as elephants), resistance to control techniques can develop as a result of a different mechanism, namely learning and transmission of learnt behaviour to other members of the species, including offspring. It is the basic economics of the evolutionary resistance to control techniques of pests and diseases that is the focus of this chapter.

To a significant extent, the progress of this resistance depends on human decisions about the frequency and duration of the use of control techniques. As a result, very complex social choice problems arise because beneficial use of the technique now has a user cost, namely the cost of the technique being less effective in the future. Therefore, current economic benefits from use of such a technique need to be balanced against its decreased future benefits. Because of market and political failures, an inferior allocation of the use of such techniques is likely to occur, as is demonstrated in this chapter.

This chapter is developed as follows. First, public concerns about the above-mentioned problem are outlined, and then reasons are given for market and political failures when this resistance problem exists. The comparative ability of monopoly and market competition to address this choice problem is considered and public policies for dealing with it are examined using some simple models. Decision problems involving the appropriate time at which to release new techniques for controlling unwanted organisms are then raised.

9.2 SOCIAL CONCERNS ABOUT THE DECLINING EFFECTIVENESS OF TECHNIQUES INTENDED TO CONTROL PESTS AND DISEASES

Public concerns about the declining effectiveness of techniques to control pests (such as insects and weeds) and diseases seem to have been most strongly expressed in the case of diseases affecting humans, such as the case of increasing resistance of diseases to the use of antibiotics and other antimicrobials. This is because many individuals, in using such techniques, pay little or no attention to the user costs involved and frequently they are used in cases where there is little benefit, or the benefit is only imagined.

One reporter, for example, expressed her concern as follows: 'If we don't stop our love affair with antibiotics – and that includes you, doctors – the chemotherapy and heart transplants that now offer an extra lease on life will soon not be an option. Why? Because our arsenal of antibiotics will be useless, unable to fight subsequent infection' because of the development of antibiotic resistance (Scott, 2014).

Due to the development of biological resistance of this type, common infections could once again become life-threatening. Furthermore, biological resistance to antimicrobials has not only developed among bacteria but has also evolved among viruses, fungi and parasites. This reduces the biological fitness of their hosts.

The problem of growing antimicrobial resistance is magnified because an 'excessive' amount of antimicrobial agents are given to domesticated animals, particularly in some developing countries which have few restrictions on their use. Antibiotics are given to some animals as supplements in their food to promote their growth. Prawns (shrimp) are sometimes reared in this way, but poultry and other livestock may also receive 'excessive' doses of antimicrobials. This can result in 'superbugs' developing, with adverse consequences for livestock and human health.

In some countries, antibiotics and other antimicrobials are readily available 'over the counter' and virtually at the cost of their production. A number of developing countries, including India, have pursued this policy in order to make these drugs readily available to the needy at a low cost. This was intended as a humanitarian action but it has accelerated antibiotic resistance.

Microbes with this antimicrobial resistance can spread internationally. For example, Scott (2014, p. 17) points out: 'A 2012 study found that swabs taken of 102 Australians before and after trips to South-East Asia, the Middle East, South America, India and China showed 44 of them returned carrying antimicrobial resistant *Escherichia coli*, or *E. coli*' not present before their journey.

The biological resistance problem also arises in the control of insects and weeds. In some cases, insects have become resistant to crops that have been genetically modified to express Bt crystals, which are toxic to some species of insects. This has been countered to some extent by the development of different forms of these crystals but the problem is an ongoing one. By maintaining refuges for non-resistant individuals in the population of the targeted insect species, it is sometimes possible to slow the development of resistance if the non-resistant trait is genetically dominant (Carrière et al., 2010). Suppliers of Bt modified seeds in many developed countries require farmers to maintain such refuges (and this is often a legal requirement) but this is not so in all countries, especially some developing countries.

Moreover, resistance has developed to some chemical insecticides and herbicides. For example, the number of species resistant to DDT eventually increased in the US during its use in the period 1946 to 1963 (Carrière et al., 2010). Furthermore, the number of weed species resistant to the herbicide glyphosphate has risen with its continuing use. Increasing resistance of pests and diseases to human developed techniques designed to control them is a widespread problem.

9.3 THE ECONOMIC NATURE OF THE PROBLEM OF INCREASED BIOLOGICAL RESISTANCE TO HUMAN CONTROLS OF PESTS AND DISEASES

The basic problem occurs when users of a technique designed to control pests or diseases have no incentive to take into account the effect of its use on collective user costs. This occurs when for each individual their own user cost is negligible but collective use results in a high user cost. The situation is like that for the prisoner's dilemma problem: each individual is only motivated to take into account his or her benefit from using the technique, and this is so whether or not others take into account user costs. Moreover, an increase in the pool of organisms resistant to the control technique can be considered to be a public bad, that is, the opposite to a public good. In some instances, this bad may be localized to particular geographical regions.

It needs, however, to be kept in mind that the amount and pattern of use of a technique (which is subject to declining effectiveness with increased use) depends not just on its users but also on its suppliers. A monopolist

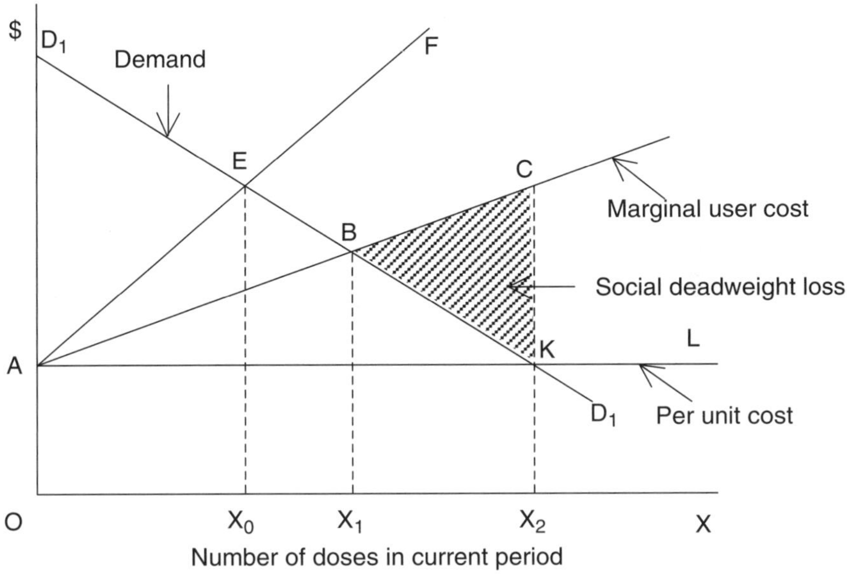

Figure 9.1 *An illustration that failure to take account of user costs results in a social deadweight loss when pest control techniques are used which decline in effectiveness with their increased use*

may restrict the supply of products required to use the technique (to limit the speed with which its effectiveness in use declines) but no such restriction will apply in a purely competitive market. Again, in some cases, a government may decide to supply the product at cost for humanitarian reasons. The consequences of the latter two cases can be illustrated by Figure 9.1. The comparative consequences of monopoly will be considered in the next section.

In Figure 9.1, the line marked D_1D_1 is the demand relationship in the current period for 'doses' of a product (for example, an antibiotic) able to combat pests or diseases. The line marked ABC shows the (collective) marginal user costs associated with the current doses, that is, their reduced economic value in the future due to increased biological resistance as a result of current use. For simplicity, the cost of supply of each dose is shown by the horizontal line AKL.

If doses of the product are supplied in a competitive market or if the government sells these at cost, X_2 doses of the pest control product will be used in the current period. Buyers will ignore user costs and socially 'excessive' use of the product will occur. The socially optimal amount of use of the product in the current period is X_1. Because user costs are ignored in this case, the social deadweight loss is equivalent to the area of the hatched triangle BKC. The faster the rate at which user cost rises, the less use should be made of the control in the current period. For example, only X_0 doses of the control are socially optimal in the current period if marginal user cost is as shown by line AEF. Failure to take account of user costs in this case results in a larger social deadweight loss than when marginal user cost is as shown by line ABC.

In the case of techniques that decline in their biological effectiveness with increased use, using them to treat diseases or control pests in those cases where the current benefit is low but a high future benefit is forgone is uneconomic. Use of the technique in such cases should be confined to high value current uses.[1] For example, in the case of antimicrobials able to treat life-threatening diseases as well as less serious diseases, it would be desirable to confine their use to the former treatments. The higher the value of the most beneficial uses compared to the less beneficial ones, the stronger the case for restricting use of these techniques to the purposes for which they are most beneficial.

Figure 9.2 provides a simple example of this proposition. It assumes a two-period model and supposes that the effectiveness of the control of pests or diseases in period 2 depends on the amount of use of the control technique in period 1. It is supposed that the control is only effective for up to $\overline{X}$ doses, that is, it is fully effective for $\overline{X}$ doses but not at all effective for a larger number of doses. Where $X(t_1)$ represents the number of doses in period 1, and $X(t_2)$ is that in period 2, the number of effective doses in period 2 is:

$$X(t_2) = \overline{X} - X(t_1) \tag{9.1}$$

where $X(t_1) \leq \overline{X}$. In other words, the number of available effective doses in period 2 is equal to $\overline{X}$ minus the number of doses already given in period 1.

Two different situations are considered in Figure 9.2. In one case, the demand function is shown by the line UD_1 and in the other case by the line VD_2. The comparative value of the doses that are in highest demand is higher in the latter case than in the former one and, therefore, excessive use of the technique in period 1 involves a greater social loss in the latter case than in the former one. This is because the user costs (or opportunity costs) in the latter case are higher than the former. It is assumed that the demand curve in period 2 is the same as in period 1.

Given that $X_2 = \overline{X}$, the optimal number of doses of the control is $X_1 = 0.5X_2$ in each period. Given the demand curve UD_1 applies to both periods and if the doses are sold at cost, consumers' surplus is then equal to the area of quadrilateral ARSU in both periods if doses are limited to

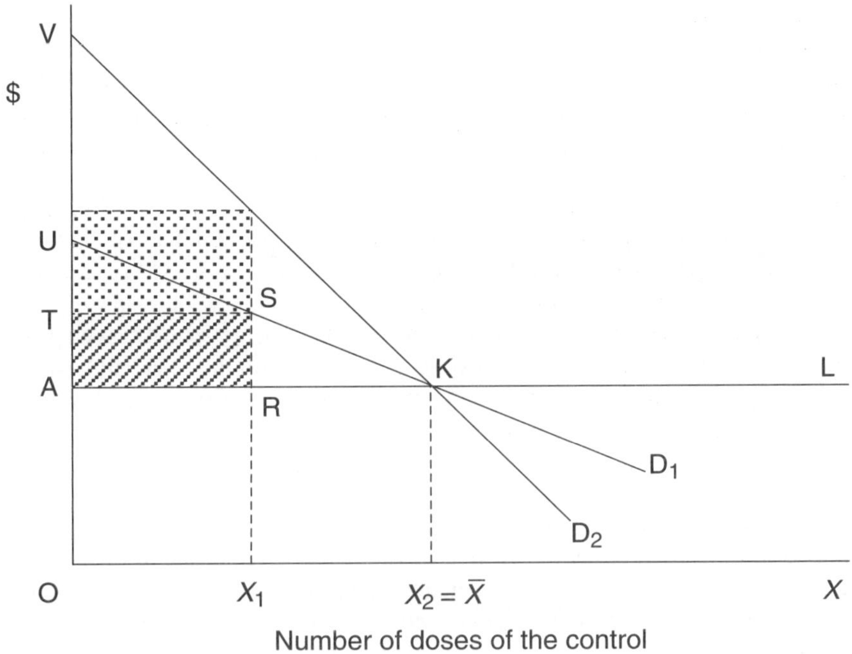

Figure 9.2 *When more valuable options are lost as a result of not restricting current use of a technique which declines in effectiveness with use, the social loss from not restricting the use of the technique increases*

$0.5X_2$ in each period and supplied only for the highest valued purposes. If, however, there is no limitation on the use of doses, X_2 doses will be administered in period 1. The surplus is then equal to the area AKU. Because the area of triangle TUS equals that of triangle RKS, the total reduction in consumers' surplus is equal to the area of rectangle ARST when the suboptimal allocation occurs compared to the optimal one. Consumers' surplus is halved in this case if it is assessed for the two periods. If the demand curve for doses is steeper, as for example is shown by the line VD_2, the total loss in consumers' surplus as a result of unrestricted use of the control is equivalent to the area of the hatched rectangle plus the dotted one. In this case, the social importance of limiting the use of doses in period 1 increases.

The inferior possible outcome illustrated by using Figure 9.2 can emerge if doses of the control technique are available in a purely competitive market or if a government decides to make the doses available at cost to any prospective buyer. Given the situation illustrated in Figure 9.2, a monopolist would restrict supplies, but whether the outcome is socially superior in this case is unclear because of the income distribution consequences of monopoly. Let us consider this matter by reference to Figure 9.2.

9.4 MONOPOLY AND THE REGULATION OF THE USE OF A TECHNIQUE THAT DECLINES IN EFFECTIVENESS WITH ITS USE

A profit-maximizing monopolist will restrict the use of a technique where it is in his/her self-interest to do so. In the particular case illustrated in Figure 9.2 (and given no discounting), the monopolist will supply $X_1 = 0.5\overline{X}$ doses in each period. This results in the optimal allocation. However, in doing so, the monopolist substantially reduces the consumers' surplus of users of the technique. If the demand curve UD_1 applies, the monopolist maximizes profit by charging OT per dose for the control and appropriates profit equal to the area of the hatched rectangle in Figure 9.2. Buyers are now only left with a surplus equal to the area of triangle TSU in each period. Their total surplus is now much less than if they had open access by purchasing the products needed to apply the technique at its cost of supply. In this case, their surplus is smaller by an amount equivalent to the area of the hatched rectangle. Users are much worse off than if doses could be sold at cost and use restricted to X_1 doses in each period. In this case, they are worse off by an amount equal to twice the hatched area in Figure 9.2. The income redistribution losses of users are larger the more inelastic is the demand for doses of the control, for example, if in Figure 9.2, the demand relationship is VD_2 rather than UD_1.

It should, however, be noted that it is a peculiarity of this example that monopoly results in the release of $0.5\overline{X}$ doses of the control in each period, the 'optimal' allocation. In many cases, monopoly will result in socially insufficient release of doses. For example, suppose that $\overline{X} = 1.5X_2$ rather than X_2 and that otherwise the above model applies but that after period 2 the control technique has no value. This may be because it becomes obsolete due to an improved control becoming available or because its effectiveness is also subject to a time constraint.

In this case, the optimal use of doses in periods 1 and 2 is 0.75 doses in each. Despite this, the monopolist will only release 0.5 doses in each period in order to maximize profit. There is a social loss.

The above theory is based on static analysis. It is possible (but not certain) that given theories of John Maurice Clark (1940) and of Schumpeter (1942), the monopolist will be motivated to invest in research and development and that in a dynamic setting, buyers will be better off as a result of new inventions than under conditions of pure competition (Tisdell, 2013, Chs 4 and 6) or in a situation in which the government supplies the product required for the control at cost.

It may also be the case where governments subsidize use of the control techniques (as in the case in some countries which have national health schemes) that governments may use their countervailing power against monopolists to keep the price of the products required for control lower than otherwise. Furthermore, in practice, it seems that many governments find it difficult to restrict the use of controls, the effectiveness of which erodes with their use. This may partly be a result of short horizons for decision-making in politics (Tisdell, 2009b, pp. 90–91). This appears to be a major problem generally for the optimal regulation of the use of depletable common-pool resources (see, for example, Tisdell, 2009a).

9.5 SOME FURTHER ISSUES

It was suggested above that, except in special cases, a monopolist is likely to restrict excessively the use of a technique which declines in effectiveness with increased use. However, there is at least one exception. If the monopoly is protected by a patent and it expires in a period in which the technique could still be effective, the monopolist may release a greater number of 'doses' of the technique before expiry than would be socially optimal.

Consider the example discussed in relation to Figure 9.2. Suppose, as before, that the technique is effective for $\overline{X} = X_2$ doses but can remain effective for three periods rather than two. Ignoring discounting and assuming that demand and supply conditions are the same in each period, social benefit would be maximized by enabling 0.33 doses of the control to be used in each of the three periods. Suppose, however, that the monopolist will lose his or her monopoly in period 3 and each dose of the control will then be sold at cost. The monopolist will then maximize profit by marketing $0.5\overline{X}$ doses in each of periods 1 and 2. The drug will, therefore, no longer be effective in period 3.

Another important issue is at what stage of development a technique for controlling population of pests and diseases should be released. Should, for example, one wait until it has high toxicity or effectiveness before its release. If it has high toxicity, it will eliminate a higher proportion of the targeted population and reduce the chances of survival of resistant members of the targeted population. Therefore, if resistance develops, it may do so more slowly where the control has high toxicity rather than a low toxicity. However, it is difficult to generalize because much depends on the nature of genetic inheritance of resistance in a pest population (Carrière et al., 2010). Carrière et al. (2010, p. 569) find in relation to Bt resistance that 'comparisons across pest species suggest that recessive inheritance and abundant refuges of non-Bt host plants are two key functions in the delay of evolution of resistance' (Tabashik et al., 2009). Referring to an article by Kruger et al. (2009), they point out that rapid development of stem borer resistance to Bt maize in a part of South Africa was a consequence of inadequate provision of refuges.

This of course raises another issue: in such cases, what is the economically optimal extent and pattern of the supply of such refuges? While the glib answer is that their provision should be such that the marginal social cost involved should be equal to the marginal social benefit of ecological resistance of the targeted pests, this is of little practical value. It is a subject requiring interdisciplinary work. Furthermore, from the point of view of economics, the provision of refuges involves an externality problem because those who provide refuges confer an unpaid for benefit on others by reducing the speed with which pests become resistant to the control.

9.6 CONCLUSIONS

Considerable public concern exists about the 'excessive' use of techniques that decline in effectiveness with greater use. As shown, market systems rarely result in a socially optimal pattern of use of such techniques and a case exists for government regulation of their use. However, governments are subject to political failures, regulated use involves cost, and many administrative systems are only able to achieve imperfect regulation of the use of such techniques. Therefore, current systems of regulating the use of these techniques remain imperfect. The role of monopolies in regulating this use can be quite varied. Depending on the circumstances, they can engage in actions that result in either excessive or insufficient social use of techniques that decline in effectiveness with use. In general, monopolists will redistribute income in their favour. There is, however, a *possibility* that they will compensate for this shortcoming by accelerating technological progress, as suggested by Joseph Schumpeter (1942).[2]

NOTES

1. However, as mentioned in Chapter 1 of this book, there is the question of how higher value use should be determined. In Figure 9.2, it is implied that this ought to be determined by the willingness (which also depends on the ability) of users to pay. This

is a standard assumption in economics. However, it is unlikely to command widespread community support in cases where the health and survival of individual human beings are at stake. Therefore, more investigation of communal responses to these issues is required.

2. This chapter draws on, extends and revises one of my earlier publications, namely C.A. Tisdell (1982), 'Exploitation of techniques that decline in effectiveness with use', *Public Finance*, **37**(3), 428–37. I am grateful to the Wim Drees Foundation for permission to draw on this earlier publication.

REFERENCES

Carrière, Y., D.W. Crowder and B.E. Tabashik (2010), 'Evolutionary ecology of insect adaptation to Bt crops', *Evolutionary Applications*, **3**, 561–73.

Clark, J.M. (1940), 'Towards a concept of workable competition', *American Economic Review*, **30**, 241–56.

Kruger, M., J.B.J. Van Rensburg and J. Van den Burg (2009), 'Perspective on the development of stem borne resistance to Bt maize and refuge compliance at the Vaalharts irrigation scheme in South Africa', *Crop Protection*, **28**, 684–7.

Schumpeter, J. (1942), *Capitalism, Socialism and Democracy*, 2nd edn, New York: Harper & Brothers.

Scott, L. (2014), 'Jagged little pills', *The Courier Mail*, 15–16 March, *Q Weekend*, pp. 16–20.

Tabashik, B.E., J. Van Rensburg and Y. Carrière (2009), 'Field-evolved resistance to Bt crops: Definition, theory and data', *Journal of Economic Etomology*, **102**, 2011–25.

Tisdell, C.A. (2009a), 'Poverty, policy reforms for resource-use and economic efficiency: neglected issues', *The Singapore Economic Review*, **54** (2), 155–66.

Tisdell, C.A. (2009b), *Resource and Environmental Economics: Modern Issues and Applications*, Singapore, Hackensack (NJ), London and Beijing: World Scientific.

Tisdell, C.A. (2013), *Competition, Diversity and Economic Performance*, Cheltenham, UK and Northampton, MA, USA: Edward Elgar Publishing.

10. Property rights in human developed genetic material: an economic assessment

10.1 INTRODUCTION

Originally, genetic material was for the most part a common pool resource but in recent decades (especially since the middle of the 1980s), newly developed genetic material has increasingly become intellectual property (often private property) as well as some of the methods for producing it. For example, plant variety rights now provide limited property rights to developers of new plant varieties produced using conventional breeding methods. Furthermore, patents bestow private property rights on those who develop genetically modified plants or develop specialized methods for producing such plants. The consequences of granting such property rights, particularly the granting of biotechnology patents (biopatents) have been subject to considerable controversy. Plant variety rights provide some market protection for plant breeders, and patents mainly protect those who are able to develop new varieties of plants or attributes in plants by genetic modification, or who invent specialized biotechnological processes for transforming plants.

Given these privatization trends, legal steps have also been taken to protect farmers' property rights in genetic material. For example, the International Treaty on Plant and Genetic Resources for Food and Agriculture (ITPGRFA), which came into force in 2004, aims to do this. This treaty's purpose is to protect the access of farmers to seed and other propagating material and to ensure that they are fairly compensated for the use by others of their genetic material, for example, by developers of new genetically modified crops using this material to produce new plant varieties. However, this aim conflicts with another aim of the treaty, namely to ensure relatively open access globally to heritage agricultural germplasm so as not to hinder the development of improved varieties of crops. As highlighted later in this chapter, this resulted in the agreed payment for access to heritage genetic material of farmers being extremely small, so small as to provide farmers with little economic incentive to conserve heritage agricultural biodiversity.

Given these policy developments, this chapter discusses property rights in new varieties of crops and methods of producing these, paying particular attention to their consequences for agricultural diversity, technological progress, biodiversity conservation, pricing and market structures. Before doing this, however, background information will be provided about how and why there has been a significant switch from a common pool approach to the accessibility of germplasm in agriculture to primarily a private property approach.

10.2 BACKGROUND ON POLICY DEVELOPMENTS WHICH HAVE INCREASED THE IMPORTANCE OF INTELLECTUAL PROPERTY RIGHTS IN FOOD AND AGRICULTURE

Feindt (2012, p. 281) points out that 'since the mid 1980s, the institutional framework for agriculture has been significantly reshaped by the extension of the key institutions of the knowledge economy, intellectual property rights and patent law, to genetic resources and breeding techniques'. There are several reasons why there has been an extension of intellectual property rights to cover various types of biological material and methods for producing these. They include:

- Substantial advances in knowledge about biotechnology and processes for transforming biological material to produce genetically modified organisms, and the promise of further advances of considerable economic value. This has made new biotechnological knowledge a very valuable economic resource.
- Increasing political support for the view that an increased proportion of research and development (including biotechnological R&D) should be undertaken by the private sector rather than the public sector, and the realization that the private sector needs economic incentives to do this. This is consistent with the view of some liberals that small government is desirable either because this is more efficient or because it is less restrictive of personal freedoms than 'big' government.
- The above view is sometimes coupled with the claim that users or beneficiaries of intellectual knowledge should pay for using it. The application of this user-pays principle has resulted in payment being increasingly required for access to new knowledge provided by public bodies as well as to justify patents for and similar private rights in intellectual property.

- Support for the extension of markets (relying on the provision of adequate property rights) has increased on the basis of claims that this is an effective and desirable method of managing the environment (including the conservation of biodiversity) and it can foster the desirable development of the stock of genetic resources. Coase (1960) argued that given adequate property rights, little or no government intervention is required in environmental policy to ensure a desirable economic outcome, and as Tisdell (1996, p. 274) points out, this market-making view has influenced the general development of contemporary science and technology policy.
- Groups of influential economists (such as the Chicago School of Economics) have altered their assessment of the comparative social and economic value of different market structures. Originally, the dominant economic view was that markets consisting of many competitive firms in which entry is easy are desirable on economic efficiency and political grounds. However, subsequently the possibility was recognized that the presence of market power and concentration of ownership in some industries could lead to their superior economic performance compared with atomistic business structures after account is taken of economies of scale and the dynamics of research, development and innovation (see, for example, Tisdell, 1996, Ch. 17; 2013, Ch. 2; Svizzero and Tisdell, 2001).

The extent to which ideas (such as the above) become a part of actual economic policies depends on realpolitik, which in turn is frequently shaped by special interest groups. For example, the type of political pressure exerted by business groups is normally motivated by their own self-interest. This interest is liable to alter as technological change and economic development occur. For example, the economic value of political lobbying for extending the scope and coverage of intellectual property rights to include biological inventions has grown because of the increased actual and potential economic value of property rights in these inventions. This has increased the willingness of the developers of such knowledge to lobby for the extension and protection of their property rights in these. The governments of nations which account for the major share of such inventions, particularly the United States, have been the main proponents for extending property rights in biotechnological resources globally, not only as a result of lobbying by relevant companies but also on the basis that this serves their national economic interest. These property rights enable countries developing new biotechnologies to appropriate increased income from abroad and this stimulates their economic growth.

Even by the beginning of this century, the dominance of the United States in biotechnology patenting was evident. At that time, the Organisation for Economic Co-operation and Development (2001) reported that almost two-thirds of all biotechnology patents issued by the United States Patent and Trademark Office were granted to United States Companies, as were almost half of those issued by the European Patent Office.

Prior to legal developments that began in the last part of the twentieth century, genetic material was basically common property, that is, an open-access resource. Some countries began to provide property rights to breeders of new plant varieties in the 1950s and in 1961, and the International Union for the Protection of New Varieties of Plants (the UPOV Convention) was established by the International Convention for the Protection of New Varieties of Plants. This convention was revised in 1991. It is less restrictive than the patent system in its granting of intellectual property rights. As in the case of the patent system, property rights are granted for a limited period of time but others are free to use a new plant variety for research purposes or to breed a new variety utilizing an existing registered variety.

As a result of scientific advances in biological research, the view emerged in the United States that the UPOV Convention did not provide sufficient protection to those responsible for inventions resulting from biotechnological research. Extensions of the patent system to cover these inventions were favoured. Furthermore, the American government came to the view that it was in the economic interest of the United States to press for a strengthening of intellectual property rights worldwide. Its main vehicle for doing this was through the GATT Uruguay Round negotiations, which established the World Trade Organization.

According to Oram (1999, pp. 1–2):

> The US even threatened to boycott these negotiations if its demands for some kind of international legislation governing intellectual property rights were not met. The US favoured inclusion of IP [intellectual property] issues in the World Trade Organization because this was the only mechanism that would ensure member states could be effectively sanctioned if they failed to comply with international IPR laws.

As a result, it was agreed that compliance with Trade-Related Aspects of Intellectual Property Rights (TRIPS) be made a requirement for members of the World Trade Organization (WTO). TRIPS requires all members of the WTO to protect new plant varieties by patents or by an effective system of a similar kind, for example, by compliance with the UPOV Convention.

In the 1980s, judicial changes also occurred, which strengthened intellectual property rights (IPR) in biological inventions. In 1980, the US Supreme Court made a landmark ruling and granted the first patent on a living organism. It decided in *Diamond vs Chakrabarty* that a newly created bacterium for digesting oil in oil spills was patentable because it was man-made by genetic engineering (Anon, 2014a). Subsequently, many other genetically modified organisms have been patented in the United States and elsewhere. However, although some naturally occurring substances and processes have been patented, the extent to which this is legally possible is still subject to debate and creates some uncertainty (see, for example, Anon, 2014b).

The Convention on Biological Diversity (CBD) also weakened global open access to genetic resources because it asserts that each nation or jurisdiction has sovereign rights over genetic resources within its borders. Consequently, this required other nations or entities from foreign nations to negotiate terms and conditions for access to these resources. Given this convention, it was necessary to negotiate bilateral agreements to obtain international access to germplasm. Therefore, the transaction costs of access to global germplasm were increased and access was restricted (Chiarolla, n.d.). This was subsequently seen as a barrier to raising the global productivity of agriculture.

In order to address the above-mentioned problems, the International Treaty on Plant and Genetic Resources for Food and Agriculture (ITPGRFA) was negotiated and then came into force in 2004. It had the twin objectives of (1) facilitating international access to agricultural germplasm developed by farmers in individual jurisdictions, by specifying standard conditions for such access, and (2) protecting the perceived rights of farmers in this germplasm, that is, their right to continue to use this germplasm and to collectively receive some monetary compensation for its use by others. Nevertheless, as discussed below, this treaty fails to resolve many of the economic issues that it was supposed to address.

Table 10.1 provides a brief chronology of major international agreements

Table 10.1 A brief chronology of milestones in the legal recognition of intellectual property rights in new genetic material

Milestone	Date
International Convention for the Protection of New Variety of Plants (UPOV)	1961 (Revised 1991)
Diamond v Chakrabarty: Decision of US Supreme Court	1980
Convention on Biological Diversity (CBD)	1992
Agreement on Trade-Related Aspects of Intellectual Property (TRIPS)	1994
International Treaty on Plant and Genetic Resources for Food and Agriculture (ITPGRFA)	2004 (in force)

and legal milestones which are associated with the strengthening of intellectual property rights in genetic material. Possibly, the major milestone was the US Supreme Court's 1980 decision in the *Diamond vs Chakrabarty* case that newly developed organisms are patentable. This was the forerunner of subsequent legal decisions in the United States that extended private property rights via the granting of patents for newly developed genetic material and processes for producing such material. This led to a similar approach being adopted in many other countries and to intensified lobbying by the United States for other countries to recognize these types of property rights.

A major problem with international legislation (agreements) affecting the conservation of biodiversity and international access to germplasm is that it is piecemeal and involves inconsistencies. Feindt (2012) attributes this to the politics of accommodating different interest groups. In addition, with the passage of time, the composition and nature of interest groups change, as do social concerns and values. New legislation and agreements are often made that are to some extent inconsistent with previous legislation and agreements. Earlier agreements are not withdrawn or completely reconciled with the later ones. Feindt (2012) describes this as involving 'layering'. TRIPS, the CBD, UPOV and ITPGRFA seem to be designed to cater for the perceived interests of different groups, and these interests partially conflict with one another.

Furthermore, the approach to biological conservation adopted in the CBD differs from that in the Convention on International Trade in Endangered Species (CITES) (Tisdell, 2009, p.338). The former stresses the importance of relying on the use-value of biological material (combined with appropriate property rights) to provide economic incentives for its conservation. The emphasis of the CBD is on promoting market-making (and other forms of use) to encourage conservation. On the other hand, CITES is designed to restrict trade in endangered species, particularly those to which there is open access or de facto open access. This sometimes leads to conflicting policy decisions. For example, while Australia's sustainable use and management of saltwater crocodiles has resulted in their products being exported from Australia under permits, attempts to develop a similar management scheme in order to utilize hawksbill turtles has been thwarted by CITES (see Tisdell, 2014, Chs 8 and 9).

Given this background information, it is now proposed to analyse the economics of private property rights (especially grants of patents) in new genetic material, and processes for its development. Consideration will also be given to the potential consequences for the conservation of biodiversity of private property rights in human developed germplasm. The economic consequences of policy measures to protect farmers' rights and national rights in heritage germplasm and at the same time facilitate global access to it, will then be examined. This examination is subsequently followed by a general discussion before concluding.

10.3 THE ECONOMICS OF PRIVATE PROPERTY RIGHTS IN NEW BIOTECHNOLOGY, PARTICULARLY GRANTING OF PATENTS

As pointed out by the Commission on Genetic Resources for Food and Agriculture (2010, p. 132), the 'issue of patenting of varieties or parts of varieties (e.g. genes or traits) [of plants] and biotechnological processes (e.g. [for their] transformation) has become a controversial matter, especially as a result of increased adherence to the TRIPS Agreement'. The TRIPS Agreement (which was strongly sought by the United States) requires WTO members to give patent protection both to new products and processes developed as a result of biotechnological research and does not permit discrimination between different fields of their application including their application in agriculture (Chiarolla, n.d.). While there is scope for national governments to exclude biotechnological processes from patenting, they must provide protection for plant varieties by patents or similar means.

The patenting of biotechnology results in increased exclusion of potential users of that technology compared with the degree of exclusion associated with plant variety rights (PBR) or UPOV. For example, 'patents are generally claimed not for a single variety, as is the case with PBR but for a whole class of varieties or even a trait within a whole species' (Commission on Genetic Resources for Food and Agriculture, 2010, p. 132). Patents can also limit the rights of others to use the patented material for research purposes, for example, to produce new plant varieties. They therefore, can restrict the rights of plant breeders. In addition, they can restrict the rights of farmers to save, use, exchange and sell farm-saved seed and other propagating material (Commission on Genetic Resources for Food and Agriculture, 2010, p. 133). Thus, although both patents and UPOV limit the rights of those not possessing intellectual property, patents involve greater restriction on the use of new biotechnology by those other than the patent-holders.

These developments raise the following questions: what is the net economic benefit (if any) of these types of property rights? What are their consequences for biodiversity and for sustainable development? Is it still the case, as was argued by Godden (1987, p. 255), that 'it is currently impossible to form a definitive overall judgment as to the net economic effects of PVR, or similar methods of establishing intellectual property rights in plant varieties'? Concluding their recent review of recent research on the economic patents, Hall and Harhoff (2012, p. 559) conclude: 'Although patents may not add much to social welfare (and are therefore, not visibly effective in aggregate), they are capable of generating large private returns in many instances'. Does this imply that political support for the system is based on private self-interest rather than its social benefits?

Knowledge about New Biotechnology as a Non-rival but Excludable Commodity

Intellectual property rights, in particular patents, convert the knowledge associated with new biotechnological inventions into an excludable and therefore marketable commodity. In the absence of such rights, the knowledge associated with this technology is a pure public good if it cannot be kept secret by its originators. Following the use of the term by Brennan and Walsh (1985), Lindner (2004a, p. 303) classifies knowledge involving new biotechnological inventions as an excludable public good. He also mentions (p. 303) that this knowledge may be classified as a club good. However, this seems to be less appropriate because the use of club goods can involve a degree of rivalry and, as a result, the optimal number of club members is usually restricted to a finite number (see, for example, Tisdell, 1984, Figure 1).

The creation of excludable public goods creates an economic dilemma. Ideally from an economic point of view, newly created intellectual knowledge should be shared by all. On the other hand, if non-exclusion is practised, discoverers of new intellectual knowledge (which has economic value) are likely to be unwilling to share it and they may be able to keep it secret. Secondly, in the absence of being able to exclusively market the results of their new intellectual knowledge, private entities have little economic incentive to engage in research and development to create new knowledge. Both these factors (lack of revelation of new knowledge and the absence of economic incentives for research and development) can retard technological progress. However, these factors are not in themselves sufficient to justify making new knowledge an excludable public good by, for example, the granting of patents. It needs to be shown that these effects do occur and are of sufficient economic benefit to offset any restrictive market behaviour made possible by granting private property rights in newly created knowledge.

The Value of Knowledge Revealed and of Accelerated Technological Progress Due to Patents for New Plant Varieties

Even though a patent system exists, discoverers of new knowledge may still have the option of keeping it secret or of only partially revealing this knowledge. Furthermore, patent restrictions on the use by others of processes for producing new germplasm can result in the revealed knowledge being of little commercial value to others and can restrict their research efforts. Despite information being revealed, this privatization of essential methods for developing new germplasm can restrict technological progress, as has been pointed out by Lindner (2004a).

It should also be kept in mind that not all technological change is socially desirable. In developing and marketing new plant varieties, private entities can be expected to consider primarily their own economic self-interest. Therefore, there is a risk that they will develop and market new plant varieties, the cultivation of which results in significant negative environmental spillovers and which undermines the sustainability of development. It is widely believed that this is a significant risk in the case of genetically modified organisms but it is also possible in cases where new germplasm is created by humans by other means. The importance of these issues has been recognized in the Cartagena Protocol in the case of GMOs.

Restrictive Market Behaviours and Biotechnology Patents

Patents and similar means of creating intellectual property can provide their owners with market power. However, initially this market power in the case of new plant varieties is limited by the availability of competitive substitute varieties. This can limit the monopoly pricing of owners of property rights of the new plant varieties. If, however, growers find it more profitable to use the new plant varieties than existing ones, many existing varieties will disappear. If the new varieties are the property of a single firm or a small group of firms, their market power will increase and they may engage in increased restrictive market behaviour. This situation (the strengthening of monopoly power over time) can eventuate either as a result of deliberate predatory pricing or merely as a result of natural market dynamics.

Market Concentration, Barriers to Entry and New Plant Varieties

Although Lindner (2004b) lists five potentially beneficial outcomes from advances in plant breeding technologies which might be obtained from privatized plant breeding and ownership, he also warns that privatization of plant breeding might result in a small number of large multinational companies crowding out other competitors and then exploiting their market power 'to capture almost all the values created by plant breeding' (Lindner, 2004b, p. 6). In fact, recent developments add weight to his concern. The Advisory Board on Biodiversity and Genetic Resources at BMELV found that 'market concentration is high in many agricultural breeding segments and has significantly increased in recent years' (Feindt, 2010, p. 19).

This advisory board found that the market for agricultural seeds in 2008 was 'dominated overall by five large companies which include Monsanto, DuPont, Syngenta and Bayer CropScience and that since 1985, the percentage market share of global sales of the four largest plant breeding companies has more than tripled' (Feindt, 2010, p. 19). Furthermore, the five largest companies in the plant breeding business accounted for over 80 per cent of genetic engineering patents granted in the US in the period 2000–2004, and a very high proportion of applications to the European Patent Office in a similar period.

From the above, it follows that there is solid evidence of substantial and increasing market concentration in plant breeding. The patent system, by adding to barriers to entry, appears to have facilitated this trend. The existence of this system and the wide content of intellectual property claims associated with some patents is a deterrent to new entrants. However, it is not the only factor involved. Developing new germplasm and methods for producing it is very costly, takes considerable time and the results of the R&D involved can be very uncertain. This is a major barrier to the entry of new firms considering supplying new genetically manipulated germplasm. Particularly in the development and supply of genetically engineered propagating material, both economies of scope and of scale are likely to be important barriers to entry.

In general, multiple factors (the patent system, legal environmental safety and product quality requirements, protection of property rights, essential expenditure on political lobbying and on information provision) as well as the nature of technical relationships involved in the supply of engineered germplasm seem to result in substantial economies of scale and scope in the supply of genetically modified seeds, thereby favouring market concentration in this area. However, this domination may also spill over in other types of seed modification (for example, the production of hybrids), which do not rely on genetic engineering.

Wasteful Duplication and Concentration of Research Effort

Lindner (2004a) stresses that research designed to produce new genetic material and to create processes for producing it can result in wasteful

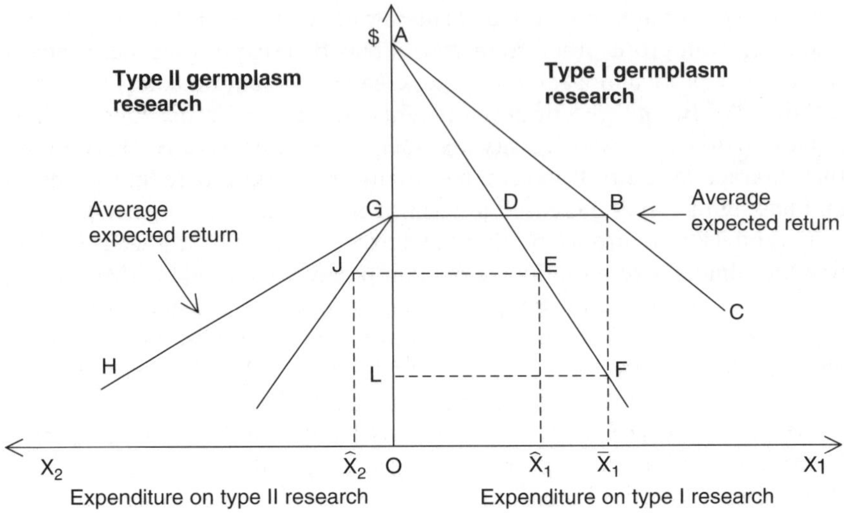

Figure 10.1 *An illustration of the misallocation of research funds due to a combination of open access to research and the granting of private intellectual property rights (for example, patents) for commercially valuable outcomes from research*

duplication of research. This is a possibility in the case of all intellectual property because capturing new knowledge (which can become private intellectual property once captured) is an open-access activity. The consequences of research activity in this case have some similarity with open access to the use of natural resources (see, for example, Tisdell, 2005, Figure 6.1). In this case, there is an over-concentration of research efforts on activities likely to yield the highest expected return on average whereas, from a social economic point of view, equalizing (social) marginal expected returns from different areas of activity is likely to be socially optimal.

Figure 10.1 provides a simple illustration of this misallocation problem. It is assumed that research funds, X, can be allocated between type I germplasm, which is expected to be very valuable, and type II germplasm, which is expected to be less valuable. In Figure 10.1, line ABC represents the expected average return from the amount of research expenditure on type I germplasm, X_1, and the corresponding expected average return relationship for the amount of research expenditure on type II germplasm, X_2, is shown by the line GH. The corresponding marginal return relationships are shown by lines AE and GJ respectively. Now suppose that the total amount of funds available for developing new germplasm is $X = \overline{X}_1$. Firms will maximize their average expected returns by allocating all of these funds to research on type I germplasm. No research is conducted on type II germplasm. However, this results in the expected marginal returns from research expenditure on developing type I germplasm being much lower than is obtainable from research expenditure on developing type II germplasm. In the former case, the marginal return is equivalent to OL, compared to OG in the latter case. By altering the allocation of the available research funds to equalize the marginal expected returns from each alternative field of research, total expected returns from genetic research can be increased and maximized. This occurs in this case when $\hat{X}_1$ of available research funds, $X = \overline{X}_1$, are allocated to research on type I germplasm and $\hat{X}_2$ to research on type II germplasm.

Note that the type of bias in the allocation of research funds illustrated by Figure 10.1 is not peculiar to biotechnological research. Whether it is likely to be more pronounced in this case than in other cases where research has prospects of delivering commercially valuable outcomes (which can become private intellectual property) is unclear.

Patent Blocking of Research, Patent Thickets, Anti-commons and Transaction Costs

Biopatents can be used to block or retard the research of potential competitors as well as the research of scientists interested only in advancing basic knowledge or science. This has been described as anti-commons activity (Buchanan and Yoon, 2000). In the case of biopatents, 'it is possible to secure temporary private ownership of existing genetic resources [unlike under plant variety laws PBR or UPOV] provided the resources are isolated in a technical process' and this monopoly may be extended by buying derivative patents or evergreening (Feindt, 2010, p. 18).

Patents for DNA sequences create particular problems. According to Feindt (2010, p. 19), 'genes are limited in number and it is extremely hard for research to sidestep gene patents; given a gene can have many functions and many patents relate to all functions, including ones yet to be discovered by others'. Furthermore, the patenting of processes for creating new germplasm can substantially block downstream research. Consequently, some limitations on the scope of such patents seem to be desirable from a social point of view. This is because they can seriously restrict access to essential plant-breeding infrastructure and result in its underutilization from a social economic point of view (Lindner, 2004a).

In addition, biopatents may be acquired, held and generated primarily with the aim of blocking potential entrants or to extract fees from them. The threat of patent litigation, given the potential costs involved, will be sufficient to deter some entrants even if they are not infringing on the intellectual property rights of others, or they may pay a fee to an alleged patentee to avoid costly and uncertain litigation. Some of the latter cases border on the use of blackmail. High transaction costs are involved both in applying for biopatents and in protecting them (Feindt, 2010, p. 22). High transaction costs in cases such as this considerably reduce the efficiency of the economic system (see, for example, Tisdell, 2013, Figure 9.5 and the surrounding discussion).

10.4 ECONOMIC ASPECTS OF THE INTERNATIONAL TREATY ON PLANT GENETIC RESOURCES FOR FOOD AND AGRICULTURE

As was pointed out above, new international agreements, changes in national legislation and court decisions have in recent decades resulted in a growing proportion of global stock of germplasm becoming private property and/or the property of sovereign states. The pool of genetic resources available for open access has declined both because of these institutional developments and because of a reduction in the stock of inherited germplasm, which has occurred for reasons outlined in Chapters 3 and 5. Furthermore, while institutional changes have strengthened the property rights of plant breeders and developers of new germplasm, virtually no recognition was given to the property rights of farmers until the International Treaty on Plant Genetic Resources for Food and Agriculture (ITPGRFA) came into force in 2004. However, this treaty does not seem to be very effective in preserving farmers' rights and in achieving some of its other objectives, such as conserving traditional plant varieties *in situ*. The reasons for the lack of effectiveness of this treaty are varied. A major reason, however, seems to be the conflict of interests of plant breeders and developers of germplasm with the interests of persons conserving heritage germplasm. The former have an interest in easy or unrestricted access to this heritage germplasm. Those nations (which appear to be the major economic beneficiaries of intellectual property rights in human developed germplasm), such as the United States, can be expected to be strongly opposed to institutional change which restricts this access. Because international treaties must be negotiated between nations (each of which is motivated by its own perceived self-interest), compromises are usually required to reach agreement. This can result in vague and qualified agreements which have little effect. Apart from this, there can also be several practical difficulties involved in adopting policies to compensate farmers for the use of the heritage germplasm which they have developed collectively.

The following questions need to be considered in assessing farmers' rights in heritage agricultural germplasm:

1. Should farmers be restricted (as a result of the granting to others of intellectual property rights in new biotechnology) in making use of their heritage germplasm in traditional ways such as the saving and replanting of seed?
2. Should farmers be financially compensated for the use of their heritage germplasm by others, for example, by those who genetically modify it to create and patent new germplasm?
3. If compensation is to be paid, how should its aggregate level be determined and how ought it be distributed to farmers?
4. To what extent is the compensation paid to farmers intended as an equity payment or as an incentive to conserve heritage germplasm? If the latter, how effective is the payment likely to be?

Avoiding Restrictions on Farmers' Rights to Use Heritage Germplasm in Traditional Ways

There is a danger that the wide range of intellectual property rights claimed for some new biotechnological discoveries could restrict the rights of farmers to use their heritage germplasm in traditional ways. Chiarolla (n.d.) points out that:

> legal and regulatory requirements may *de facto* limit the freedom of farmers to continue their traditional agricultural practices, such as the saving of seeds and tubers from consumption and subsequent use. At the national level, the restriction of the rights that farmers currently enjoy and the lack of initiatives to create appropriate incentives may affect negatively the conservation and development of plant genetic resources for food and agriculture, and increase food security.

Although an ostensible aim of ITPGRFA is to protect farmers' rights to use their heritage germplasm in traditional ways, close examination of the treaty indicates that these rights may still be restricted. Although the preamble to the treaty proclaims these rights, they are not specifically stated within the treaty itself even though Article 9.3 states: 'Nothing in this article shall be interpreted to limit any rights that farmers have to save, use, exchange and sell farm-saved seed/propagating material, subject to national law and as appropriate'. However, what is 'appropriate' is not specific and some national laws can restrict farmers' rights. Furthermore, this treaty states that it does not annul the rights and obligations of the consenting parties agreed to under other international agreements, such as UPOV and TRIPS. The latter do restrict the traditional rights of farmers to use improved varieties in the same way as heritage varieties of crops. A major concern, however, is that intellectual property rights in new plant varieties be not so widely encompassing as to restrict trade in heritage varieties or farmers' re-use of the propagules of these varieties. Such restrictions would be indefensible on economic efficiency grounds and could give rise to an unjustifiable restrictive practice.

Should Farmers be Compensated When Others Use Their Heritage Germplasm to Create New Plant Varieties?

It is assumed in the ITPGRFA that farmers have a right to equitably participate in sharing benefits arising from the utilization of plant genetic resources by others. However, it is possible to argue that income received in this way constitutes a windfall gain because their heritage germplasm was developed and conserved for their own benefit alone. Furthermore, there is no clearcut guide to what would constitute an equitable share for farmers of any benefit arising from new plant varieties developed using heritage varieties. In addition, it could be argued that the extra benefit is purely due to the ingenuity of those who devise new plant varieties using existing genetic material. In many cases also, it is farmers' ancestors who developed these heritage varieties, not the present generations of farmers.

If Compensation is to be Paid, How is its Aggregate Amount to be Determined and How Ought it be Distributed to Farmers?

Estimating the contribution of landraces to the development of improved plant varieties is difficult because such developments often rely on genetic resources obtained from many different sources of 'elite breeding lines several generations removed from farmers' varieties' (Brush, 2007, pp. 1510–11). In such cases, there is no obvious way to determine the contribution of any individual landrace to the economic benefit obtained from the new variety. In fact, according to Brush (2007, p. 1511): 'Estimating the contribution of a single landrace or collection to the value of a modern variety has not been accomplished and is likely to become more difficult as pedigrees become more complex'.

While it might be thought that it could mostly be farmers in developing countries that would lose out in the absence of being paid for the use of their landraces in other countries for breeding material, this might not be the case in practice because foreign landraces provide important resources for breeding in developing countries, such as India and Nigeria (Brush, 2007, p. 1511).

There is also the problem of determining to which farmers any compensation should be paid because the development of landraces depends on the efforts of many different individuals and earlier generations. One cannot really determine to what extent the presence of an existing landrace is due to a particular individual or set of individuals. Therefore, the appropriate rules for sharing any payments for use of landraces are indeterminate.

Is the Payment to be Made on Equity Grounds or as an Incentive to Conserve Agrobiodiversity?

The responsibilities of the consenting parties to the ITPGRFA include promoting efforts by farmers and local communities to manage and conserve on-farm their plant genetic resources, but the way in which they should do this is not specified. Brush (2007) argues that only a meagre amount of funds are likely to be generated by payments to use the genetic resources contained in landraces and these will provide insufficient funds if paid to farmers to entice them to continue to conserve landraces. Brush (2007, p. 1511) considers that a weakness of ITPGRFA 'is that it does not give proper emphasis to the obligations of industrial countries and developing countries alike to support the conservation of resources beyond funds raised in connection to commercializing improved crop varieties'.

Sovereign Rights of National Governments over Plant Genetic Resources

The treaty affirms the sovereign rights of national governments in their plant genetic resources. Consequently, individual states are responsible for the management of their plant genetic resources. Individual state governments collect any funds due to farmers for benefit sharing of their genetic resources, and they decide how funds will be disbursed to farmers. As a result, dissimilar policies may be pursued by different governments in disbursing funds to farming communities. Furthermore, the amount of funds available to farmers is likely to be considerably reduced if the agency costs of governments are deducted from these funds.

Facilitated Access to Plant Genetic Resources

The ITPGRFA, by its intent to facilitate multilateral access to landrace plant varieties, erodes property rights which farmers are recognized as having under the treaty. For example, the treaty requires that 'access shall be accorded expeditiously, with a minimal or no charge at all' (Hasan, 2004, p. 4). This being so, only a small amount of funds are likely to be collected from benefit sharing. Furthermore, the treaty specifies that 'no IPR [intellectual property rights] so as to limit access shall be claimed over PGRs [plant genetic resources] or their genetic parts or components in the form received from the multilateral system' (Hasan, 2004, p. 4).

In effect, the facilitated access conditions in the treaty almost re-establish multilateral open access to landrace varieties of plants. Open access to genetic resources was restricted by a clause in the Convention on Biological Diversity assigning sovereignty to individual states over these resources. It became apparent that this clause would restrict international access to landraces genetic resources because international access would depend on bilateral agreements, which in all likelihood would entail substantial transaction costs and impede access. Furthermore, it was argued that easier access globally to heritage germplasm would benefit both developed and less developed countries: it would be mutually advantageous (Brush, 2007, p. 1507). Nevertheless, countries such as the United States, which are in a superior position to genetically modify this germplasm and patent the results, might be the major beneficiaries of easy access to heritage germplasm.

10.5 DISCUSSION

The content of the ITPGRFA illustrates problems that can arise in developing serially economic policies in a piecemeal fashion, each of which is based on principles fashionable at the time and designed to take account of particular interest groups on each occasion. This partly involves what Feindt has described as 'layering'. International agreements resulting in plant variety rights (UPOV), the Convention on Biological Diversity and TRIPS (reinforced by extended applications of the patent system to biological discoveries) supported the principle that strengthening private and national property rights in genetic resources would have a desirable collective economic impact on the conservation and development of genetic material, including agrobiodiversity. These developments appeared to be in the economic interests of those nations possessing genetic resources of economic value and those nations and corporations able to develop improved genetic material. However, possible negative consequences of such a system were overlooked.

Subsequently, it was realized that such a system could block international access to existing genetic resources and/or result in high transaction costs to obtain access to them. There was a possibility that this would impede economic growth based on the utilization of existing genetic material. On the other hand, it was realized that open-access to heritage genetic material would provide no incentive to the guardians of this material to conserve it, and that not to compensate these guardians for its use, might be (arguably) inequitable. The problem posed was how to support almost open access to this material and provide some economic incentive to its guardians to conserve it. The resulting compromise, resulting in ITPGRFA, provides little economic incentive for conserving heritage agrobiodiversity but allows almost open access to heritage agrobiodiversity. It is favourable to developers of agrobiodiversity using modern scientific methods, such as genetic engineering and hybridization, particularly those that are able to obtain actual private property rights (for example, via patents) in genetic developments or de facto rights via secrecy and other methods of exclusion. Consequently, it appears to provide greater economic benefit and incentive to developers of genetic material and little incentive to farmers (and others) to conserve heritage genetic material. The appropriate economic balance between conserving existing heritage genetic material and developing new genetic material has yet to be found, as has been pointed out by Brush (2007).

The extension of the patent system to cover a wide range of biological discoveries is of major economic and social concern. There are several reasons why this system (which creates an excludable public good) is not fully efficient from an economic point of view. The system relies on market restrictions to achieve its usually intended result of accelerating technical progress in order to improve living standards generally. The net benefit of the system depends on the social economic benefits from greater technical progress exceeding the economic drawbacks caused by restrictive behaviour by the patentee and by the transaction costs associated with the system. Another problem is that the system, by accelerating technical progress in developing new germplasm, might seriously erode the existing genetic stock and eventually result in unsustainable economic development.

That raises the question of what, if any, would be a better system from an economic point of view. Furthermore, how can the existing patent system in relation to its application to biopatents be improved to increase its social economic benefits?

A possible alternative to motivating private enterprise to generate biotechnological advances by granting intellectual property rights in new biological discoveries (including patents) is to rely on public bodies to supply these advances. However, because these bodies would have to be funded by taxes, the amount of funds available for public research of this type may be less than in the case of private provision combined with grants of intellectual property rights. Furthermore, the separation of public bodies from the market and commercial activities could result in research results of lower economic value than when there is reliance on private enterprises. In addition, public bodies are not subject to the discipline of market competition and this could result in greater X-inefficiency in their research than when R&D is conducted by private enterprises. Consequently, both private and public provision of R&D to generate agricultural genetic improvements are subject to economic shortcomings. Comparing imperfect systems is difficult. However, a social economic role could exist for both public and private provision of research and development of biotechnology. For example, public provision would be desirable in cases where social economic returns are likely to be high but private returns are low. Furthermore, the question needs to be considered of whether or not fees should be charged for access to results from research and development funded by the public and, if so, when these fees should be levied and how the charges for access should be determined. Also under what circumstances should public bodies patent their inventions and limit open access to these?

A further issue that needs to be addressed is whether it is possible to make some changes in the patent system in order to reduce its imperfections as a means for developing and supplying new biotechnology. In particular cases, it may be desirable to limit the wide range of genetic material over which intellectual rights can sometimes be claimed and/or reduce the length of time for which the patent applies for particular applications. Property rights in new processes for producing new germplasm can restrict the research endeavours of non-patentees and hinder biotechnology research.

According to a survey of agricultural scientists conducted by Lei et al. (2009), 'scientists believe that contrary to the current consensus, proliferation of IP protection has a strongly negative effect on research in their disciplines' by restricting their access to intellectual property. Restrictions of this type should be considered for relaxation. Such relaxation of intellectual property rights will, of course, reduce their economic value for those hoping to claim such rights. The extent to which this will reduce their research effort is not known, and the economic gains of easier access to the intellectual knowledge generated have not been quantified. Therefore, this subject needs more investigation.

As pointed out by Raidt (2014), many other shortcomings in the patenting of biotechnology exist. These include patent backlogs, a shortage of qualified patent examiners and legal uncertainty due to 'overly broad or ill-defined intellectual property (IP)', and 'excessive litigation, including actions of patent trolls created to buy patents for the purpose of claiming infringements and extracting fees, rather than using IP to market useful goods' (Raidt, 2014, p.1). Furthermore, uncertainty remains about the degree of alteration required to naturally occurring biological material to qualify for a patent and to what extent the alteration must be described.

10.6 CONCLUDING COMMENTS

Major changes occurred in the closing decades of the last century in scientific procedures for supplying new plant varieties and, in turn, these have also resulted in significant changes in intellectual property rights in new germplasm and procedures for producing it. These are now often private intellectual property whereas in previous times, open access to germplasm was the general rule. Furthermore, an increasing proportion of R&D activity designed to supply new germplasm is now undertaken by companies rather than by public bodies. There are several reasons for these institutional changes. They include a strengthening of the view that market provision of commodities based on private property rights is a more effective (economic) means of their provision than their supply by governments or public bodies, as well as the fact that it became apparent that substantial profits can be obtained from new biotechnology given intellectual property rights in it. The latter provided a strong motive to lobby politically for such rights by companies able to generate profitable biotechnology inventions.

However, the current system of granting intellectual property rights in biotechnology inventions has some serious imperfections from a social economic point of view, as has been highlighted in this chapter, and needs reform. On the other hand, relying only on public provision of biotechnology also has its drawbacks, as was pointed out. Furthermore, there are several social conflicts that are difficult to reconcile. For example, the attempt to reconcile farmers' interest in property rights in germplasm (by means of the ITPGRFA) with those of plant breeders (who want easy access to this germplasm) has not been entirely successful, and current economic mechanisms for conserving heritage plant germplasm remain weak. Furthermore, the disappearance of existing stock of this heritage germplasm is likely to be accelerated by the successful development of new germplasm using biological engineering and other techniques. In the end, this could limit the scope for further genetic advances and threaten food sustainability.

REFERENCES

Anon (2014a), 'Biological patents in the United States', Wikipedia. Available at: http://en.wikipedia.org/w/index.php?title=Biological_patents_in_the_United_States. (Accessed 7 October, 2014).

Anon (2014b), 'Biological patent', Wikipedia. Available at: http://en.wikipedia.org/w/index.php?title=Biological_patent. (Accessed 7 October, 2014).

Brennan, G. and C. Walsh (1985), 'Private markets in (excludable) public goods: A re-examination', *Quarterly Journal of Economics*, **100**, 811–19.

Brush, S.B. (2007), 'Farmers rights and protection of traditional agricultural knowledge', *World Development*, **35**, 1499–514.

Buchanan, J.M. and Y.J. Yoon (2000), 'Symmetric tragedies: Commons and anti-commons', *Journal of Law and Economics*, **43**, 1–14.

Chiarolla, C. (n.d.), 'FAO International Treaty on Plant Genetic Resources and Farmers' Rights', United Nations University. Available at: http://archive.ias.unu.edu/sub_page.aspx?catID=35&ddlID=191. (Accessed 2 October, 2014).

Coase, R.H. (1960), 'The problem of social cost', *Journal of Law and Economics*, **3**, 1–44.

Commission on Genetic Resources for Food and Agriculture (2010), *Second Report on the State of the World's Plant Genetic Resources for Food and Agriculture*, Rome: FAO.

Feindt, P.H. (2010), 'Biopatents – A threat to the use and conservation of agro-biodiversity?' Bonn: Advisory Board on Biodiversity and Genetic Resources at The Federal Ministry of Food, Agriculture and Consumer Protection (BMELV), Federal Office for Agriculture and Food (BLE). Translation from the original German paper.

Feindt, P.H. (2012), 'The politics of biopatents in food and agriculture, 1950–2010: Value conflict, competing paradigms and contested institutionalisation in multi-level governance', *Policy and Society*, **31**, 281–93.

Godden, D. (1987), 'Plant variety rights: Framework for evaluating recent research and continuing issues', *Journal of Rural Studies*, **3**, 255–72.

Hall, B.H. and D. Harhoff (2012), 'Recent research on the economics of patents', *Annual Review of Economics*, **4**, 541–65.

Hasan, S.R. (2004), 'ITPGRFA and the protection of farmers' rights', *Policy Brief*, No. 9, Kathmandu: SAWTEE.

Lei, Z., R. Juneja and B.D. Wright (2009), 'Patents versus patenting: Implications of intellectual property protection for biological research', *Nature Biotechnology*, **27** (1), 36–40.

Lindner, B. (2004a), 'Privatised provision of essential plant breeding infrastructure', *The Australian Journal of Agricultural and Resource Economics*, **48**, 301–21.

Lindner, B. (2004b), 'Economic issues for plant breeding in public funding and private ownership', *Agribusiness Review* [Online] **12**. Available at: http://www.agrifood.info/review/2004/Lindner.html. (Accessed 30 September, 2014).

OECD (2001), *Biotechnology Statistics in OECD Member Countries: Compendium of Existing National Statistics*, Paris: OECD.

Oram, J.A. (1999), 'The TRIPS agreement and its implications for food security', [Online] Cork: International Famine Centre, University College. Available at http://www.recrea.org/rrf/biopatents.htm. (Accessed 2 October, 2014).

Raidt, J. (2014), *Patents and Biotechnology*, Washington, DC: US Chamber of Commerce Foundation.

Svizzero, S. and C. Tisdell (2001), 'Concepts of competition in theory and practice', *Revista Internazionale di Science Economiche e Commerciali*, **48** (2), 145–62.

Tisdell, C.A. (1984), 'The provision of wilderness by clubs', *Rivista Internazionale di Scienze Economiche e Commerciali* (*International Review of Economics and Business*), **31** (8), 758–61. Reprinted in C.A. Tisdell (2002), *The Economics of Conserving Wildlife and Natural Areas*, Cheltenham, UK and Northampton, MA, USA: Edward Elgar Publishing.

Tisdell, C.A. (1996), *Bounded Rationality and Economic Evolution*, Cheltenham, UK and Brookfield, VT, USA: Edward Elgar Publishing.

Tisdell, C.A. (2005), *Economics of Environmental Conservation*, 2nd edn, Cheltenham, UK and Northampton, MA, USA: Edward Elgar Publishing.

Tisdell, C.A. (2009), *Resource and Environmental Economics: Modern Issues and Applications*, Singapore, Hackensack (NJ), London and Beijing: World Scientific.

Tisdell, C.A. (2013), *Competition, Diversity and Economic Performance*, Cheltenham, UK and Northampton, MA, USA: Edward Elgar Publishing.

Tisdell, C.A. (2014), *Human Values and Biodiversity Conservation: The Survival of Wild Species*, Cheltenham, UK and Northampton, MA, USA: Edward Elgar Publishing.

PART III

Natural biological capital: wild genetic resources and natural ecosystems

11. The conservation and loss of wild biodiversity and natural ecosystems: basic economic issues

11.1 INTRODUCTION

Whereas the previous part of this book focused primarily on human managed biodiversity and human developed ecosystems, this part deals mainly with the economics of the conservation and loss of wild biodiversity and natural ecosystems. The discussion in this chapter extends that in Chapters 2 and 3, by paying particular attention to the basic economics of the loss of wild biodiversity, including natural ecosystems. However, it needs to be recognized that there is no distinct division between wild (unmanaged) biodiversity and human managed biodiversity because the extent to which human beings influence biodiversity forms a continuous spectrum.

Nevertheless, because of bounded rationality, it is not practical to analyse all possibilities on the multidimensional spectrum of biodiversity. It is only practical to analyse some of the 'ideal' or selected types in this spectrum. By doing so, we can obtain an improved appreciation of the whole spectrum and can develop our selection of economic techniques to assess these different situations. In cases where biodiversity is highly controlled by human beings, private goods (the consumption of which is rivalrous and excludable) are mainly produced. By contrast, in cases where wild (unmanaged) biodiversity is a prominent feature of an ecosystem, the commodities (services) supplied usually contain a large public good or quasi-public good component, even though a private good component may also be present. For example, products, such as wild fish, involving rivalrous consumption may be extracted from unmanaged or slightly managed ecosystems.

In the latter case, mixed goods are supplied by the natural ecosystem. Although market and political failures can occur both in relation to human managed ecosystems and natural ecosystems, the problem of economic failure seems to be more pressing in the latter case. This might be why in recent decades economists have given greater attention to the economic value of conserving wild biodiversity than to the conservation of biodiversity that has evolved primarily due to anthropocentric management. However, both types of biodiversity are potentially of considerable economic value and therefore studies of the economics of biodiversity conservation and loss are both needed.

This chapter is presented in the following sequence. First, there is a discussion of the economic valuation of wild biodiversity and the services provided by natural ecosystems. This is followed by an examination of the role of opportunity costs in the evaluation of wild biodiversity and in deciding whether or not natural ecosystems should be conserved. Subsequently, socio-economic drivers of wild biodiversity loss are identified and policy implications of this analysis are outlined.

11.2 OBSERVATIONS ON THE ECONOMIC VALUATION OF WILD BIODIVERSITY AND NATURAL ECOSYSTEMS

Most studies of the economic value of wild biodiversity concentrate on estimating the value of particular living natural objects (for example, individual species) or collections of these (for example, in national parks and protected areas and ecosystems). Few focus on estimating the economic value of the variety or diversity of such objects per se. This is so despite considerable attention being given to the consequences of biological diversity for the way in which ecosystems function, for example, to the view that greater biodiversity increases the resilience or stability of an ecosystem, a view which is itself subject to qualifications.

The failure of most economic studies of biodiversity to consider the economic value of diversity (as such) was already commented on by Pearce and Moran over two decades ago. They state that most economic studies measure the economic value of particular biological resources rather than their diversity and point out that 'diversity valuation requires some idea of the WTP [willingness to pay] for the *range* of species and habitats rather than the specific biological resources they happen to support' (Pearce and Moran, 1994, p. 94). It is still the case that most studies of biodiversity focus on valuing particular biological resources rather than their range and variety.

Although both of the above-mentioned aspects of biodiversity are important, it seems to be more urgent to give increased attention to the analysis of what constitutes diversity in nature, how this should be valued, and to determining how much it is valued by individuals. Therefore, in the following five sections, particular consideration is given first to issues involved in the measurement and specification of the value of the diversity of living things per se. This is followed by a brief discussion of standard economic methods of estimating the value of natural objects and then some views about the comparative economic value of natural versus modified ecosystems are examined.

11.3 THE ECONOMIC VALUATION OF WILD BIOLOGICAL DIVERSITY PER SE (VALUATION OF NON-MATERIAL BENEFITS)

Biological diversity may be valued for a variety of reasons. It may, for example, be valued for its own sake (its existence value) and for the enjoyment of seeing, hearing about, and experiencing diverse living things. It may also be valued for the different products that it supplies or that it is believed to be able potentially to supply, such as ingredients for medicines, and germplasm for improving the genetic attributes of crops.

Although individuals differ in the value they place on being able to observe, experience or learn about diverse living things, for many, the opportunity to visit and view environments containing diverse wildlife is an attraction which contributes positively to their utility. For example, in a survey of 622 visitors to a subtropical rainforest (Lamington National Park in Queensland), 332 (53 per cent) of respondents said that the presence of a considerable amount of biodiversity in Lamington National Park was very important to them and a further 225 (36 per cent) said that it was important. Only 43 of those responding to this survey said that the presence of a considerable biodiversity was unimportant to them. Furthermore, the fact that the ecosystem in the Lamington National Park was considered to be rare was also highly rated by respondents.

The relevant statistics for these responses are summarized in Table 11.1. The same group of respondents were also asked about the importance to them of various attributes of birdlife in Lamington National Park. Both the large variety or diversity of birds present and their rarity were regarded as important or very important by most respondents. The relevant statistics are summarized in Table 11.2. Note that the majority of the respondents considered the local diversity of birds to be more important than their rarity even though both aspects were highly rated.

While a variety of factors can influence individuals' valuations and decisions to visit a natural area, the extent of the biodiversity present and the rarity of the ecosystem is likely to be an important influence. Nevertheless, these attributes are not important for everybody, as is apparent from the

Table 11.1 The importance (as stated by a sample of visitors) of the biodiversity present and the rarity of the ecosystem in the environment of a subtropical rainforest, namely Lamington National Park, Queensland[(a)]

Response	Considerable biodiversity present		Rare ecosystem	
	No.	%	No.	%
Very important	332	53	344	55
Important	225	36	217	35
Unimportant	43	7	41	7
No response	22	4	20	3
Total of respondents	622	100[(b)]	622	100[(b)]

Notes:
(a) Data extracted from Tisdell and Wilson (2012, Table 6.8, p. 187).
(b) Percentages rounded to the nearest whole number.

Table 11.2 The importance (as stated by a sample of visitors) of the diversity of birds and their rarity in Lamington National Park, Queensland[(a)]

Response	Large variety or diversity of birds		Presence of rare birds	
	No.	%	No.	%
Very important	353	57	324	52
Important	236	38	217	35
Unimportant	27	4	66	11
No response	6	1	15	2
Total of respondents	622	100[(b)]	622	100[(b)]

Notes:
(a) Data extracted from Tisdell and Wilson (2012, Table 6.12, p. 191).
(b) Percentages rounded to the nearest whole number.

examples shown in Table 11.1 and 11.2. Furthermore, areas in which rare species occur but which are not biodiverse can be highly valued.

Note that the percentage of the general public who rate biodiversity as unimportant is likely to be higher than that indicated by the results reported from the survey in Lamington National Park. First, those who do not visit national parks or protected areas (or who do so less frequently) are likely to be less interested in nature and in its diversity than those who do visit, or do so frequently. Visitors are unlikely to be completely representative of the total population. Secondly, those who responded to the above-mentioned survey may have had a greater interest in the fate of nature and its diversity than those who did not, and therefore might have been more highly motivated to respond.

The economic valuation of natural biodiversity for viewing and for recreation is likely to vary with its geographical location and with the density of the biodiversity in a locality. Other things being held constant, low-cost access (low travel costs) to wild biodiversity, and its close proximity to major population centres, can be expected to increase its economic value for tourism and recreation.

Both the presence of local biodiversity and of global biodiversity is valued by many individuals, but possibly most individuals place greater weight on conserving biodiversity that is closest to where they live. The more distant is a natural attraction away from where individuals reside, the less concerned they are likely to be about its conservation, other things being held constant. Therefore, a distance-related drop-off effect in valuing biodiversity may occur. Although empirical testing for this drop-off effect has not (it seems) been specifically done, there is some evidence of a relationship of this kind. For example, a survey of visitors to the Jourama Falls Section of the Paluma National Park in northern Queensland was conducted by Tisdell and Wilson (2012, Ch. 7). A part of this survey enquired about the stated willingness of respondents to contribute to the conservation of the mahogany glider (*Petaurus gracilis*). This is an endangered gliding marsupial (Tisdell, 2014, Ch. 6). It was found that respondents who lived outside Queensland (either living overseas or in another Australian state) were less likely to state that they would donate to its conservation than were Queenslanders (Tisdell and Wilson, 2012, p. 231). Furthermore, it is commonly stated in relation to benefit transfer analysis that willingness-to-pay values tend to decline the further individuals reside from the site where environmental benefits are generated (Bateman et al., 2000).

This should not be taken to imply that no economic value is placed by individuals on the existence of distant biological objects, the survival of which helps to maintain global biodiversity. Nevertheless, a distinction ought to be made between how much biodiversity is valued in local areas and its valuation globally. It is possible for local extinction of many wild species to occur and their global number to remain unchanged. Even though global biodiversity (as measured by number of species) is unchanged, human well-being may be reduced because local biodiversity declines. Economic valuation of biodiversity is subject to spatial influences.

Note that moral sentiments influence individuals' valuations of biodiversity. For example, when a sample of Brisbane residents were asked whether they supported the survival of different reptile species, 80 per cent of respondents (as reported in Tisdell, 2014, Ch. 10) said that they favoured the survival of the least liked species, a very venomous type of snake, the taipan (*Oxyuranus* spp.). The most frequent reason given for this response was that all species have a right to exist (Tisdell, 2014, p. 220). A number of respondents to the same survey also supported the survival of taipans on the basis that all species have an ecological role to play in ecosystems. However, as discussed in Tisdell (2014, Chs 7 and 10), moral or ethical sentiments are shaped (to a significant extent) by social and cultural influences and are liable to alter with the passage of time.

Although biodiversity is a complex type of economic good, in some situations its supply can be modelled as a pure public good or as an open-access quasi-public good not subject to congestion or similar constraints. The former attribute may arise, for instance, from the presence of the existence and bequest attributes of biodiversity, and the latter attribute can occur when wild biodiversity is used for non-consumptive tourism and recreation.

The amount of the wild biodiversity supported by wild habitats is likely to increase with the size of ecosystems or protected areas, and the less they are disturbed by human activities. Furthermore, the probability of such biodiversity continuing to exist can be expected to increase with the size of a protected area and the less it is disturbed by human activities. In principle, the optimal economic size of such an area can be determined if the Kaldor–Hicks (potential Paretian improvement) criterion is adopted.

However, this approach has several limitations. First, it is based on partial analysis. Therefore, the values used to determine the social optimum are influenced by what is assumed about other things of relevance, for example, how much biodiversity is conserved elsewhere. Second, it ignores the income distribution consequences of policies affecting the conservation of biodiversity. Therefore, important ethical issues are side-stepped. This can be illustrated by the following simple example.

Suppose that, as illustrated in Figure 11.1, an existing natural (or near natural) area has a size of x_2 km^2. Those who value the conservation of biodiversity in this area place a collective marginal valuation on the size of this natural area, as shown by line ABC. Those who want to use the natural area to supply material goods experience a marginal net loss from its protection which is indicated by the line OBD. This is the marginal opportunity cost to this group of protecting the area in order to conserve

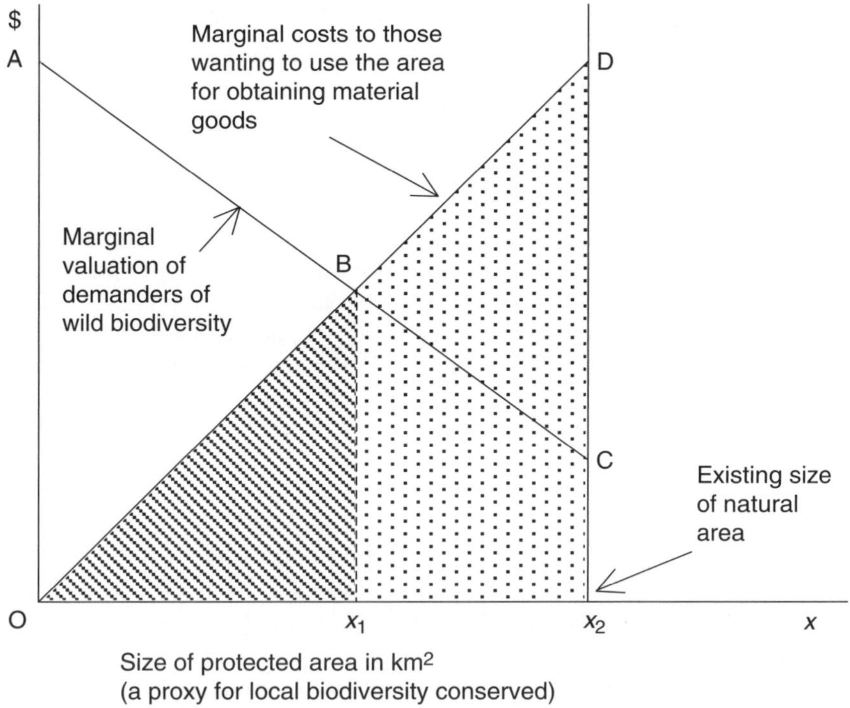

Figure 11.1 An illustration of a Kaldor–Hicks optimum for protecting local wild biodiversity

its biodiversity. Using the Kaldor–Hicks criterion, it would therefore be socially optimal to protect x_1 of this natural ecosystem and make $x_2 - x_1$ of it available for the supply of material goods.

However, the application of the potential Paretian improvement principle involves at least two policy problems. If losers from a social (environmental) change are not adequately compensated for their loss, it does not resolve social conflict. Furthermore, even if losers are compensated for their loss, social conflict may remain about whether the payment of compensation is just.

In the case illustrated in Figure 11.1, an amount equal to the hatched triangle would need to be paid to compensate those who lose the opportunity to use x_1 of the protected area to supply material goods. Whether this should be paid might depend on whether those wanting to use the area for material purposes have already been using it in this way and therefore, have de facto property rights in its use. This may, for example, be a tribal group.

Further considerations might include how poor they are and what alternative opportunities they have for their livelihood if deprived of the material use of this natural area.

On the other hand, if those wanting to use the area have not been using it and did not previously have property rights to use it, no moral case would seem to exist for the payment of compensation. Nevertheless, even in the previous case, compensation may not be paid in practice or inadequate compensation may be paid. In the past, some tribal groups have been forced off the land they have traditionally used for their subsistence to create protected areas (or to enable the 'development' of their land) without compensation being paid or adequate compensation being paid to them. Furthermore, as discussed later, market and political failures can also result in failures to conserve wild biodiversity in a socially optimal manner and could result in none of the natural area considered in Figure 11.1 being conserved.

11.4 VALUATION OF MATERIAL BENEFITS OF CONSERVING WILD BIODIVERSITY FOR NEW COMMERCIAL PURPOSES

The above discussion relates primarily to the non-material attributes of wild biodiversity. However, wild biodiversity may also be valued because of the new material benefits (commercial goods) it might create. As Craft and Simpson (2001, p.1) state: 'one argument frequently made is that biodiversity is a source of new industrial, agricultural and particularly pharmaceutical products. Natural organisms it is argued, are great repositories of genetic information', which can be potentially used to supply valuable new commercial products. Ten Kate and Laird (1999) provide many specific examples of the commercial use of biodiversity. Craft and Simpson (2001) also point out that several natural scientists (they mention Wilson, 1992; and Rubin and Fish, 1994) claim that the prospects of discoveries of valuable commercial products by utilizing existing wild biodiversity establish a strong case for biodiversity conservation.

Some economists have also adopted a similar point of view. For example, Ciriacy-Wantrup (1968) argued that given the potential future value of species in relation to the cost of their conservation, conserving a minimum viable population of all species has economic merit. Other economists (for example, Krutilla, 1967; Arrow and Fisher, 1974) have shown that conserving biodiversity to keep future economic options open for its use can be economically worthwhile and therefore it is wise to err in favour of the conservation of biodiversity. However, the extent to which this is economic in practice remains rather elusive. This is not surprising given the amount of uncertainty involved in evaluating the commercial potential of biodiversity and the different criteria which individuals may adopt for valuing biodiversity conservation.

The basic approach adopted by economists to estimate the value of wild biodiversity (wild germplasm) for the potential supply of a new commercial product is to multiply the estimated probability of this germplasm supplying such a product by its predicted value (Simpson et al., 1996, p. 164). For example, experts may be asked the probability of a specified wild area or type of biota being able to supply organic material for the development of new types of drugs, and economic values may be estimated taking account of the economic value obtained from similar drugs in the past. Although this approach is logical, many challenges are involved in applying it.

Experts often are unsure or have very different opinions about the above-mentioned probability. According to Pearce and Moran (1994, pp. 106–107), when Principe (1989; 1991) asked different drug company experts about the probability of wild plant species giving rise to a potentially valuable new pharmaceutical drug, he obtained a wide range of estimates. These probabilities ranged from 1 in 10 000 plant species resulting in a new valuable drug to 1 in 1000. This indicates that the probabilities of new commercial discoveries based on wild germplasm are very uncertain. So also are the expected economic values of such discoveries. These will vary with the 'effectiveness' of the new product that is developed and with the basis (the method) used for determining economic values (see, for example, Pearce and Moran, 1994, pp. 106–12). For example, in the case of humans, the impact of new drugs both on the length of life (the statistical value of life, Pearce et al., 1992) and the quality of life might be taken into account. Market value approaches (estimated retail values calculated by Principe, 1991) seem less satisfactory. Furthermore, the appropriate basis for economic valuation is likely to vary with its purpose and the type of products obtained. For instance, the economic valuation of new products obtained from wild germplasm to treat diseases of livestock or plants is likely to be based on a different set of considerations from the valuation of those products used to treat diseases in human beings. Changes in consumers' surplus and producers' surplus are likely to be relevant in the former case but not whole of life considerations.

Simpson et al. (1996, p. 183), after reviewing the literature on the expected value of wild biodiversity for supplying new pharmaceuticals, came to the conclusion that the private value of conserving a marginal hectare of threatened wild habitat to supply new pharmaceuticals (its rental value for this purpose) is fairly small, even under relatively favourable conditions. Therefore, because this value is negligible, relying on the payment of this value to those able to conserve wild biodiversity will, in their view, be ineffective as a means for conserving wild biodiversity. This rental value may also be low for the development of other potential new products of commercial value other than pharmaceuticals. Nevertheless, the economic value of a marginal hectare of natural habitat for enabling the discovery of new products will undoubtedly be higher when the full range of the types of new commercial products that could be produced by it is taken into account. Still, after the transaction costs involved in any payment to landholders are taken into account, the potential value to landholders of conserving wild germplasm for potential commercial uses may be quite low.

11.5 TOTAL ECONOMIC VALUATION OF NATURAL AREAS/ECOSYSTEMS

The economic values of conserving natural areas do not depend solely on their potential for supplying new commercial products. Their provision of multiple services (as well as some possible disservices) needs to be taken into account. Some of the important factors that need to be accounted for when determining the size of a natural area (ecosystem) to be conserved in a particular locality can be grasped by reference to Figure 11.2. There line ABC represents the marginal opportunity costs of conserving an increased amount of the natural area. This is assumed to be constant for simplicity. Line DBF represents its marginal expected value for potentially valuable new commercial products as a function of the size of the protected area. Line GCH indicates marginal changes in the total economic value of the protected area as a function of its size. The difference between line GCH and DBF is its marginal value for supplying public and quasi-public goods. The public goods supplied could be global public goods, such as CO_2 sequestration, or local public goods, such as improved flood control. Quasi-public goods can be expected to reflect the use of the protected area for recreation and tourism.

From Figure 11.2, it can be seen that if holders of a natural area are paid the sum of the marginal values of its potential to produce commercial products, they will only have an incentive to conserve x_1 km^2 of it. They obtain a rental income equivalent to the area of triangle ABD. However, using the Kaldor–Hicks criterion, it is socially optimal to conserve x_2 of the natural area. Therefore, relying only on this type of use value (the potential to produce new commercial products) to determine whether to conserve natural areas can be expected to result in the area conserved being undersupplied. In some localities, it is also possible for line DBF to be located entirely below the marginal opportunity cost line. In that case,

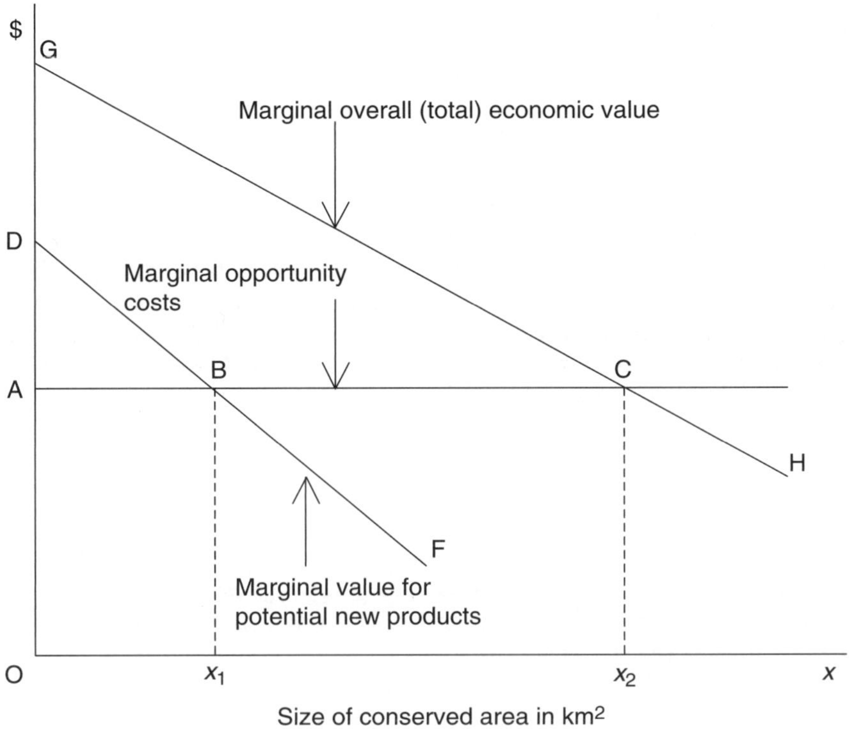

Figure 11.2 An illustration that relies only on the potential of natural areas to supply new commercial products in order to assess their social worth will undervalue those areas

payment of the potential economic use value of the area for supplying new commercial products fails to provide sufficient economic incentive to landholders to protect any of the natural area.

Although the illustration in Figure 11.2 relies on partial static analysis, it can provide some comparative insights. First, other things being held constant, the line DBF will be higher the less likely other protected areas are able to supply the new commercial products (or substitutes) which might be supplied by the focal area. In general, the substitutability of the commodities supplied by other natural sites needs to be considered both in relation to use values and non-use values. In completing the analysis, one needs to make assumptions about the alternative future sizes of other sites. Secondly, the socially optimal size of the protected area (if the Kaldor–Hicks criterion is adopted) will be larger, other things remaining constant, the lower are marginal opportunity costs of protecting it.

The potential economic value of a natural area for producing new commercial products cannot be judged on the amount of diversity of germplasm present, although it may be an influence. Areas with little diversity of species could be very valuable for their potential to produce new products (including pharmaceuticals of economic value) if just a few species could result in highly valuable commercial products. For example, while many temperate natural areas contain less natural species diversity per unit area than tropical natural areas, they may have a high conservation value because of their potential to supply new commercial products, including remedies for diseases. Areas containing the wild relative of some cereals (grasslands) may show little diversity but potentially could be very valuable for helping to sustain cereal yields in the case of the development of new diseases of cereals. To be specific: the expected economic value of a wild area for producing new commercial products is equal to:

$$E[V] = \sum_{i=1}^{n} \rho_i V_i \qquad (11.1)$$

where ρ_i is the probability of the *i*th species producing a commercially valuable new product, V_i is its estimated value and n represents the number of different species in a wild area. Clearly, $E[V]$ cannot be determined merely by considering the size of n.

11.6 METHODS FOR EVALUATING THE NON-USE VALUES OF WILD BIODIVERSITY

The major portion of the total economic value of wild biodiversity often consists of its non-use value and/or its non-consumptive use value for recreation and tourism. It is very difficult to elicit these economic values accurately, no matter whether revealed preference methods (such as travel cost methods) or stated preference methods (such as contingent valuation) are adopted. These limitations have been extensively catalogued in the relevant literature and so I'll not repeat the details here. Rather I'll concentrate on considering what stated preferences for non-use values of biota might really measure, the stability of these stated preferences and knowledge deficiencies that limit their value.

As pointed out by Salles (2011), a difference of opinion exists in the relevant literature about whether procedures for obtaining stated preferences uncover existing preferences or whether these procedures are instrumental in the formation of preferences. In fact, in many cases, elicitation methods used to obtain stated preferences may partly reveal existing preferences and may partly create new ones. As Salles (2011, p. 473) states, 'it is widely acknowledged that stated preference methods are design-dependent, especially when related to non-use values'. Furthermore, 'created' or elicited stated preferences may not be stable. Consequently, the standard assumptions of economists that 'preferences pre-exist and need to be [can be] uncovered by a process involving practice, repetition and experience' and that the uncovered values are stable (Salles, 2011, p. 473), are likely to be violated. Both theoretical and empirical evidence for such violations are reported in Tisdell (2014, see especially pp. 28–34 and Ch. 6) and several aspects of their violation are outlined in Tisdell (2007; 2013), Tisdell and Wilson (2012) and Tisdell et al. (2008).

Once it is accepted that the values and stated values associated with many types of goods are subject to construction, are not stationary and are not independent of the processes of their elicitation, interpretations of the relevance of stated values for policy purposes become problematic, as do statements about willingness to pay to conserve biota or to act in other ways to promote its conservation.

Some of these complications can be illustrated by reference to Figure 11.3. Imagine that the relationship ABCDE represents the 'true' willingness of an individual to contribute (pay) towards the conservation of a wild species, a set of species, or a natural area as a function of the passage of time. Prior to t_1, the individual has only limited information

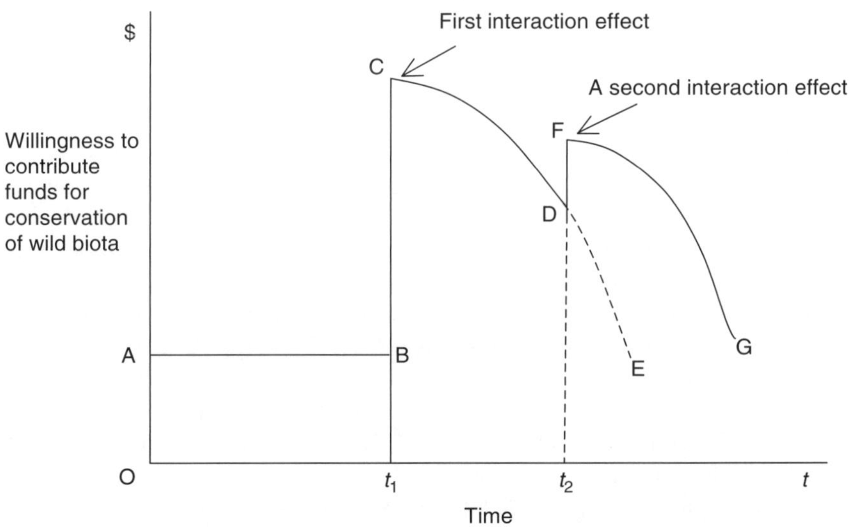

Figure 11.3 *An illustration of complications involved in the valuation in the case of the conservation of wild biota having the attributes of public good or quasi-public good*

and no experience with the wild biota (which for expositional convenience is assumed to be a single wild species) but is 'truly' willing to pay OA towards its conservation. Suppose that at t_1, the individual has a favourable emotional experience with the species and also obtains more information about it, for instance, it is more endangered than the individual previously thought and/or its continuing existence has positive ecological effects. The individual's true willingness to contribute to its conservation then jumps to point C. However, it then begins to decline (a drop-off effect occurs) in the absence of an individual obtaining further positive information about the species and experiencing additional favourable interaction with it. For example, path CDE is followed. A part of the reason for the drop-off effect is forgetting and crowding out by other events, both consequences of individuals having bounded rationality. These phenomena are discussed in Tisdell et al. (2008)

Although it was assumed above that both the emotional experience of being in contact with a species and increased information about it elevated the willingness of individuals to pay at time t_1, this elevation can occur if only one factor is present. Furthermore, the relative importance of these factors in shaping preferences is liable to differ between individuals (see, for example, Tisdell and Wilson, 2012, Ch. 6).

In the case illustrated in Figure 11.3, it can be seen that, following the experience of the individual at time t_1, the individual's true willingness to contribute to the conservation of the focal wild species does not remain stationary. Eliciting this stated payment after t_1 will yield a higher figure the smaller is the duration of time between t_1 and the subsequent elicitation. Furthermore, the elicitation process itself is unlikely to be neutral even if it does not provide new information or an extra emotional experience. This is because it forces the respondent to focus attention on the object to be evaluated, and in this process the individual's focus on other objects can be expected to be reduced. Where the object is viewed favourably, this is liable to elevate its stated value and where the opposite is the case, it is likely to depress its stated value. In the situation illustrated in Figure 11.3, elicitation at t_2 elevates the willingness to pay to an amount corresponding to F and subsequently a path such as FG might be followed. This effect is similar to the Heisenberg principle in quantum theory (Heisenberg, 1930). Therefore, it can be difficult to decide which values from t_1 onwards should be selected for valuation purposes.

There are also other complications. Not all citizens are well informed about the functions of species and natural ecosystems, and therefore their values. Consequently, choice that relies on the preferences of the general public may fail to yield a socially desirable outcome. At least, these preferences ought not to be blindly accepted for policy purposes. They ought to be supplemented, for example, by the opinions of experts. However, even experts can disagree and policy-makers also have to exercise judgement in accepting their advice. Rational scepticism always needs to be exercised.

11.7 DIVERGENT VIEWS ABOUT THE COMPARATIVE ECONOMIC VALUE OF NATURAL AND MODIFIED ECOSYSTEMS

Various estimates are available of the economic value of services provided by different types of natural ecosystems as well as of the global value of ecosystem services. However, their validity have been questioned, for instance by Toman (1998). Costanza et al. (2014) update some of these estimates and review relevant contributions.[1] Despite this, estimates of the economic value of modified natural ecosystems are patchy and no comprehensive empirical estimates are available of trade-off functions for the economic value of services provided by modified versus natural ecosystems. This is probably because of the mammoth amount of data required for such a task and the complex calculations that would be needed to construct these trade-off functions. Nevertheless, from a human perspective, a world in which all ecosystems were natural (or nearly so) would be less than ideal and unable to support the current level of human population. Past modifications of natural environments have enabled both the level of human population and human well-being to increase, even though many past changes to natural ecosystems have been less than ideal. Furthermore, legitimate concerns exist about the extent to which the continuing transformation of natural ecosystems will, in fact, increase human well-being.

Figure 11.4 shows a hypothetical relationship, ABCDE, between the level of human well-being and the extent to which natural ecosystems are modified (see also Figure 11.1 in Tisdell, 2005). The extent of modification of natural ecosystems is measured on a qualitative scale from zero to unity, where zero corresponds to no modification and unity to total modification. Relationship ABCDE in Figure 11.5 is intended to indicate that while the marginal benefit to humans of altering natural ecosystems as a function of time initially rises, eventually it tapers off with continuing alteration and becomes negative. Contemporary civilization may now have reached the stage where economic gains from its further modifying or altering natural ecosystems will result in little increase in human well-being or even reduce it. The opinions of scholars, however, differ about whether this is so.

The illustrations in Figure 11.4 and 11.5 should be regarded as purely hypothetical. The relationship in Figure 11.5 suggests that later civilizations

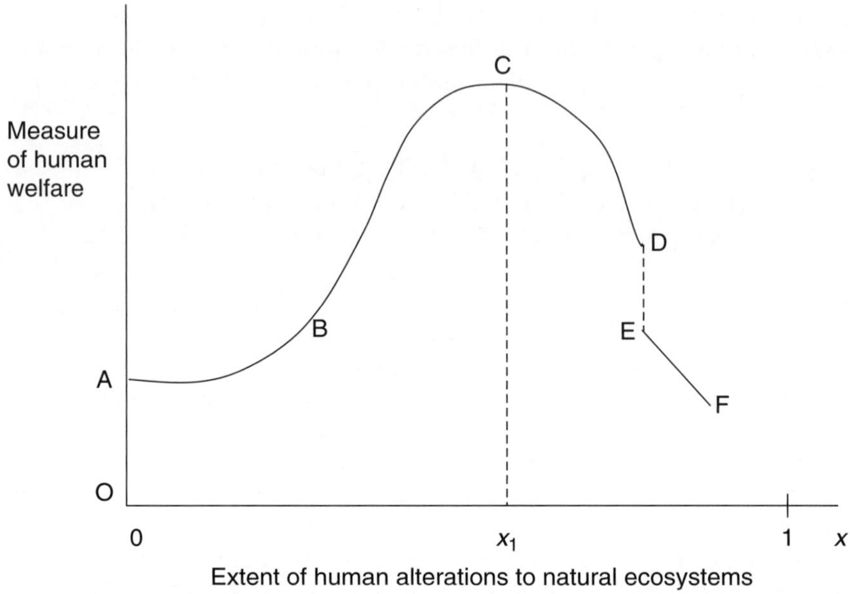

Figure 11.4 Hypothetical relationship between the extent to which human beings alter natural ecosystems and the level of their well-being

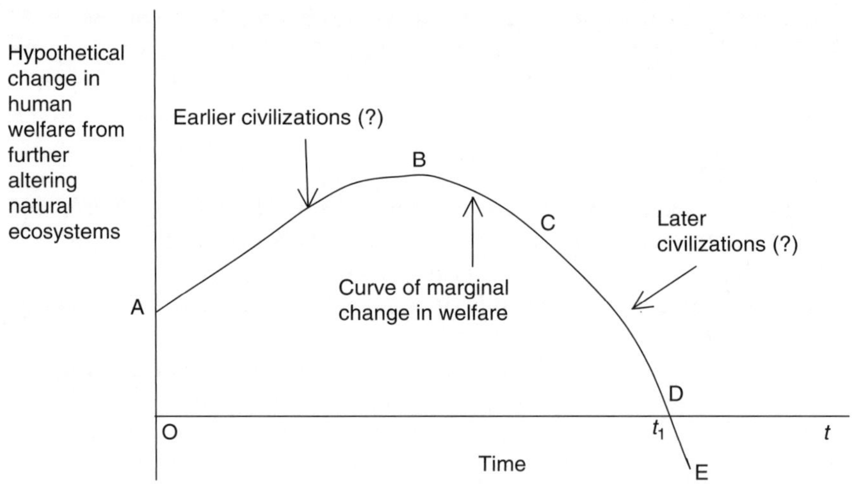

Figure 11.5 An illustration of the possible scope for increasing human welfare by altering natural ecosystems given previous alterations to these: the conversion process is assumed to be subject to eventually diminishing returns

need to pay greater attention to the welfare consequences of altering natural ecosystems than earlier ones. Eventually a time, t_1 in Figure 11.5, may be reached where any further reduction in natural ecosystem services results in no increase in human welfare, and in fact, further loss of such services reduces human welfare. There are concerns that contemporary civilizations may now be approaching such a situation.

Although general perspectives on the value of natural ecosystems and of their transformation can help in understanding their comparative value, their ability to do so is limited. This is because optimal patterns of transformation of ecosystems or land use are usually complex. The same degree of transformation of an ecosystem is not optimal everywhere and varied degrees of transformation are usually needed even for the same end-purpose. For example, Braat and ten Brink (2008) suggest that natural ecosystems provide maximum value for recreation and tourism when they are lightly used (modified a little) and they suggest general relationships between the value of other types of services provided by ecosystems and the extent of their modification using the classification of the Millennium Ecosystem Assessment (2005). Their graph of these relationships is reproduced in Salles (2011, p.478, Figure 2).

There are some limitations, however, to this perspective. First, the degree of optimal transformation of an ecosystem to serve a particular purpose will vary within the type of ecosystem. Secondly, a range of different degrees of transformation of an ecosystem is likely to be ideal for supplying a particular ecosystem service. In the case of recreation and tourism, for example, it may be optimal to set aside some areas, such as national parks, with little modification and to intensively change other areas to supply urban public parks or playing fields to cater for recreation. Consequently, a mixture of different degrees of transformation is likely to be ideal. The same applies to other categories of services. Furthermore, interdependence exists between the valuation of the availability of different types of environmental services. This interdependence further complicates the valuation problem even in a static setting. If dynamics is considered, this adds to the challenge of determining optimal use of the biosphere to maximize the value to human beings of environmental services. Therefore, we can only expect, given our current knowledge and abilities, limited (yet useful) insights into the important issues involved in determining the optimal transformation of the biosphere by human actions.

11.8 THE RELATIVE NATURE OF ECONOMIC VALUATION AND COST-EFFECTIVENESS ANALYSIS AS A BASIS FOR ECONOMIC DECISION-MAKING

Economic valuation is relative in nature. Economic methods of valuation do not determine absolute values. Consequently, the application of techniques for economic evaluation only provides a limited guide to the desirability of conserving biodiversity and ecosystems. Furthermore, while cost-effectiveness or cost-efficiency analysis can in some circumstances reduce the amount of information about economic values required to solve a resource allocation problem, it does not eliminate economic valuation requirements entirely, especially if social costs are to be minimized to achieve a particular objective. Consider each of these matters in turn.

The Relative Nature of Economic Valuation and Conservation Choices

The relative nature of the economic valuation can be illustrated by an example. For simplicity, suppose that the sole value of a near natural area is for recreation and tourism and consider the depiction of this problem using Figure 11.6. Given the status quo, the net demand for visits to this

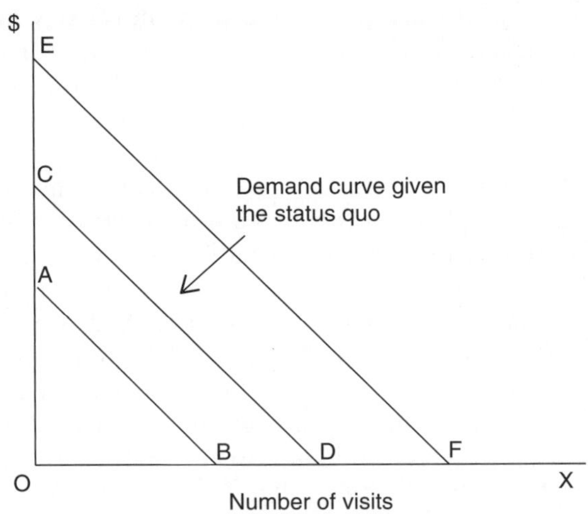

Figure 11.6 A diagram to illustrate the relative nature of economic valuation: in a deteriorating environment, the ascribed economic value of an environmental facility which degrades, but shows relatively less degradation than others, can increase

area might be as indicated by the line CD. Therefore, its imputed economic value is equal to the area of triangle ODC. Now suppose that this site becomes degraded but everything else remains unchanged. The relevant demand relationship for visits may fall to AB and the imputed economic value of the area declines to an amount equal to the area of triangle OBA. On the other hand, if alternative substitute natural sites record greater degradation than the focal site (or some or all of these disappear) and other things remain unaltered, the demand for using the focal site for recreation and tourism may in fact increase. For instance, in the illustration, the demand relationship may shift up to EF, and consequently the ascribed economic value of the focal site increases to an amount equal to the area of triangle OFE. Even though (as a result of degradation) the quality of the recreational and tourism experience in the relatively natural focal area is reduced and the absolute value of that experience declines, the relative value of the focal area rises. Similarly, the non-use economic values attributed to a relatively natural area may rise when it registers environmental degradation if its substitutes become even more degraded. Furthermore, if the wild biodiversity in an area is reduced, the economic value of conserving the area to sustain this reduced biodiversity can increase, even though the remaining biodiversity has less absolute value than when it was more abundant.

Cost-effectiveness Analysis and Opportunity Costs

Policy-makers may sometimes be asked how a specified biological conservation objective can be achieved at minimum economic cost. For example, the problem posed may be to advise on how land use should be allocated to conserve the minimum viable population of a species (or mixture of species) at minimum economic cost. This case is analysed in the next chapter. Or again, the problem posed may be to minimize the cost of preserving a specific area of a particular biotype. The primary aim is usually a political decision and is often the reaction to the concerns of pressure groups. It may also be a response to expert opinions.

Because the primary aim does not require the elicitation of individual preferences, this reduces the amount of information which needs to be collected to provide advice in these cases. Nevertheless, the decision-making problem is still complex if the objective is to minimize social economic opportunity costs. If that is so, multiple environmental values will need to be taken into account, many of which are unlikely to be reflected in market relationships. Depending on the extent to which individual preferences are to count, these preferences will need elicitation, for example, by techniques for obtaining stated preferences.

Sometimes the economic opportunity costs to be taken into account may be politically determined. For instance, the economic costs to a particular lobby group may be the main political concern. For example, the political aim may be to obtain a particular conservation objective subject to the income forgone of foresters being minimized. A case of this type is considered in the next chapter.

11.9 SOCIO-ECONOMIC REASONS FOR SOCIALLY DETRIMENTAL ALTERATIONS TO NATURAL ECOSYSTEMS AND WILD BIODIVERSITY

Natural changes and anthropocentric influences can alter wild biodiversity and natural ecosystems. Direct alteration of natural ecosystems by human initiatives have become the prime cause of wild biodiversity loss. Other anthropocentric causes include the over-harvesting of selected species and the destruction of wild species regarded as pests. In general, these actions are motivated by the prospect of private economic gain. Nevertheless, they may prove to be socially detrimental, or even though socially beneficial, the benefit obtained may be much lower than could have been achieved by a different pattern of change. There are several reasons why this is so. Consider the following ones.

Ignorance or lack of knowledge about the consequences of altering natural ecosystems are frequently the cause of unwanted changes to these systems. For example, early European settlers in Australia arrived in an unfamiliar environment and could not accurately predict the environmental consequences of changing its land use by introducing agriculture and livestock. So some changes were made, which in retrospect were not socially ideal. This also happened in the Americas and elsewhere. Today, the environmental effects of applying many new technologies (including new biological technologies, such as those relying on GMOs) cannot be predicted perfectly. Therefore, unwanted and unpredicted environmental consequences of attempts to alter ecosystems to benefit humankind still continue to be a problem. Unfortunately, some of these consequences are irreversible or too costly to reverse once they happen.

Decisions about altering ecosystems also depend on attitudes to uncertainty and the bearing of risk. Societies in which there is greater willingness to take risks are more likely to modify ecosystems for potential economic gain. However, if widely different attitudes to risk-taking exist in a society, decisions by risk-takers to alter natural ecosystems (and in some cases, other stable ecosystems) may result in conflict with risk-avoiders if potential negative spillover effects from the decisions of the former impact negatively on the welfare of the latter group.

The time preference for personal gains of those able to change ecosystems is also a behavioural influence on alterations to ecosystems. Those having a high preference for present versus future income may choose to irreversibly alter an existing ecosystem for immediate gain even though the type of alteration substantially lowers the ecosystem's long-term economic benefits. Furthermore, they will be more inclined to do this if they think that technological progress will to some extent offset the environmental damage done.

Individual behaviours are influenced by the nature of the organization or administration of societies, and in particular by the nature of their economic and political systems. Although market systems of economic organization are pervasive in contemporary societies, a considerable amount of decision-making about resource use depends on administrative decisions within companies, as well as choices by politicians, public servants and by those managing NGOs. No contemporary society depends on market systems alone to administer its use of resources, including ecosystems. Unfortunately, all of these components of social organization (including markets) are subject to economic and other failures resulting in less than ideal alterations in ecosystems and changes in biodiversity. Let us briefly consider in turn market failures, political problems, shortcomings in public administration and in decision-making by NGOs.

Market systems are limited in their ability to achieve an ideal economic allocation of resource use (and non-use) both in terms of minimizing scarcity and in ensuring a just distribution of income. In this context, potentially avoidable economic scarcity exists if, in principle, it would be possible by some administrative change to make someone better off without making another worse off, that is, if Paretian economic efficiency is not achieved. Both the absence of markets or incomplete markets are commonly identified as reasons for human wants being less fully satisfied than they could be in principle. Reasons for such failures include the presence of public goods and environmental and other externalities. However, possibly the main factor limiting the economic extension of market systems is the transaction costs involved in the enforcement of property rights and in exchange. The inability of holders of natural areas (containing wild biodiversity) to market all the benefits they may generate for others limits their economic incentive to conserve these areas and the biodiversity they contain. Therefore, a case exists for their provision by governments. However, it is also evident that government provision of such resources is subject to political and administrative failures.

Several possible sources of political failure occur in conserving natural and heritage areas as well as biodiversity. These can all be described as failures in governance, and occur both in higher and lower income countries. Their importance varies with the structure of government, for example, the nature of political constitutions, institutions and political representation. They are influenced by existing customs and codes of conduct in society. Examples of political failures (as judged by Pareto's criterion) include decision-making about resource use based on majority voting (Tullock, 1959; Tisdell, 2005, p. 80). However, this failure can be moderated by political compromises (Tullock, 1959). Arrow (1951) has shown that many rules or methods for collective decision-making fail to achieve a social optimum and can result in inconsistent outcomes. In addition, Arrow (1951) has proven that it can be impossible to derive social welfare functions which satisfy sets of criteria considered by many to be socially desirable.[2] Furthermore, in all democracies (and probably in most political systems) political lobbying by interested parties (able to gain from a particular political decision) plays a significant role in influencing political decisions. Lobbying may take many different forms. Politicians may be provided with 'information' by lobbyists in an attempt to influence their decisions, they may be promised personal or party financial support in the event of a favoured decision, or publicity campaigns may be mounted to sway public opinion and increase the social acceptability of a wanted political decision or social rejection of an unwanted one. However, the ability of different social groups in society to successfully lobby varies considerably and this variation can result in development versus conservation decisions that are far from socially optimal. The playing field is not level.

Marcus Olson (1965) has identified several social situations in which this is so. For example, consider a natural area, the existence of which benefits many, each by a small amount and for which the aggregate benefit of its conservation is high. For instance, suppose that 100 million persons belong to this group and that each benefits by $5 from the area's conservation so that their aggregate benefit is $500 million. On the other hand, members of a small group of developers each stand to benefit by a large amount from transforming the natural area. However, their aggregate benefit is significantly less than the total loss which would be experienced by those favouring the conservation of the natural area. For example, five developers might stand to gain $20 million each by developing the area. They have an economic incentive to lobby government for development if there is a reasonable probability of their campaign being a success relative to its cost. The large numbers of those benefiting from the conservation of the natural area may not be able to mount an effective counter political campaign for two reasons. Because the conservation of the natural area is likely to involve a significant public good component, free-riding can be expected. Secondly, because of the large numbers involved, the transaction costs involved in organizing the collection of finance for a political campaign to conserve the natural area can be a large burden on initial organizers, and no individuals may be prepared to accept this burden. Consequently, political decisions about conservation may fail to promote the general interest.

Also account needs to be taken of the fact that even when politicians are well intentioned in their discussions, their policies may be faulty, or they may be swayed by popularism to subscribe to false hopes and unachievable aspirations.

Public (civil) servants also play a key role in governance. It cannot be assumed that they are always well informed and efficient and so they can sometimes propose environmental changes that are not in the public interest. Furthermore, Niskanen (1971) has argued that public servants have an incentive to act in their own self-interest. They may, for example, compete for power and influence in the government. Different parts of the public service may both be supported and influenced by 'client' groups and, in extreme cases, may be subject to 'capture' by influential special interest groups. For example, farming groups may significantly influence public officials in departments concerned with agriculture as may manufacturers supplying products used by farmers. To some extent, the relative power and size of government departments may reflect the potential power of different political pressure groups in society. In turn, this can be expected to affect the pattern of development and type of conservation decisions made by governments. Moreover, some civil servants may (sometimes in cooperation with politicians) accept bribes for bending or ignoring rules and regulations. Swarna Nantha (2014) provides examples of this in relation to permits to use rainforests for growing oil palm in South-East Asia. In addition to this, the size and scope for sustained funding of the civil service depends upon public revenues. Marketed goods provide a basis for taxation but not those that are unmarketed, such as the supply of pure public goods. This may lead powerful public servants to favour the supply of the former rather than the latter, thereby creating a bias in favour of development.

NGOs play an important role in many societies in influencing the conservation of natural areas and wild biodiversity. They can act as political lobbyists or as direct providers of conservation services. Many do both. They fill gaps in ensuring or promoting the supply of wanted commodities in societies but for various reasons can fail to do this in a socially optimal manner. First, in order to carry out their mission and remain economically viable, NGOs need to raise funds. This may result in their over-concentration on conservation activities that have the greatest emotional appeal to potential donors in order to raise funds and in the wasteful duplication of fundraising efforts (Tisdell, 2009a). Secondly, agency problems can arise in the case of large NGOs. For example, decisions by executives of NGOs may be little influenced by the wishes of the donors who are non-members, and in cases where an NGO has many members, members may have little influence on the NGO's executive decisions. In addition, NGOs are not subject to the type of market disciplinary forces that regulate the behaviour of commercial business enterprises, although some develop commercial business units, in order to boost their revenues or to increase their economic viability.

The managers and employees of an NGO have an interest in ensuring its economic sustainability and are often keen to increase its size and hence its influence. This can result, for example, in considerable expenditure by NGOs on advertising and promotion to attract funds to support their causes. In some cases, government funding and provision of the services provided by an NGO would be a more economic option than their supply by an NGO. However, given the institutional structure of societies, the most economical option for provision of ecological services, such as the supply of protected areas, is unlikely to occur. This is so even though it seems to be socially desirable for nature conservation services to be provided both by governments and NGOs (see, for example, Tisdell and Wilson, 2012, pp. 373–4).

11.10 POLICY ISSUES RAISED BY THE ABOVE DISCUSSION

The implications for the above analysis for policy recommendations about the conservation of biodiversity and ecosystem services depend on the way in which policy decisions are made. Traditionally, economists have assumed that rational policy advice is prepared on the basis of well specified ends (that is, on the basis of a well specified objective function) to be optimized subject to available means. The procedures involved for determining optimal policies, therefore, rely on a static rational approach to decision-making. However, this is not the only possible form of decision-making. Frequently, political decision-making involves dynamic interactive processes which can result in altered objectives as well as changed assessments of available means and their relationships to the achievement of objectives. Furthermore, objectives may be incompletely specified and various types of satisficing approaches to policy-making and iterative procedures for policy formulation are possible. While examining these alternative approaches to policy promotion would be worthwhile in this context, the focus here will be on the traditional optimizing approach to policy formulation.

Even if the traditional economic approach to providing policy advice is adopted, it should be borne in mind that the relevance of economizing arises for a wide range of different social objective functions. This is clear from the economic analysis of Bergson (1938).

While the type of social objectives specified in neoclassical welfare economics, such as adopting policies to bring about Paretian-type gains (Kaldor–Hicks gains) with the ultimate objective of achieving Paretian optimality, can be pursued, the economics of satisfying other types of objectives is also relevant for policy purposes. For example, it may be decided by politicians that selected wild species should be saved from extinction or that a 'representative' amount of natural ecosystem should be conserved. A pertinent problem is how to achieve these objectives most economically. Policy options are often limited by political and institutional constraints and this restricts the practical choices available to policy-makers (Hagedorn, 1993; Tisdell, 2009b). Therefore, in practice, the most economical solutions need to be determined taking into account such constraints.

Neoclassical welfare economics has been criticized as being 'Nirvana' economics. This is partly because of its failure to take account of bounded rationality, agency and transaction costs. However, a further reason is that it has resulted in the development of policy proposals without taking into account the institutional and political environment in which policy is to be implemented. This may help to explain why, as stated by J.-M. Salles (2011, p. 479), 'in real decision-making [public policy for conserving biodiversity and ecosystems], particularly in France, economic valuation appears to have little influence on final [political] choices'.

Nevertheless, there is a case for assessing policy possibilities for conservation from varied points of view, including the neoclassical perspective. Moreover, the most economic conservation choices subject to political institutional and social constraints often need to be found. Although it may be desirable to change political and institutional constraints and social attitudes affecting policy outcomes, it is normally beyond the ability of an economist to alter these. Nevertheless, social scientists can bring attention to desirable changes in political structures, institutions and social attitudes and explain why they believe these changes to be desirable.

The services provided by most natural ecosystems or natural areas are varied and numerous, unlike those supplied by most human dominated ecosystems. This limits rational choice about their desirable conservation and transformation options, as does poor knowledge about the biodiversity present in them and their supply of ecosystem services.

Because many heterogeneous components of natural ecosystems have economic values (some positive and some possibly negative), rationally planning their use and conservation is a major challenge. This is because of the existence of bounded rationality and the complexity of interdependence of environmental variables. Therefore, in making social decisions about land use, there is a temptation in determining policy to focus on one or a few of the ecosystem service variables involved. For example, there may be a single focus on carbon sequestration of vegetation in an area with the area's contribution to the conservation of biodiversity being ignored, as well as its role in supplying other valuable ecosystem services. Consequently, in cases where subsidies are given for conserving natural areas or near natural areas, payment to landholders may only be based on one or a very limited range of services provided by these areas. Hence, an optimal balance between conservation and economic development will only be achieved by chance. However, the optimality of different forms or patterns (mosaics) of land use depend on the extent to which it is optimal for individual parcels of land to provide multiple ecosystem services or a single such service. This is discussed in Chapter 17 of this book.

A variety of policy instruments and procedures are available to influence the conservation of wild biodiversity and natural areas. The chosen policy may be influenced by:

- its social and political acceptability; and
- its economic efficiency, taking into account such factors as the agency and transaction costs involved in its implementation as well as its ability to achieve the goals desired.

The range of possible policy instruments for conserving biodiversity and supplying ecosystem services include the following (the list is not exhaustive):

1. Subsidies or payments to agents who maintain biodiversity and the supply of wanted ecosystem services.
2. Taxes payable by those agents who reduce biodiversity or the supply of wanted ecosystem services.
3. The use of tradable permit and environmental offset systems.
4. Direct maintenance by governments of ecosystem services, for example, by the public provision of protected areas.
5. Zoning may also be used to manage the supply of ecosystem services.
6. Laws and regulations designed to limit the loss of particular ecosystem services. Application of these may be subject to judicial (court)

decisions. Hence, in these cases, their application does not depend solely on administrative decisions.

7. In some instances also, extensions of property rights in wild species and wild genetic resources may help to foster their provision.

Consider the main features and consequences of these different policies.

Observe that their income distribution implications differ. Subsidies for or payments for the provision of ecosystem services usually redistribute income in favour of the providers of those services, whereas taxes or restrictions on landholders' use of 'natural' ecosystems usually reduce their incomes. However, in the latter case, some (or in extreme cases, all) of the cost of restricted use of ecosystems can be passed onto buyers of commodities supplied by these ecosystems. For example, measures which restrict the clearing of trees on grazing properties could reduce beef supplies and increase beef prices. Therefore, the economic incidence of a restriction on ecosystem use needs to be considered. This incidence will depend on several factors, including the elasticities of supply and demand of marketable goods affected. Similarly, the incidence of payments for provision of ecosystem services needs to be explored.

If government restrictions on the use of ecosystems do not change the market price of any marketed produce affected (or do so by very little), the cost of restrictions on the use of ecosystem services are likely to be borne completely (or almost completely) by those supplying these services. In the case of payments for these services, their suppliers benefit fully from the subsidy in these circumstances, but the subsidy may not fully compensate them for their loss.

Government payments for the provision of ecosystem services may be less than the cost to landholders of providing these if they are forced by government regulation to provide these services. For example, saltwater crocodiles, *Crocodylus porosus*, are a protected species in Australia but landholders in the Northern Territory of Australia are able to obtain payment from crocodile farmers for collecting crocodile eggs on their land. The allowable quantities of the eggs which can be collected in a region are regulated by the Northern Territory Government. However, the payments received by graziers do not fully compensate them for their livestock losses and only in exceptional cases does this result in graziers adopting land management favourable to the population of saltwater crocodiles (Tisdell, 2014, Ch. 9). Nevertheless, this scheme goes some way to making policies for the conservation of saltwater crocodiles more acceptable to graziers.

Whether or not payments are made to landholders for providing ecosystem services or compensation is given for restricting their use of ecosystems can depend, as was mentioned above, on the social acceptability of these alternative policies. If very poor landholders, for example, are restricted in their land use, then justice may require that they be compensated for loss of income if they have inadequate alternative economic opportunities to compensate for lost income.

Social attitudes to payments for provision of ecosystem services may also depend on whether their supply by landholders is voluntary or compulsory. Nevertheless, political systems sometimes operate in ways that seem to be inequitable or lacking in general social acceptability. For example, policies may be adopted that result in the poor bearing the burden of the loss of natural ecosystem services or the poor may be burdened with the cost of the provision of ecosystem services in ways that further reduce their meagre levels of income.

In the remaining chapters of this book, further consideration is given to the type of policies mentioned above.[3] In the next chapter, attention is given to determination of land use to minimize the cost of conserving wild species *in situ*, assuming that the focal species must be conserved in a completely protected area. However, this assumption is relaxed in Chapter 17 where the possibility of mixed land-use is considered, that is, the commercial use of land combined with spatial and other constraints on this use to conserve some wildlife species or to ensure the continuing provision of particular ecosystem services. Chapter 13 examines the potential (and limitations) of property rights (including global rights) in non-captive species and in wild genetic resources as a means to conserve biodiversity. This discussion is extended in Chapters 14 and 15. In Chapter 14, additional economic measures to conserve wildlife on private land (as well as further consequences of the private property solution) are considered, and in Chapter 15 the relationship of the rate of interest with the conservation of wild species is re-assessed, taking into account macroeconomic relationships.

11.11 CONCLUDING COMMENTS

Since this is a lengthy chapter, some recapitulation of its main points is desirable. It was noted that economists have given comparatively little attention to the valuation of biodiversity as such and to specifying the comparative value of transformed and natural ecosystems or alternative forms of biosphere use. For example, estimates of the economic trade-off values that can be generated by modifying natural ecosystems are sparse. Reasons for this include the complexity of these types of valuation, their large data requirements, the limited validity of the available data and the limitations of current methods of economic valuation.

Although wild biodiversity and natural ecosystems and natural areas have economic value, their exact values are uncertain. Based on statements by samples of tourists and recreationists visiting natural areas, evidence was presented that the value they place on wild biodiversity and natural ecosystems varies considerably. The wider community's valuation of the existence of wild biodiversity and the presence of natural areas for their non-material benefits is likely to be lower than that of frequent visitors to natural areas.[4] Consequently, valuations of wild biodiversity for its natural benefits by members of the public can be expected to show greater variation than in the above-mentioned cases.

The conservation of wild biodiversity can also result in material economic benefits. The stock of wild germplasm is a source for the potential genesis of new commercial products. Considerable attention has been given in the economic literature to estimating its likely economic value for the potential supply of new pharmaceuticals. Estimates of its expected economic value in this regard vary considerably but the consensus seems to be that these (marginal) values are low per hectare of natural area conserved. The variability in estimates of the expected economic value of conserving natural areas for their potential supply of new pharmaceuticals is due to differences in the economic methods of valuation adopted by researchers and differences in the data sets utilized by them.

It should be observed that the value of natural areas for supplying new commercial pharmaceuticals only takes account of their expected or average values for this purpose. However, for some types of decision-making, the probability distribution (range) of these values is important. Furthermore, the potential of different natural areas to supply ingredients for new drugs can differ significantly, and this may not depend solely on the extent of biodiversity present in them. The value of conserving a natural area depends not only on its potential economic value for supplying ingredients for new drugs. Its potential for supplying components for other commercial new products must also be taken into account, as well as additional types of potential economic benefits such as the value of its environmental or spillover benefits, its passive economic values and its worth for recreation and tourism.

Economic methods for estimating non-marketed economic values associated with wild biodiversity conservation (and nature conservation generally) rely on stated valuations by members of the public and these are subject to several basic limitations. Many members of the public are ill-informed about the worth of conserving wild biodiversity or components of it. Methods for trying to elicit stated values are often 'preference-forming' because gaps exist in the pre-existing preference functions of individuals. Existing preferences of individuals for conserving some natural objects are absent or hazy. In such cases, the valuations formed by respondents are much influenced by the information provided to them by interviewers or by their experiences. Furthermore, biases in responses arise from arousal effects and attention focusing. Moreover, drop-off or erosion effects occur with the passage of time in stated values and, presumably, in the felt values of individuals in a non-intrusive state. Stated values can only be elicited by intruding on the mental processes of individuals and consequently it is often impossible to elicit the underlying values that trigger their behaviours.

These problems raise the question of the extent to which it is wise to rely on elicited individual relations when making decisions about nature conservation. Alternatives were discussed, but an ideal method for specifying values (that is for specifying objectives for nature conservation) has yet to be found. Consequently, in practice, different approaches are often employed for specifying objectives for nature conservation.

Scholarly views about the relative value of natural versus modified ecosystems were considered. It was speculated that with the passage of time, the extra economic value of altering natural ecosystems at first increased, but that the transformation process is subject to eventually diminishing returns. It is possible that the marginal returns for such transformation may eventually become negative. This could partly explain why, in recent times, increased concern has been expressed about the sustainability of economic growth and development, particularly when it is achieved at the expense of the conservation of natural ecosystems. Although some scholars have hypothesized general relationships between the extent of modification or transformation of natural ecosystems and the value of the ecosystem services they provide, the relationships posited are problematic. This is because these values vary with different types of ecosystems, the nature of the modifications and their geographical location of the modification. Furthermore, the demand for some ecosystem services is diverse. Consequently, the optimal supply of a particular ecosystem service can require varying (heterogeneous) modifications to an ecosystem, as was illustrated by ecosystem modification for satisfying the demand for recreational and tourism services.

Economic valuations, as a rule, focus on relativities or comparisons. As was illustrated, these valuations fail to take account of the reduced absolute economic value of ecosystem services. Therefore, it is inappropriate to rely entirely on this type of analysis in determining environmental policy. Given the difficulty of determining some types of economic values, there is a temptation to assume that this problem can be avoided by relying on the application of cost-effectiveness analysis in which the prime objectives are independently specified. While this can reduce the burden of valuation, it may not eliminate it entirely. This is obviously so if the aim is to achieve the conservation objective at minimum social cost. This issue is considered in the case considered in the next chapter.

In outlining reasons for socially detrimental alterations in natural ecosystems and in wild biodiversity, it was emphasized that these are not only due to market failures. Political and public administrative failures, which occur in every country but which vary internationally in their importance, are also important. Examination of historical records provides many examples of such failures, and they are ongoing occurrences. Furthermore, as was pointed out, the influence of NGOs on the supply of ecosystem services is less than ideal, even though their influence may be preferable to no such influence.

The selection of policies to conserve biodiversity and to manage ecosystem transformations or conservation needs to take account of social, institutional and political constraints. This reduces the leeway for adopting economic policies found to be socially desirable on theoretical economic grounds, such as those based on the application of neoclassical welfare economics. While there is value in exploring the application of traditional welfare economics to conservation policies, economic policy formulation should be widened to take account of social political and institutional constraints, even though this is intellectually very challenging. Such an approach seems to be especially important in devising conservation policies that incorporate economic considerations. Moreover, rather than formulate a policy based on a single perspective, it is worthwhile, in my view, to outline and explore alternative policies based on a variety of different perspectives.

NOTES

1. One reason why estimates of the global value of ecosystem services should be considered cautiously is that many are essential for human survival, and presumably humankind would be prepared to make an extreme sacrifice for them not to disappear (Salles, 2011, p.478). This aspect is not captured in economic valuations of essential ecosystem services. Nevertheless, economic estimates of the comparative historical value of global flows of ecosystem services can be qualitatively indicative of trends and can provide a useful basis for assessment of global environmental change. For example, Costanza et al. (2014, p.156) estimated that in real terms the annual aggregate value of ecosystem services between 1977 and 2011 decreased by US$70.2 trillion. While increased economic values were recorded for cropland, grass/rangelands and urban biomes (human dominated ecosystems), these were greatly outweighed by losses in the economic value of ecosystem services supplied by other, more natural biomes. This is concerning. However, on the basis of these estimates, it would be unwise to conclude that human well-being declined globally between 1997 and 2011 due to a reduction in the real supply of ecosystem services. Further consideration of this matter is called for. It is given some further attention in Chapter 16 of this book.

2. Arrow's impossibility theorem shows that in many cases it is impossible to rationally base social welfare functions on individual values.
3. Various aspects of the use of these policy instruments are also discussed in Tisdell (2005, Ch. 3; 2009c, Chs 4 and 5).
4. There is some evidence that frequent visitors to natural areas have a more positive attitude towards nature conservation than non-visitors or infrequent visitors (see for example, Bandara and Tisdell, 2003).

REFERENCES

Arrow, K.J. (1951), *Social Choice and Individual Values*, New York: John Wiley.

Arrow, K.J. and A.C. Fisher (1974), 'Environmental preservation, uncertainty and irreversibility', *Quarterly Journal of Economics*, **88**, 312–19.

Bandara, R. and C.A. Tisdell (2003), 'Use and non-use values of wild Asian elephants', *Sri Lankan Economic Journal*, **4** (2), 3–30.

Bateman, I.J., I.H. Langford, N. Nishikawa and I. Lake (2000), 'The Axford debate: A case study illustrating different approaches to the agregation of benefit data', *Journal of Environmental Planning Management*, **43** (2), 291–302.

Bergson, A. (1938), 'A reformulation of certain aspects of welfare economics', *Quarterly Journal of Economics*, **52**, 310–34.

Braat, L. and P. ten Brink (eds) (2008), *The Cost of Policy Inaction (COPI) in The Cost of not Meeting the 2010 Biodiversity Target*, Alterra, Wageningen: European Commission. Available at: https://archive.org/details/costofpolicyinac08braa. (Accessed 24 March, 2015).

Ciriacy-Wantrup, S.V. (1968), *Resource Conservation: Economics and Policies*, 3rd edn, Berkeley, CA: Division of Agricultural Science, University of California.

Costanza, R., R. De Groot, P. Sutton, S. Van der Ploeg, S.J. Anderson, I. Kubiszewski, S. Farber and R.K. Turner (2014), 'Changes in the global value of ecosystem services', *Global Environmental Change*, **26**, 152–8.

Craft, A.B. and R.D. Simpson (2001), 'The value of biodiversity in pharmaceutical research with differentiated products', *Environmental and Resource Economics*, **18**, 1–17.

Hagedorn, K. (1993), 'Institutions and agricultural economics', *Journal of Economic Issues*, **27**, 849–80.

Heisenberg, W. (1930), *The Physical Principles of Quantum Theory*, C. Eckart and F.D. Hoyt [trans.], Chicago: Chicago University Press.

Krutilla, J.V. (1967), 'Conservation reconsidered', *American Economic Review*, **57**, 777–86.

Millennium Ecosystem Assessment (2005), *Ecosystems and Human Well-being: A Synthesis*, Washington, DC: Island Press.

Niskanen, W.A. (1971), *Bureacracy and Representative Government*, Chicago: Aldine.

Olson, M. (1965), *The Logic of Collaborative Action*, Cambridge, MA, USA: Harvard University Press.

Pearce, D.W. and D. Moran (1994), *The Economic Value of Biodiversity*, London: Earthscan.

Pearce, D.W., C. Bann and S. Georgiou (1992), 'The social costs of fuel cycles',

London: Centre for Social and Economic Research on the Global Environment, London University College. Mimeo.

Principe, P. (1989), *The Economic Value of Biodiversity Among Medicinal Plants*, Paris: OECD.

Principe, P. (1991), 'Monetizing the pharmacological benefit of plants', Washington DC: US Environmental Protection Agency. Mimeo.

Rubin, S.M. and S.C. Fish (1994), 'Biodiversity prospecting: Using innovative contractual provisions to foster ethnobotanical knowledge, technology and conservation', *Colorado Journal of International Environmental Law and Policy*, **5**, 23–58.

Salles, J.-M. (2011), 'Valuing biodiversity and ecosystem services. Why put economic values on Nature?', *Comptes Rendus Biologies*, **334**, 469–82.

Simpson, R.D., R.A. Sedjo and J.W. Reid (1996), 'Valuing biodiversity for use in pharmaceutical research', *The Journal of Political Economy*, **104**, 163–85.

Swarna Nantha, H. (2014), 'The financial and political opportunity costs of orangutan conservation in the face of oil-palm expansion', *Economics, Ecology and the Environment*, Working Paper no. 195, Brisbane: School of Economics, The University of Queensland.

Ten Kate, K. and S.A. Laird (1999), *The Commercial Use of Biodiversity: Access to Genetic Resources and Benefit-Sharing*, London: Earthscan.

Tisdell, C.A. (2005), *Economics of Environmental Conservation*, 2nd edn, Cheltenham, UK and Northampton, MA, USA: Edward Elgar Publishing.

Tisdell, C.A. (2007), 'Knowledge and the valuation of public goods and experential commodities: information provision and acquisition', *Global Business and Economics Review*, **9** (2/3), 170–82.

Tisdell, C.A. (2009a), 'Institutional economics and the behaviour of conservation organizations: Implications for biodiversity conservation', in K. Ninan (ed.), *Conserving and Valuing Ecosystem Services and Biodiversity: Economic Institutional and Social Challenges*, London, UK and Sterling, VA, USA: Earthscan, pp. 175–91.

Tisdell, C.A. (2009b), 'Complex policy choices regarding agricultural externalities: Efficiency, equity and acceptability', in V. Beckmann and M. Padmanabhan (eds), *Institutions and Sustainability. Political Economy of Agriculture and the Environment – Essays in Honour of Konrad Hagedorn*, Heidelberg: Springer, pp. 83–106.

Tisdell, C.A. (2009c), *Resource and Environmental Economics: Modern Issues and Applications*, Singapore, Hackensack (NJ), London and Beijing: World Scientific.

Tisdell, C.A. (2013), 'Ecotourism experiences promoting conservation and changing economic values: The case of Mon Repos turtles', in R. Ballantyne and J. Packer (eds), *International Handbook on Ecotourism*, Cheltenham, UK and Northampton, MA, USA: Edward Elgar Publishing.

Tisdell, C.A. (2014), *Human Values and Biodiversity Conservation: The Survival of Wild Species*, Cheltenham, UK and Northampton, MA, USA: Edward Elgar Publishing.

Tisdell, C.A. and C. Wilson (2012), *Nature-based Tourism and Conservation: New Economic Insights and Case Studies*, Cheltenham, UK and Northampton, MA, USA: Edward Elgar Publishing.

Tisdell, C.A., C. Wilson and H. Swarna Nantha (2008), 'Contingent valuation as a dynamic process', *Journal of Socio-Economics*, **37** (4), 1443–58.
Toman, M. (1998), 'Why not to calculate the value of the world's ecosystems and natural capital', *Ecological Economics*, **25**, 57–60.
Tullock, G. (1959), 'Problems of majority voting', *Journal of Political Economy*, **67**, 571–9.
Wilson, E.O. (1992), *The Diversity of Life*, Cambridge, MA: Belknap.

12. Allocating land use to minimize the opportunity cost of conserving wild species in their natural habitats

12.1 INTRODUCTION

It was argued in the previous chapter that constrained economic optimization problems are often relevant in formulating policy solutions for biodiversity conservation as well as for decision-making about protecting particular or selected sets of wild species and portions of ecosystems. An example of this problem for a particular wildlife species is as follows: find a contiguous natural area of a size which (if protected) will minimize the opportunity cost of sustaining the minimum viable population of the focal species or some other specified (targeted) level of population of this species. This is the type of policy problem given most attention in this chapter.

In practice, both the biological objectives to be pursued in these types of policy choices may be politically prescribed as well as the types of opportunity costs to be taken into account. Although, from an idealistic point of view, social opportunity costs should be considered, government policy may be based on a narrower perception of opportunity costs. For example, a government may be mainly concerned about the private opportunity costs imposed by conservation decisions on particular economic groups, for example, farmers or foresters, or in some countries, those wanting to establish oil-palm plantations. The solutions to such problems are not self-evident even though these problems all have a similar mathematical structure. That these solutions are not self-evident, is underlined by the fact that in the ecological literature, recommendations for reserving different types of land for conservation have been based on comparisons of either the absolute economic returns they generate if converted to commercial use, or on differences in the density of a species they support. These approaches are shown to be deficient because they ignore *relative* trade-offs between populations of species conserved and economic gains from conversion of natural habitats at alternative sites. A model is developed to demonstrate this deficiency and is applied to the case of the conservation of the orangutan.

The importance of opportunity costs and comparative (relative) opportunity costs for the optimal allocation of resources has been emphasized in the economic literature since the early nineteenth century (Ricardo, 1817). In discussing the implications of comparative opportunity cost, Ricardo emphasized that differences in the relative opportunity costs of resource use rather than differences in absolute opportunity costs are relevant for determining the optimal economic use of resources. Although Ricardo applied this principle to the determination of optimal specialization in international trade, his theory of comparative advantage can be applied to other resource allocation problems, including the allocation of land use so as to achieve conservation goals at minimum economic cost. However, the significance of Ricardo's theory has not been adequately recognized in the literature focusing on the choice of ecological policies (see Cullen, 2009). This can lead to errors in recommendations for the best allocation of available land for the conservation of wild nature and for commercial use such as for agricultural development.

The World Conservation Strategy (IUCN–UNEP–WWF, 1980) and its successor *Caring for the World* (IUCN–UNEP–WWF, 1991) recommended that the most productive land for agriculture or for similar commercial purposes be set aside for agriculture rather than be conserved. This implies that land having a low productivity from a commercial point of view should be reserved for conservation purposes. However, as pointed out in Tisdell (1991, pp. 29–31; 2005, 34–8), this ignores the principle of comparative advantage, that is, the comparative (or relative) productivity of land when used for alternative purposes. It relies on a comparison of absolute opportunity costs (economic returns forgone) rather than relative opportunity costs as the basis for determining the purpose for which land is used. Making this distinction is important because the absolute-cost principle does not ensure that conservation goals are achieved at minimum economic costs. Minimizing economic costs makes conservation more attractive or palatable to decision-makers; more habitats or a larger population of a species could be conserved for a given amount of resources or for less monetary sacrifice.

Models have been proposed for optimizing absolute economic returns and species richness over a given landscape (for example, Polasky et al., 2008). The purpose of this chapter, however, is to show how the economic principle of comparative advantage can be applied to decisions about conserving different types of land and their associated habitats so as to achieve the probable survival of a single species' population at minimum economic cost. The principle is illustrated by taking the survival of the orangutan in South-East Asia as a particular case. After discussing conceptual issues involved in this analysis and its sometimes counterintuitive results, general obstacles to conserving the orangutan are outlined. The economics of reserving different types of land areas for the conservation of the orangutan in Borneo, such as peat swamps and other land types, is then explored. The scope for the general application of this analysis is discussed.

12.2 RELEVANT CONCEPTUAL ISSUES

One aim of this chapter is to determine how to allocate completely protected land for the *in situ* conservation of a minimum viable population of a species so as to ensure that this is achieved at the lowest possible economic cost, where the economic cost is opportunity cost. Opportunity cost in this case is the maximum economic benefit forgone by setting aside the land for the conservation of the species rather than using it for alternative purposes. For example, in Borneo and Sumatra, the private opportunity costs of conserving land and its associated habitat for the conservation of the orangutans are often the returns forgone from not growing oil palm on this land. The social opportunity costs would include environmental externalities in addition to this.

Expressed somewhat differently, an intended outcome of the analysis in this chapter is to determine the pattern of land use that conserves a minimum viable population of a focal wild species but which does so at minimum opportunity cost. However, it is important to recognize in advance that the concepts of a minimum viable population of a species and that of opportunity cost have limitations. Consider each of these concepts in turn.

First, no conservation strategy can ensure the survival of any species forever. Therefore, strictly speaking, there is no completely safe minimum population for ensuring the continuing survival of a species. The best we can achieve are strategies that preserve or increase the chances of survival of a species (Hohl and Tisdell, 1993). In these circumstances, a workable objective is to determine the level of population needed to ensure that a species will survive with a minimum acceptable probability, for instance, 95 per cent, for at least a specified period of time. One can then attempt to minimize the economic cost of achieving this objective. This is the approach adopted here in considering the fate of the orangutan and similar wild species.

There are also further issues that arise in aiming to conserve a minimum viable population of a single species. For example, to what extent is it desirable to conserve populations of sub-species (there are three sub-species of the Bornean orangutan) or a meta-population of a species? If the former, then some preference ordering (utility analysis) is required of the combinations of sub-species that it is desired to conserve. However, obtaining collective social agreement on such a preference ordering is likely to be difficult. Furthermore, some criteria which can be applied to decisions to conserve individual species (such as ensuring a minimum viable population of these) become unworkable when the conservation of some species must be abandoned to conserve others due to limited resources being available for protecting wild species (Tisdell, 1990). For example, resource constraints can be expected to rule out policies to ensure that a minimum viable population of every species is maintained. Nevertheless, there will be substantial social support for protecting some wild species in order to ensure that a minimum viable population of them continues to exist. This support is likely to be reinforced when these species are umbrella species, that is, a species the protection of which *in situ* incidentally results in the conservation of many other species. The orangutan is considered to be such a species. However, as pointed out below, the orangutan requires a very large area of connected tropical forest to maintain a minimum viable population and the opportunity costs of conserving it *in situ* can be very high when account is taken of the economic returns forgone from alternative land use, for example its value for growing oil palm.

Turning to the economics side of the decision problem, opportunity costs are not always easy to estimate, and the question needs to be addressed of whether private opportunity costs or social opportunity costs should be used in solving this optimality problem. The private opportunity cost of protecting land for the conservation of species is often regarded as being the amount of profit forgone by being able to use the land commercially. This is assumed to be the case in the analysis developed here. Social opportunity cost is the collective economic benefit forgone of adopting a particular type of land use rather than alternative uses; in this case, it is collective economic gain forgone by protecting land from commercial use in order to conserve a wild species. In practice, it is useful to consider both types of cost. Private opportunity costs are important because private economic returns from land conversion are the main driving force politically for alterations of natural habitats. But they need also to be considered as a part of social opportunity costs. Usually, more reliable information is available about this component of social opportunity cost than about its other components (for example, environmental costs). Therefore, estimation and consideration of the private opportunity costs of conserving a species is an important first step in taking into account the social opportunity cost of its conservation. This chapter concentrates on this first step. This is also done because estimation of social opportunity cost can be complex and often requires controversial value judgements.

The private opportunity cost of conserving a species is highest (other things being held constant) when, simultaneously, each member of its population requires a large home range for its survival and the available land gives a high return when used for alternative purposes. This appears to apply, for instance, to the allocation of fertile tropical lowlands for conserving the orangutan. In such circumstances, it can be very difficult to justify, in terms of cost, the goal advocated by Ciriacy-Wantrup (1968) that (at least) one safe minimum level of the population of each species be conserved, and it is unclear whether all subspecies should be conserved at this level. Furthermore, since the opportunity cost of conserving such a species is high, significant economic gain may be made by ensuring that the land area(s) set aside for its conservation is such that the aggregate opportunity cost of reserving this land is minimal.

Although the focus in this chapter is on the conservation of a single species, in the process of protecting land for its conservation, other valued species may also be conserved. Thus the benefit from conserving a single species may be of greater value (or come at a lower cost) than ensuring its own survival, as is especially evident in the case of umbrella species. For example, conserving areas to protect the orangutan also helps conserve the distinctive proboscis monkey (*Nasalis larvatus*) (Tisdell and Swarna Nantha, 2008). Ideally, relevant synergies and associations of this type should be taken into account. At this stage, however, this analysis will focus on the conservation of a single species.

The question arises of whether the conservation of a species should occur in more than one area. Other things held constant, the greater the number of separate areas in which viable populations of a species are conserved, the more likely the species is to persist for a specified period of time, and possibly the larger would be the supply of ecosystem services provided by natural areas (Luck et al., 2003). However, opportunity costs can be expected to rise as the number of protected areas increase. The analysis presented here initially assumes that only one area is to be set aside for conserving the focal species. Subsequently, consideration is given to the possibility of conserving separate populations of the species.

Bearing in mind the above qualifications, consider some of the economic and related challenges involved in conserving the orangutan, and then, drawing on available scientific evidence available in Husson et al. (2008), let us apply the theory of comparative economic advantage (comparative opportunity costs) to identify types of land which, if conserved, are likely to minimize the opportunity cost of protecting the orangutan.

12.3 BACKGROUND ON ORANGUTAN CONSERVATION IN BORNEO AND SUMATRA

The orangutan has been selected to illustrate applications of the concepts outlined in the previous section and as a case to which the analysis developed in the next section can be applied. It is also because the economic trade-offs required to conserve it *in situ* appear to be very high and it is in danger of extinction. There are two species of the orangutan: the Borean orangutan (*Pongo pygmaeus*) and the Sumatran orangutan (*Pongo abelii*). The IUCN Red List of Threatened Species lists the latter as critically endangered (Singleton et al., 2008) and the former as endangered (Ancrenaz et al., 2008).

The orangutan is a highly specialized hominid which relies primarily on fruits from tropical trees for food (Rijksen and Meijaard, 1999). This is supplemented by the eating of young shoots, barks and insects. Because of its comparatively specialized diet, each orangutan requires a relatively large land area of diverse flora to provide it with food. Unlike other great apes, it is not gregarious, and this appears to be a consequence of the large forested area that each orangutan requires to provide it with sufficient food.

Another factor that obligates orangutans to live in natural forests is that they find it easiest to travel through a closed forest canopy rather than along the ground. Therefore, they normally occupy the forest canopy and sleep in beds (nests) built high in the canopy. The fact that orangutans are usually solitary and are normally present high in the forest canopy makes them difficult to locate and view. This results in their potential for nature tourism in the wild being much less than for other hominid species, such as the gorilla.

There has been a decline of orangutans of more than 50 per cent in the last 60 years in Borneo and an 80 per cent decline in Sumatra in the last 75 years (IUCN, 2009). Although the orangutan once had a wide distribution in Asia, its remaining populations are now confined to the aforementioned islands, where less than 50 000 and about 6500 orangutans are found in Borneo and Sumatra, respectively. In Borneo, a large minimum viable population of the orangutan only exists in Kalimantan (Indonesia) and Sabah (Malaysia). Sarawak (Malaysia) has a border area with Kalimantan that contains orangutans and provides a minimum viable population of orangutans when combined with an adjoining area in Kalimantan. There are no orangutans in Brunei.

A problem in conserving the orangutan is that it has the slowest birth rate of all mammals. Orangutans reach reproductive maturity only after a period of 8 to 15 years and produce one offspring every 7 to 8 years (Payne and Prudente, 2008). Consequently, the survival of the orangutan is vulnerable even to the relatively low rates of population loss which may occur because of hunting or its capture for the pet trade or similar purposes.

The greatest threat to the continuing survival of the orangutan is believed to be the conversion of forested land to oil palm plantations (Swarna Nantha and Tisdell, 2009; Wich et al., 2008). There are also threats from land conversion for other purposes as well, such as for timber plantations and also from large-scale intensive logging that is often illegally done. In many cases, logging is a forerunner to the establishment of oil palm plantations or other types of agriculture. Profits from the initial logging of a forested area can add to the financial viability of converting these lands into oil palm plantations (Clay, 2004). This they can do by reducing initial cash flow problems and providing investible funds. Other things being held constant, land areas that give a high and rapid return from logging add to the financial attractiveness of starting oil palm plantations. These large returns from logging have also resulted in cases where lands are abandoned once timber gains are made (Holmes, 2002).

Fires are also a threat to the survival of the orangutan. Fires are often lit after logging to dispose of remaining vegetation and residual forest wastes and to return any fertilizing nutrients contained in them to the soil. These fires can escape and create forest fires during dry periods. Orangutans are often burnt to death and their habitat destroyed.

As mentioned above, the hunting of orangutans and their capture is also a threat. As well as this, orangutans *in situ* have limited potential to support ecotourism (Tisdell and Swarna Nantha, 2008). Consequently, the economic returns from this form of non-consumptive use value are low. All these factors add to the threats to their survival.

12.4 A MODEL FOR THE APPLICATION OF THE THEORY OF COMPARATIVE ADVANTAGE TO THE CONSERVATION OF A SINGLE SPECIES

For ease of application and as a first approximation, a linear model is proposed as the basis for determining the most economic allocation of land in order to conserve, *in situ*, a minimum viable population of a species, X. This population is assumed to be sufficient to ensure the survival of the focal species. The problem is to find the allocation of land between its reservation for conserving a species, X, and its use for commercial purposes (such as agriculture) such that this allocation minimizes the economic cost (that is, the opportunity cost) of conserving the species. In the next section, this model will be applied specifically to the conservation of the orangutan.

Suppose that $i = 1, \ldots, n$ separate areas of land of varying types of natural habitat are available and that they are individually suitable for conserving a minimum viable population, $XMVP_i$, of species X. The minimum viable population (MVP) of the focal species could vary between land areas. For example, other things being equal, the MVP of a species may be higher in regions where its food supply fluctuates more than elsewhere.

Assume that in location i, a natural area (single parcel) of land of $\overline{L}_i$ km^2 is available (and suitable) to support species X, at an average density of λ_i per km^2 and that L_i^a km^2 of this land may be allocated to the growing of crops (or other forms of commercial production), with the remainder being left in a natural state. Commercial economic production is assumed to provide an average private economic benefit (per year) of Π_i per km^2. In the area of land where commercial production occurs, species X is assumed to disappear. There is, therefore, a trade-off between the level of population of X in location i and the use of this land for commercial production. The relationship can be expressed as:

$$X_i = \lambda_i(\overline{L}_i - L_i^a) = \lambda_i L_i^p \qquad (12.1)$$

where X_i is the population size of the focal species and L_i^p is the area of protected land at site i. Private economic benefit, R_i, from converting land to commercial production in location i can be expressed as:

$$R_i = \Pi_i L_i^a \qquad (12.2)$$

where Π_i is the average private economic benefit (for example, profit) per km^2 from converted land. Consequently, the trade-off function between the population of species X conserved at site i and private economic returns can be found. From equation (12.2):

$$L_i^a = R_i / \Pi_i \qquad (12.3)$$

Substituting in equation (12.1), the trade-off function is:

$$X_i = \lambda_i \overline{L}_i - \lambda_i R_i / \Pi_i \qquad (12.4)$$

Letting $\lambda_i \overline{L}_i = Xmax_i$ and $\lambda_i / \Pi_i = b_i$, this can be re-expressed as:

$$X_i = Xmax_i - b_i R_i \qquad (12.5)$$

The intercept of this line with the vertical axis, $Xmax_i$, indicates the maximum population of species X that site i is able to sustain in the absence of land conversion and b_i represents the rate at which the size of the population of X at site *i* is reduced in order to obtain an increase in private benefits from commercial production. It is the rate of trade-off between conservation of the population of X and private economic benefit from land conversion. Its inverse represents the extent to which private economic benefits from land conversion have to be forgone to elevate the size of the population of X. The intercept of the line represented by equation (12.5) with the R_i axis is:

$$R_i = Xmax_i / b_i = \Pi_i \overline{L_i} \tag{12.6}$$

It represents the total private return that can be obtained by converting all the available suitable land at site *i* to commercial production. Note that not all the available land area may be suitable for commercial production and therefore Π_i needs to be adjusted accordingly by a scaling variable. For example, if only 80 per cent of the land is suitable for commercial use, this variable would be 0.8.

Given this approach, knowledge of only two numerical values is needed to estimate the trade-off function. These are the maximum population of species X that site *i* can sustain and the maximum private economic benefit to be obtained from total economic conversion of land at this site. These values are respectively, $Xmax_i = \lambda_i \overline{L_i}$ and $\Pi_i \overline{L_i}$.

This aspect is illustrated in Figure 12.1. There line ABC represents the trade-off between the level of population of species X at site *i* and the private economic benefit to be obtained from converting the land to commercial production. Point A corresponds to the maximum population of X that can be sustained at the site and point C corresponds to the maximum private economic benefit that could be obtained by land conversion. If at site *i*, the MVP of species X is $XMVP_i$ = OF, and this constraint can be met at point B. Hence, a private economic benefit of OD can be obtained by converting some land to commercial production at site *i* and leaving enough unused land to conserve the MVP of X.

The amount of land that needs to be conserved to achieve point B and thereby conserve the MVP of X can be found by substituting $XMVP_i$ into equation (12.1) and rearranging. It is:

$$XMVP_i = \lambda_i(\overline{L_i} - L_i^a) \tag{12.7}$$

and rearranging,

$$\overline{L_i} - L_i^a = XMVP_i / \lambda_i, \tag{12.8}$$

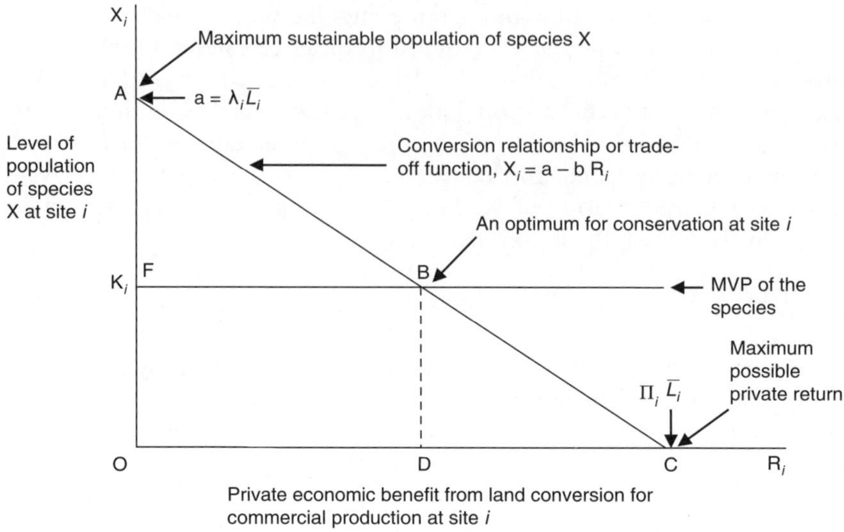

Note: This figure is also used to illustrate the solution of an optimization problem outlined in the text.

Figure 12.1 An illustration of the trade-off between conversion of land for private economic gain and the population size conserved of a species

The remaining area of land,

$$L_i^a = \overline{L_i} - \text{XMVP}_i = \lambda_i \tag{12.9}$$

can be set aside for commercial production.

In order to solve the problem of minimizing the total private opportunity cost of conserving a MVP of species X (when only a single site is set aside for its conservation), it is necessary to consider the level of private economic benefits forgone at each of the available sites in order to achieve the MVP of species X at each of these sites, that is XMVP_i. This can be found by taking the difference between R_i when $X_i = 0$ and R_i when $X_i = \text{XMVP}_i$. This is the difference between the maximum private returns (profit) when all the available land at site *i* is utilized commercially with no attention being given to the conservation of species, and that when just sufficient land is fully protected to achieve the MVP of species X at this site with the remaining land being used commercially. The site should be singled out for conserving the species for which this difference is lowest.

This will minimize private opportunity cost. The solution to this opportunity cost problem also maximizes private returns (profit) from land use subject to the condition being satisfied that a single connected natural site of sufficient size be reserved to ensure that the minimum viable population of species X is conserved.

The approach of minimizing opportunity cost in order to solve this problem can be detailed mathematically in the way that follows. When $X_i = 0$, and therefore no consideration is given to the conservation of species X, equation (12.5) implies that:

$$X_i = 0 = \text{Xmax}_i - b_i R_i \tag{12.10}$$

Rearranging this equation, maximum private profit at site *i* is:

$$R_i = \text{Xmax}_i \,/\, b_i \tag{12.11}$$

Let this value be represented by $\hat{R}_i$. When $X_i = \text{XMVP}_i$ and commercial use of the land is just sufficiently restricted to ensure the minimum viable population of X at site *i*,

$$\text{XMVP}_i = \text{Xmax}_i - b_i R_i \tag{12.12}$$

Rearranging this equation, profit from commercial land use of site *i* when XMVP_i is satisfied is:

$$R_i = (\text{Xmax}_i - \text{XMVP}_i) \,/\, b_i \tag{12.13}$$

Let this value be represented by $\overline{R_i}$. Consequently, when just sufficient land is set aside at site *i* to conserve XMVP_i of species X, the private economic benefit forgone, C_i, is equal to the difference between expressions (12.11) and (12.13), that is:

$$C_i = \hat{R}_i - \overline{R_i} \tag{12.14}$$

$$C_i = \text{Xmax}_i \,/\, b_i - (\text{Xmax}_i - \text{XMVP}_i) \,/\, b_i \tag{12.15}$$

$$= \text{XMVP}_i \,/\, b_i \tag{12.16}$$

where $b_i = \lambda_i/\Pi_i$. This result is of central importance for decision-making in this conservation case because the site should be selected for conservation of species X that minimizes XMVP_i/b_i if a single minimum viable population is to be conserved so as to minimize opportunity cost.

Some important implications of expression (12.16) should be noted. First, the location for which $XMVP_i/b_i$ is lowest is the one that minimizes the opportunity cost of conserving a single viable population of species X. If more than one separate viable population of the species is to be maintained, the private opportunity costs are minimized by adding the conservation of the species at other available sites in ascending order of $XMVP_i/b_i$. Observe that since $b_i = \lambda_i/\Pi_i$, the value of this expression does not depend on the absolute densities of species X which can be supported at a site nor does it depend on the average level of economic returns per km^2 yielded by commercial 'development' of a site. Only the relative marginal opportunity costs of conserving species X at a site and the required size of its MVP are relevant in this optimization problem.

It can be seen that, other things held constant, a rise in the required MVP for X at a site increases the opportunity cost of conserving it at that site, and also an increase in marginal opportunity cost does likewise. If the MVP for X is the same at all sites, the site for which b_i is largest (the one for which a reduction in R_i results in the largest increase in population of X) is the one that minimizes the opportunity cost of conserving species X. If two sites have a similar value for b_i, the one with the lowest value of $XMVP_i$ minimizes the private opportunity cost of conserving species X.

The computational procedures involved in finding the solution to the above model are straightforward. They are:

- eliminate all land areas that are of insufficient size to satisfy $XMVP_i$;
- for the remaining areas, calculate $XMVP_i / b_i$;
- from this set, select the area with the lowest value of $XMVP_i / b_i$ to conserve the focal species.

Because both the number of alternative sites to be taken into account is likely to be low and the number of variables to be estimated is low, this approach is practical.

It is true that the linear model used involves considerable simplification. Actual systems are likely to be non-linear and judgement is required about how satisfactory a linear approximation is. Secondly, the model is static, whereas actual systems are dynamic. The relevant model of Ricardo (1817) was also linear and static. Introducing dynamics would complicate the modelling greatly and is unnecessary to establish the essential point that differences in relativities rather than absolute values are germane to the solution of this problem.

12.5 DISCUSSION AND EXTENSION OF THE ABOVE MODEL

First, it should be noted that the trade-off function in the above model is assumed to be continuous. However, if the MVP for the species X represents a threshold (as it is supposed to) then in reality, the trade-off function has a discontinuity. In Figure 12.1, for example, once the extent of land conversion is such as to yield a private economic return marginally in excess of OD, the probable survival of species X falls below the targeted level and its population is liable to disappear. Thus, its population may fall to zero for private commercial returns in excess of OD. However, this complication does not invalidate the procedure outlined above for solving the optimization problem specified.

Observe that there is also a qualitative dimension to the above problem. Although homogeneity (on average) at each site is assumed, differences in densities of the focal species and profits from commercial use of the land can occur between sites. This is allowed for in the model. In other words, the trade-off function ABC shown in Figure 12.1 can differ between sites. It is also possible that the MVP of the focal species will be higher at some sites than others, for example, at sites when there is greater variation in food availability, and this can be allowed for in this model. Therefore, the area of land that will be needed to protect the focal species, for example the orangutan, can vary between sites. This is allowed for in this model but heterogeneity at any particular site is not. Furthermore, the shape and habitat connectivity of the land set aside for conservation can also be important in land use decisions.

A further limitation of the model is that it does not allow for uncertainties about the MVP of the focal species or for uncertainty about the profitability of land conversion. The extent of these uncertainties can vary between sites. This adds to the complexity of the decision-making problem but they can be allowed for in principle.

An additional problem is that the optimal solution presented in the model is unlikely to be achieved because of political pressures. Other things being held equal, the greatest political pressure is likely to be applied to allow conversion of natural areas to commercial use that gives the largest absolute private economic gains. Their relative productivity for conserving a focal species is likely to be given scant consideration by commercial interests. As a result, the aggregate private opportunity cost of conserving a species may not be minimized.

This can be illustrated by Figure 12.2. Consider two sites, I and II. Compared to site I, site II is assumed to conserve more of species X per km^2 (for example, it has a higher density of orangutans on average, λ_i) if left

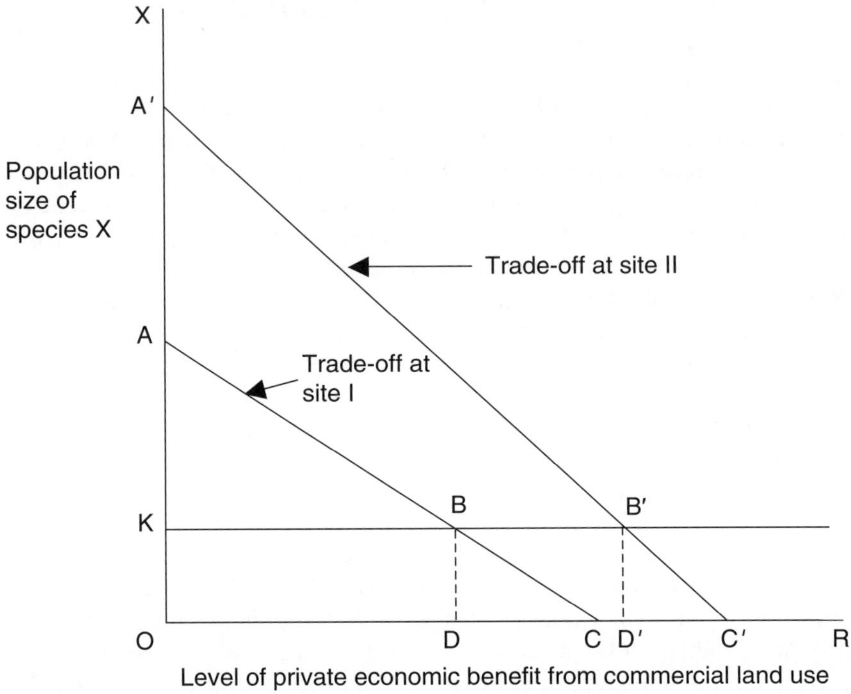

Figure 12.2 Sometimes conserving a species on land that has the greatest commercial value minimizes opportunity costs as is explained

in its natural state. It is also assumed to give greater private profit per km^2 if converted to commercial use. There will, therefore, most likely be greater commercial pressure to develop site II than site I. For example, in Figure 12.2, let ABC represent the trade-off function between the level of population of species X and private profit at site I and A′B′C′ represent that at site II. Given that OK is the MVP of species X at both sites and therefore, is the same at both sites, private opportunity costs are minimized by conserving the species at site II rather than site I. The opportunity cost of conserving a MVP of the species at site I is equal to DC whereas it is only D′C′ at site II. The latter is smaller in amount because A′C′ declines more sharply than AC. The rate at which profit has to be forgone at site II to increase the population of X is less than at site I. On the other hand, a higher absolute amount of profit per km^2 is forgone if site II is used to conserve the species. Such a choice is likely to generate considerable political opposition by commercial interests, because it is likely that each commercial agent is guided only by its absolute profit forgone.

The above exposition indicates that even if overall private economic gains are the only consideration, social pressures can result in land conversion that does not minimize the private opportunity costs of conserving a species. Furthermore, economic failure is even more likely if the goal is to conserve the focal species in a way that minimizes social opportunity costs. The social economic gains from land conversion are likely to vary with the location of the land and its properties, and in most cases, these differ from private economic benefits. Ideally, the social opportunity costs of conserving a species should be minimized, although it needs to be noted that social opportunity costs are difficult to measure and there can be considerable differences in opinion about their magnitude. However, the concept is important. For example, if there are two different areas where a MVP of species X can be conserved for a similar private opportunity cost, socially one should not be indifferent about the area to select for their conservation. For example, the negative externalities generated by commercial production in one of these areas may be greater than in the other. Therefore, other things being held constant, it would be socially desirable to conserve the species in the area that would generate the least negative spillover when used commercially. This type of issue arises in setting aside areas for the conservation of the orangutan, as will become evident in the next section of this chapter.

Note also that, as explained earlier, in many cases it is not just the size of an area reserved for the conservation of a species that is important but also its spatial characteristics. Areas conserved for the orangutan basically need to be compact and continuous and not disjointed and scattered, that is, not fragmented. In addressing the above issue, the spatial requirements of a species cannot be disregarded and it may add to the opportunity costs of their protection. For example, orangutans cannot be effectively conserved if only small separated pockets (islands) of forested land are reserved for their protection. Landscape features may also be important. For example, if the land reserved for the conservation of orangutans is dissected by a major river, this constitutes a barrier to their movement because orangutans do not swim. Will the conserved area on each side of the river be enough to conserve a minimum viable population of this species in this area?

Furthermore, the model presented here is a linear one and more complexity could be introduced into it, such as curvilinearity to reflect changes in comparative cost, variations in thresholds relating to the type and intensity of land use, the biological sensitivity of the species to habitat modification, and curves of varying slopes to reflect financial dynamics (for example, the rate of return may decline as large amounts of land are converted from a natural state for commercial production causing, for example, a supply-side glut). However, these factors are not modelled here because to do so is likely to obscure the fundamental point illustrated in this chapter.

Note that each of the potential sites for conserving species X are assumed to be sufficiently separated from one another to make movement of the species from one site to the other impossible. Furthermore, it is supposed that the species cannot survive for any length of time outside of its protected habitat. This is approximately true of orangutans but does not apply to all species; for example, it is not true of all marine species and many of these show movement outside the boundaries of their reserves (Grafton et al., 2005).

12.6 APPLICATION OF THE MODEL TO SELECTING PROTECTED AREAS TO CONSERVE ORANGUTANS

The development of the above model was partly motivated by its possible application to the selection of protected areas to conserve the orangutan. The theory implies that areas should be selected for the conservation of the orangutan for which their density is highest relative to profits forgone per unit of land as a result of this protection. This is so if the objective is to conserve the orangutan at minimum private opportunity cost. When this condition is satisfied, the private opportunity cost of conserving each orangutan reaches its lowest level. Put differently, $b_i = \lambda_i / \Pi_i$ reaches its lowest value when this decision rule is followed. This can be clarified by a numerical example applicable to the orangutan. Singleton et al. (2004) suggest a single population of around 500 orangutans would be sustainable in the long run. The area required to sustain such a population varies with the carrying capacity of the habitat. While in general orangutan densities per km^2 are low, they can vary considerably (Husson et al., 2008). Densities of 1–2 orangutans per km^2 are not uncommon. The profitability of the commercial area of land also varies considerably. The annual profit from growing oil palm in Malaysia has been in the range of US\$528–\$790 per km^2 (Swarna Nantha and Tisdell, 2009), that is US\$52 800–\$79 000 per km^2 if all the land is suitable for growing this crop. In many cases, not all of the land will be commercially usable so the profit per km^2 would be lower in such cases.

It is assumed that the minimum viable population of orangutans at a site is 500. For simplicity, suppose that there are two sites, Site I and Site II, where the MVP for the orangutan of 500 can be achieved. Also consider two cases. In case (1), the density of orangutan at Site I is 1.5 per km^2

Table 12.1 Examples of the (annual) opportunity cost (profit forgone) in order to conserve an MVP of 500 orangutans

Cases	Orangutans per km² λ_i	Profit per km² Π_i	Conserved area in km² needed for MVP, L_i^p	Total profit forgone to ensure MVP, C_i
Case 1				
Site I	1.5	$50 000	333.3	$16.7m
Site II	2.0	$80 000	240.0	$20.0m
Case 2				
Site I	1.5	$50 000	333.3	$16.7m
Site II	4.0	$80 000	125.0	$10.0m

and profit forgone per km² (per year) from conserving the site is $50 000. Therefore, the cost of conserving each orangutan is $33 333.3. At Site II, the density of the orangutan is 2 per km² and the profit forgone per km² is $80 000 from conserving the site. Therefore, the cost of conserving each orangutan at Site II is $40 000. The least cost option is to conserve the orangutan at Site I. This will require 333.3 km² of land to be set aside for the purpose at Site I. The approximate profit forgone, C_1, is $16.7 m. If Site II were selected, the profit forgone, C_2, would be 250 × $80 000 = $20m in order to conserve 500 orangutans.

In case (2), suppose the same as in case (1) except that Site II is able to support four orangutans per km². In this case, each orangutan can be conserved at Site II at a cost of $20 000. The relative cost (comparative cost) of conserving one orangutan at Site II is now lower than at I. The profit forgone, C_2, by conserving the MVP of orangutans at Site II is now $10 m and, therefore, lower than at Site I where it, C_1, is $16.7 m. Therefore, the most economical decision is to establish the MVP of 500 orangutans at Site II rather than Site I.

The details of these two cases are summarized in Table 12.1. Case 1 shows that it is not necessarily most economical to achieve an MVP of a species (for example, orangutans) by protecting an area where their densities are highest. Case 2 demonstrates that reserving land for commercial use (for example, agriculture) which gives the highest return per unit of area and using land of lesser commercial value for conserving a species is not necessarily the most economical way to conserve an MVP of the species. This is because in both cases comparative values involved in trade-offs are ignored.

Some ecologists have claimed that 'sites with the highest density of orangutans should be prioritized for conservation' (Husson et al., 2008, p. 96).

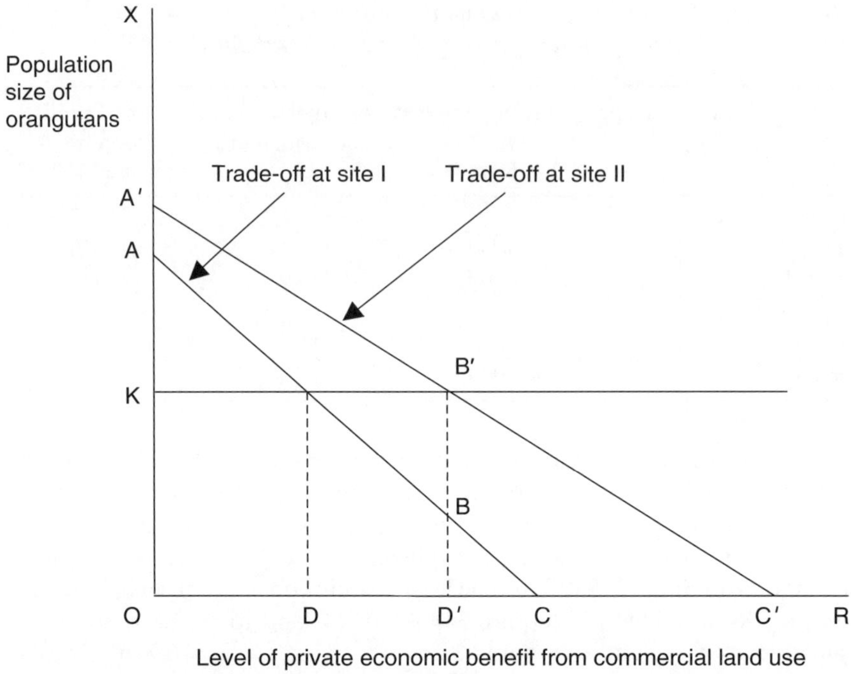

Figure 12.3 An illustration that the criterion of Husson et al. (2008, p. 96) can fail to minimize the economic cost of conserving the orangutan

But as the above example indicates, this is not necessarily so from an opportunity cost point of view. This can also be illustrated by a diagram similar to those shown earlier.

Using the same approach as above, Figure 12.3 illustrates the point that the criterion of Husson et al. (2008, p.96) does not necessarily minimize the opportunity cost of conserving the orangutan. Once again, assume that there are two alternative sites, I and II, where the orangutan can be conserved and that site II can support the highest density of orangutans. Suppose also that a higher rate of profit per km from conversion can be obtained at site II than at site I but that the rate of reduction in the population of orangutans at site II for increased profit is less at site II than site I. Also assume that the MVP at both sites is H. If it is decided to conserve an MVP of orangutans at site II rather than site I (as the Husson et al. criterion recommends), the opportunity cost would be higher than if the opposite choice was made. A line like ABC in Figure 12.3 can be drawn to represent the trade-off at site I and A′B′C′ can be drawn to illustrate that of site II. Should just enough orangutans be conserved at site II to satisfy XMVP = OK, the opportunity cost is equivalent to D′C′, whereas if it is done at site I, it is DC. The latter amount is less than the former one because the trade-off function ABC is steeper than A′B′C′. This drives home the point that differences in the absolute densities of the population of a species able to be supported by dissimilar areas and differences in the profitability of converting these areas to commercial use cannot be relied on to determine the reservation of land to minimize the cost of conserving a species.

It is difficult to decide a priori which type of habitat is likely to minimize the opportunity cost of conserving a minimum viable population of orangutan species. This is because there is still some uncertainty about the ability of different types of habitat to conserve orangutans and the annual profits that they are likely to yield when converted to commercial use, for example the growing of oil palm. Nevertheless, Husson et al. (2008) have made progress in identifying differences in orangutan densities supported by different types of landforms. On the whole, they found that densities were highest in descending order in mosaic landscapes (peat and dry forests), peat swamps, dry forests, and karst forests. However, they state that while they found that peat swamp forests support higher mean densities of orangutans than dry forest habitat, the difference was not statistically significant, contrary to what they had initially predicted (Husson et al., 2008, p.93).

If we assume that these densities are much the same, and we consider two areas (one consisting of peat swamps and the other of dryland forest), both of which are capable of sustaining a MVP of orangutan, the economic choice of which habitat to select for a reserve will hinge on the relative opportunity cost of providing the reserve. The private economic returns from growing oil palm on areas of peat swamp are likely to be somewhat lower per unit area of land (and for a given density of planting) than from the cultivation of palms on dryland forests with mineral soils. Palms grown on peat require heavier rates of fertilizer application (for example, phosphate, copper and zinc) and it can be more difficult to sustain yields (Andriesse, 1988, p.99; Mutert et al., 1999; Singh, 2008). Higher densities of palm per hectare may need to be planted on deep peat, which could mean additional production costs. Therefore, reserving peatlands for orangutan conservation is likely to involve lower marginal private opportunity costs (marginal profit forgone) than if areas of dryland forest are used for this purpose.

Furthermore, from several perspectives, the use of peatland rather than dry forest habitat for growing oil palm involves higher negative externality costs. Overall, taking this into account as well as private profitability, the relative social cost of converting peatland to oil palm production seems likely to be higher than for dry forest conversion. Agricultural use of peatland is associated with greater levels of release of CO^2 to the atmosphere and a greater frequency and duration of wild fires. Moreover, exploiting only a part or certain sections of a peat swamp for oil palm planting could still jeopardize overall peat swamp ecological and hydrological stability leading to its systemic collapse (FAO, 2008). These negative externalities suggest that the social economic benefit of reserving peatland for conservation purposes is higher than for dryland forest.

The relative opportunity costs of preserving karst forests to conserve orangutans are unclear. While karsts host significant endemic species, they support low orangutan densities and are not considered an important focus for orangutan conservation (Marshall et al., 2007). Although most karst forests do not have rich and productive soils suitable for agriculture, some can be used for rice and oil palm production. They are most valued economically for mining and are quarried for cement, marble and lime for agriculture (for example, to reduce the acidity of peat soils when used for agriculture) (Schilthuizen et al., 2005). The necessary trade-offs between orangutan populations and the returns from the development of agriculture on karst lands are unknown at this time.

More research is urgently required to estimate accurately the opportunity costs, (both private and social) of conserving different types of habitat and land areas for orangutans. This will improve decision-making about the selection of sites for the conservation of this species. In order to improve choices about the conservation of species, greater attention should be given to the approach of ecological economics, which combines ecological and economic considerations.

12.7 CONCLUSION

This chapter has combined ecological and economic considerations to demonstrate the applicability of the theory of comparative advantage (sometimes called comparative cost theory) to choices about the protection of land and habitat designed to ensure that the minimum viable population of a wild species is conserved at minimum private opportunity cost. It was shown that the relative trade-off between the population of the focal species and private profit from land conversion is the critical consideration involved in making the optimal choice. The view was rejected that areas where densities of the species are highest should be prioritized for conserving them. So was a similar view that the areas where private profits are highest should be reserved for commercial development. Each of these views takes account of only one variable in the conservation equation and relies on differences in their absolute values to draw a conclusion. However, both the relevant ecological and the economic variables need to be simultaneously considered and it is the differences in their relative values (see equation (12.15)) and not their absolute values that are critical to making the best decision.

The application of a simple linear model to protecting land areas to conserve orangutan species was considered. This added relevance to the point that relative trade-offs are important in deciding which land area should be protected to conserve the species. This application also highlighted some of the difficulties that are likely to arise in practice in applying the basic model developed in this chapter. For example, reliable data are not readily available on orangutan densities in all habitats and areas where they occur (although Husson et al., 2008, have made progress in collecting such data) and there are insufficient data on the private returns to be expected from land conversion in many areas.

Additional limitations of the model were noted. It is, however, impossible to capture all complexities of the conservation problem in a simple model. Furthermore, very complex models may be undesirable if they obscure a simple point, such as the one made in this chapter. Models that capture the nature of things are often more desirable than those that include most of the complexities of reality.

It has been assumed in this chapter that the focal wild species can only be conserved in a fully protected area or that this is the primary choice even if it is able to survive (possibly in reduced population densities) in somewhat disturbed areas, for example, in areas subject to selective light logging. When the latter possibility exists, the question arises of whether it is more economical to conserve the species in a fully protected area or for it to be conserved allowing controlled commercial use of an area which might, for example in the case of the orangutan, involve selective (light) logging. The optimal spatial conservation of the orangutan and other wild species can be regarded as a management issue involving the optimal supply of ecosystem services. It belongs to a class of problems for which decisions are needed about whether mixed or separated areal (spatial) use of ecosystems is economically optimal. This problem is considered in Chapter 17, where more consideration is given to the conservation of the orangutan as an example.[1]

NOTE

1. This chapter contains several excerpts from a previously published article of which I am the principal author, namely C.A. Tisdell and H. Swarna Nantha (2011), 'Comparative

costs and conservation of wild species in situ e.g. orangutans', *Ecological Economics*, **70**, 2429–39. I am grateful to Elsevier for permission to use material from this article.

REFERENCES

Ancrenaz, M., A. Marshall, B. Goossens, C. van Schaik, J. Sugardjito, M. Gumal and S.A. Wich (2008), '*Pongo pygmaeus*', in *IUCN 2010. IUCN Red List of Threatened Species Version 2010.4*. Available at: http://www.iucnredlist.org. (Accessed 9 May, 2011).

Andriesse, J.P. (1988), *Nature and Management of Tropical Peat Soils*, Rome: FAO.

Ciriacy-Wantrup, S.V. (1968), *Resource Conservation: Economics and Policies*, 3rd edn, Berkeley, CA: Division of Agricultural Science, University of California.

Clay, J. (2004), *World Agriculture and the Environment: A Commodity-by-Commodity Guide to Impacts and Practices*, Washington, DC: Island Press.

Cullen, R. (2009), 'Is economics being outcompeted in environmental decision-making?' Connections Paper no. 98. Available at: http://www.agrifood.info/connections/2009/Cullen.pdf. (Accessed 3 February, 2010).

FAO (2008), 'Oil-palm estate development in Southeast Asia: Consequences for peat swamp forests and livelihoods in Indonesia', FAO, *Scoping Agriculture–Wetland Interactions: Towards a Sustainable Multiple-response Strategy*, Rome: FAO, pp. 81–6.

Grafton, R.Q., T. Kompass and V. Schneider (2005), 'The bioeconomics of marine reserves: A selected review with policy implications', *Journal of Bioeconomics*, **7**, 161–78.

Hohl, A. and C.A. Tisdell (1993), 'How useful are environmental safety standards in economics? The example of safe minimum standards for protection of species', *Biodiversity Conservation*, **2** (2), 168–81. Reprinted in C.A. Tisdell (2002), *The Economics of Conserving Wildlife and Natural Areas*, Cheltenham, UK and Northampton MA, USA: Edward Elgar Publishing.

Holmes, D.A. (2002), 'Where have all the forests gone?', *Environmental and Social Development, East Asia and Pacific Region Discussion Paper Series*, Washington, DC: The World Bank.

Husson, S.J., S.A. Wich, A.J. Marshall, R.A. Dennis, M. Acrenaz, R. Brassey, M. Gumal, A.J. Hearn, E. Meijaard, T. Simorangkir and I. Singleton (2008), 'Orangutan distribution, density, abundance and impacts of disturbance', in S.A. Wick, S.S.U. Atmoko, T.M. Setia and C.P. Scheik (eds), *Orangutans: Geographical Variation in Behavioural Ecology and Conservation*, Oxford: Oxford University Press, pp. 77–96.

IUCN (2009), *The IUCN Red List of Threatened Species*. Available at: http://www.iucnredlist.org. (Accessed 4 January, 2010).

IUCN–UNEP–WWF (1980), *The World Conservation Strategy: Living Resource Conservation for Sustainable Development*, Gland, Switzerland: International Union for the Conservation of Nature and Natural Resources.

IUCN–UNEP–WWF (1991), *Caring for the World: A Strategy for Sustainable Living*, Gland, Switzerland: World Conservation Union.

Luck, G.W., G.C. Daily and P.R. Ehrlich (2003), 'Population diversity and ecosystem services', *Trends in Ecology and Evolution*, **18**, 331–6.

Marshall, A.J., L.A. Salas, S. Stephens, E.L. Nardiyono, E. Meijaard and

S.A. Stanley (2007), 'Use of limestone karst forests by Bornean orangutans (*Pongo pygmaeus morio*) in the Sangkulirang Peninsula, East Kalimantan, Indonesia', *American Journal of Primatology*, **69**, 212–19.

Mutert, E., T.H. Fairhurst and H.R. von Uexkull (1999), 'Agronomic management of oil palms on deep peat', *Better Crops International*, **13**, 22–7.

Payne, J. and C. Prudente (2008), *Orangutans: Behaviour, Ecology and Conservation*, London: New Holland Publishers.

Polasky, S., E. Nelson, J. Camm, B. Csuti, P. Facker, E. Lonsdorf, C. Montgomery, D. White, J.R. Arthur, B. Garber-Yonts, R. Haight, J. Kagan, A. Starfield and C. Tobalske (2008), 'Where to put things? Spatial land management to sustain biodiversity and economic returns', *Biological Conservation*, **141**, 1505–24.

Ricardo, D. (1817), *The Principles of Political Economic and Taxation* (reprinted 1955), London: Dent.

Rijksen, H.D. and E. Meijaard (1999), *Our Vanishing Relative: The Status of Wild Orangutants at the Close of the Twentieth Century*, Dordrecht: Kluwer.

Schilthuizen, M., T.-S. Liew, B.B. Elahan and I. Lackman-Ancrenaz (2005), 'Effects of karst forest degeneration on pulmonate and prosobranch land snail communities in Sabah, Malaysian Borneo', *Conservation Biology*, **19**, 949–54.

Singh, G. (2008), 'Sustainable development of peatland for oil palm', paper presented at *The International Symposium and Workshop on Tropical Peatland*, Kuching, Sarawak Malaysia, 19–22 August 2008.

Singleton, I., S.A. Wich and M. Griffiths (2008), '*Pongo abelii*', in *IUCN 2010. IUCN Red List of Threatened Species. Version 2010.4* [Online]. Available at: http://www.iucnredlist.org. (Accessed May 9, 2011).

Singleton, I., S.A. Wich, S.J. Husson, S.S.U. Atmoko, M. Leighton, N. Rosen, K. Traylor-Holzer, R. Lacy and O. Byers (2004), *Orangutan Population and Habitat Viability Assessment: Final Report*, Apple Valley, MN, USA: IUCN/SSC Conservation Breeding Specialist Group.

Swarna Nantha, H. and C.A. Tisdell (2009), 'The orangutan–oil palm conflict: Economic constraints and opportunities for conservation', *Biodiversity Conservation*, **18**, 487–502.

Tisdell, C.A. (1990), 'Economics and the debate about preservation of species, crop varieties and genetic diversity', *Ecological Economics*, **2** (1), 77–90. Reprinted in C.A. Tisdell (2002), *The Economics of Conserving Wildlife and Natural Areas*, Cheltenham, UK and Northampton, MA, USA: Edward Elgar Publishing.

Tisdell, C.A. (1991), *Economics of Environmental Conservation*, 1st edn, Amsterdam: Elsevier Science Publishers.

Tisdell, C.A. (2005), *Economics of Environmental Conservation*, 2nd edn, Cheltenham, UK and Northampton, MA, USA: Edward Elgar Publishing.

Tisdell, C.A. and H. Swarna Nantha (2008), 'Conservation of the proboscis monkey and the orangutan in Borneo: Comparative issues and economic considerations', in V.K. Gupta and A.K. Verma (eds), *Perspectives in Animal Ecology & Reproduction*, 5, Delhi: Daya Publishing House, pp. 225–50.

Wich, S.A., E. Meijaard, A.J. Marshall, S.J. Husson, M. Ancrenaz, R.C. Lacy, C.P. Van Schaik, J. Sugardjito, T. Simorangkir, K. Traylor-Holzer, M. Doughty, J. Supriatna, R. Dennis, M. Gumal, C.D. Knott and I. Singleton (2008), 'Distribution and conservation status of the orang-utan (*Pongo* spp.) on Borneo and Sumatra: How many remain?', *Oryx*, **42**, 329–39.

13. Property rights in non-captive wildlife and biodiversity conservation

13.1 INTRODUCTION: IMPORTANCE AND CONTEXT

In order to reduce the rate of human induced biodiversity loss of wild species, it has become increasingly important to stem this loss on private land. Some writers believe that granting landholders commercial property rights in wildlife will be effective in dealing with this matter and will result in the preservation of biodiversity. This chapter explores this view using economic theory. In doing so, it takes into account the total economic valuation concept. While the granting of commercial property rights is found to be effective for conserving some species, it is predicted to be a complete failure as a means of conserving other species. Furthermore, its potential for success in different regions of the world varies. Here, particular attention is given to the economics of utilization and conservation of non-captive fugitive (or mobile) wildlife.

Considerable concern has been expressed in recent decades about continuing human induced biodiversity loss in the wild (May et al., 1995; Ehrlich, 1995) as well as a loss of genetic diversity in domesticated and cultivated species (Tisdell, 2003a; Tisdell, 2014; see also Chapters 5 and 6 of this book). Several factors have contributed to human induced biodiversity decline of non-captive biota. These factors include appropriation of wildlife habitats by humans, and their alterations to more productively serve human economic goals (Swanson, 1994a; 1994b; 1995), competition of humans with other species for food resources and their other means of subsistence (for example, harvesting of animals or plants by humans at the lower end of the food chain, thereby depriving higher order species of food) and the direct over-harvesting of wild species by humans. Other factors include the introduction of invasive alien species by humans and human induced climate change.

Most of the world's land area is now in private hands. For example, Benson et al. (1999) point out that two-thirds of the land in the USA is privately owned. Most of the globe's natural area has been subject to some conversion by humans. Consequently, the continuing survival of many species, especially those that have a large home range or are migratory, depends on their ability to use private or tribal lands without undue molestation or loss of habitat (Bandara and Tisdell, 2002). If a species is harvested on such land, its continuing survival requires its rate of harvesting to be sustainable.

It might also be noted that tribal lands are significantly large in some parts of the world, although not in most. For example, the major proportion of land in Northern Australia belongs to Aboriginal tribes, their share being especially high in the Northern Territory, the far north of Western Australia and Cape York in Queensland. Similarly, in the north-east of Canada, most land belongs to Inuits. Comprehensive consideration of the impact of property rights, attenuated or otherwise, on the conservation of wildlife requires attention to both private and tribal decisions about land use. However, here I would like to concentrate principally on the analysis of decisions by private landholders and assume (unless otherwise stated) that the primary aim of their commercial activity is one of private profit maximization.

With increased support for market-based liberalism in recent decades, claims have become more common that wildlife located on private land and private property is an effective means of conserving wildlife and biodiversity. Proponents point to the apparent conservation successes that have arisen from the granting of private property rights to private landholders in southern Africa. For example, Muir-Leresche and Nelson (2000) state that 'the evidence seems to be overwhelming that the existence of private-property rights to wildlife is advancing the goal of conservation of biodiversity in Zimbabwe and other southern African nations'. Seasholes (2002) claims that:

> the creation of property rights in southern Africa has been a stunning success story, a real win–win situation in which landowners and wildlife have both benefited. After all, this makes sense. Property rights theory tells us that the creation of (private) property rights tends to lead to a better world.

It does seem that the granting of private property rights to landholders in southern Africa has been successful in increasing populations of some species of wildlife. The black rhinoceros is an often given example. However, most of the wildlife concerned is enclosed in one (or in some cases more than one) private property. Therefore, it can be regarded as captive rather than non-captive and to some extent, farmed. It also appears that the wildlife involved includes the species for which considerable use value can be commercially appropriated by private landholders. However, one should be cautious about transferring the southern African experience to other countries and regions.

This is because the cost of enclosing wildlife species varies with the species and it may be more costly in some regions than others to do this because of the nature of the wildlife involved. For example, large species of kangaroos in Australia can jump over very high fences. But probably more significantly, the use value of wild species (and the amount of their total economic value) that can be appropriated by landholders, if granted property rights, varies between regions. For example, many of Australia's native mammals are costly to enclose, are considered to be of limited value for game hunting,[1] and are relatively small and nocturnal, thereby reducing their value in open areas for viewing.

One might also be led to ask if the southern African experiment of granting private property rights to wildlife is so successful, why are national parks and protected areas in southern Africa not being privatized? One reason could be that there are still significant market failures present that cannot be eradicated. Failures of this type have been identified by Hall et al. (2000), Erickson (2000), Freese and Trauger (2000) and Isaacs (2000).

In the light of seemingly exaggerated claims for the ability of private property rights to conserve wildlife and biodiversity, it is important to clarify the position as far as non-captive wildlife is concerned by taking into account the implications of established economic theory and its modern developments for this issue. The matter is more complex than it might appear to be at first sight and it becomes apparent that some economic theories (Swanson, 1994b; Skonhoft, 1999) advocating a private property approach to wildlife conservation require qualifications.

The economic analysis of the utilization and conservation of wildlife on private lands and the question of whether private property rights are likely to be effective in conserving non-captive wildlife, have been given due consideration by economists for around half a century. Ciriacy-Wantrup (1968) suggested that assignment of private property rights to landholders in fugitive wildlife species is likely to be ineffective in ensuring their conservation and optimal utilization. He expected failures to occur similar to those that arise in the case of open access resources because of the mobility of the species involved.

Colin Clark (1973; 1976) discovered that even when wildlife or natural biota are immobile or non-fugitive, for example natural stands of timber, private commercial use could be expected to eliminate their stock if the growth in their net economic value is less than the rate of interest.[2] Clark (1973) demonstrates that if a landholder aims to maximize his/her present discounted worth, extinction of a species can occur if its price exceeds the per unit cost of harvest for all the levels of the stock of a wild species on a property and the rate of interest is sufficiently large. The net present value of the business can be increased by realizing all the farm stock of the species and investing the realized economic gain in the capital market. However, elimination of a wild species on private land can even occur when the rate of return from commercial private utilization of a species exceeds the rate of interest because there may be an even more profitable use of the land requiring the extinction of this species. In other words, opportunity costs (alternative levels of return) forgone by a landholder as a result of conserving a wildlife species need to be considered by a landholder intent on wealth maximization.

Nevertheless, commercial use of relatively immobile wildlife species (of which captive wildlife is a special case) can, on occasions, result in their conservation because it is the most profitable form of land use and the return on this use is not less than the rate of interest. In such cases, failure to give property rights to landholders for commercial use of species can result in elimination of these species on private land. Conversely, granting private property rights to such wildlife is likely to result in their conservation. The South African situation accords with this case.

In considering the extent to which private property rights of landholders in non-captive wildlife can contribute to the conservation of wildlife, the following issues will be investigated in turn: the total economic value of wildlife and its conservation prospects, variations in the patterns of mobility of wild species, basic consequences for conservation of the private appropriation of non-consumptive use rights in wildlife, and similarly, the consequences of consumptive use rights in wildlife on private land. A brief critical discussion of relevant economic articles is followed by a discussion of global rights in wild genetic resources and concluding comments.

13.2 TOTAL ECONOMIC VALUE OF WILDLIFE AND BASIC PROSPECTS FOR CONSERVATION AS A RESULT OF COMMERCIAL USE OF WILDLIFE: IMPLICATIONS OF USE AND NON-USE VALUES

The view has become widely accepted that wildlife and natural areas can be considered to have a total economic value consisting of use values and non-use values. Proponents of the concept include Albani and Romano (1998). Economic use value is generally divided into direct and indirect use value. Indirect use value may, for example, involve the use of wildlife for illustrating books and television documentaries.

Direct use of wildlife may be consumptive (as in the case of recreational hunting or fishing, or the capture of animals for meat and other products) or non-consumptive, as in the case of the use of animals for viewing by tourists. In most cases, use values are marketable because exclusion is possible and maybe profitable.

By contrast, most non-use values are not marketable. For wildlife species, these include existence, bequest and option values. These attributes have the characteristics of pure public goods. Hence, many wildlife species are mixed goods in the sense that they have attributes making them private goods as well as attributes that make them pure public goods.

The relative importance of these attributes varies according to the type of wildlife species considered. The total economic value of some fish species, for example mullet, seems virtually to be accounted for entirely by their direct consumptive use value. On the other hand, most of the value of some species consists of their non-use value, for example, whale species in Australia, tree kangaroos (Tisdell and Wilson, 2004), and in Sri Lanka, the Asian elephant (Bandara and Tisdell, 2003).

In principle, under ideal conditions, it would be possible for landholders to market all the use value of wildlife species. The question can then be posed: in the absence of mobility of wildlife, would there be an economically optimal degree of conservation of wildlife taking account of the total economic value of species of wildlife? If the Kaldor–Hicks approach to socio-economic optimality is accepted for the purpose of this analysis, the answer is that economic optimality will not be achieved for the conservation of all species if landholders are only able to appropriate their use value. Market failure is, therefore, a problem. This can be illustrated by Figure 13.1. The line ABCDF represents marginal use generated by the stock of a species or focal wildlife species on a property, and curve NRSCF represents the marginal value of the total economic value of benefits from the stock of the species on the property. The difference between curve NRSC and line ABC represents the marginal non-use value of the stock of the species.

In the absence of human use of the land area, assume that the population of the relevant wildlife species is x_5. Suppose that on this land the marginal opportunity cost of maintaining the population of the species is as shown by line GDH. Then, given private rights to use this species, profit maximization would result in a reduction of its stock from x_5 to x_4. This is optimal in this particular case from a socio-economic point of view because non-use benefits are infra-marginal (Tisdell, 2005a, pp. 58–62).

However, if the marginal opportunity cost of conserving the population of the species on this property is higher (say corresponds to JBK), socio-economic optimality is not achieved. Only x_1 of the population of

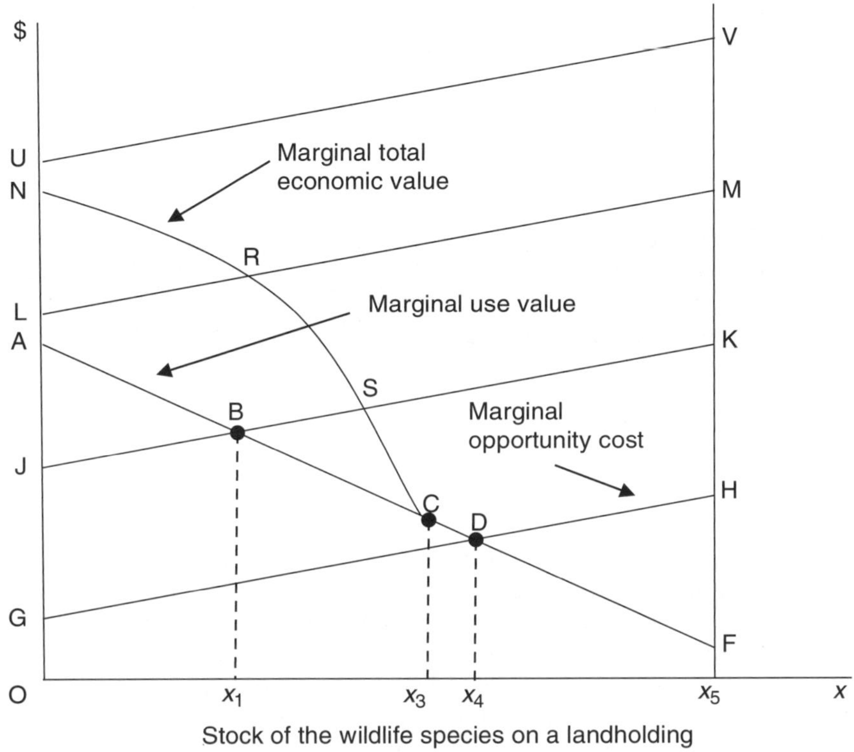

Figure 13.1 *When non-use economic values are important, commercial rights for landholders in wildlife are unlikely to result in the optimal level of conservation of wildlife from an economic point of view: in that respect, one should consider whether non-use values are infra-marginal or not*

the species is conserved, whereas the population corresponding to S, x_2, is optimal from a collective viewpoint. Nevertheless, the species survives, *provided* that the aggregate level of populations of the species surviving on all properties is sufficient and their interdependence is such as to ensure the conservation of at least a minimum viable population of the species.[3] If instead, the marginal opportunity cost is as shown by line LRM, the population of the species is wiped out on the property even though a population corresponding to point R would be Kaldor–Hicks optimal. On the other hand, if the marginal opportunity cost curve is UV, extinction of the species on the property is optimal from a strictly socio-economic viewpoint. This is based on the assumption that economic willingness to pay or the total economic benefit obtained from a species, as judged by humans, should be the arbiter of the continuing existence of species. This ethic is not, however, acceptable to everyone.

We can observe from Figure 13.1 that, keeping other things equal, the lower is the marginal use value relative to marginal non-use value of the wild species, the greater the likelihood of market failure. If a species has substantial non-use value, market failure will occur unless private use results in a level of population for which its non-use value is infra-marginal. It is clear that species with little or no use value have little chance of survival on private land even when they may have a very high non-use value and therefore a high total economic value.

13.3 THE NATURE OF THE MOBILITY OF WILDLIFE AND ITS ECONOMIC CONSEQUENCES

The geographical mobility of wildlife can take different forms and patterns, and some species are more mobile than others. The pattern of mobility can influence the commercial utilization of wildlife on private land.

Some natural biota are sedentary in some stages of their life and display dispersion or movement geographically, only in some part of their life. This is true for many plants. For example, only their seeds may be spread by wind, water or other means. Otherwise, they remain fixed where they grow. A similar pattern exists for some molluscs, such as oysters, and for coral species. These species are only mobile at one stage in their lifecycle.

Most wild species are, however, relatively mobile during their whole lifetime. This is true, for example, of most mammals. Nevertheless, mammals differ significantly in the range of their movements and the pattern of these. The smaller the range of movement of a group of animals in relation to the size of a private property, the more likely are their user costs and economic benefits to be internal to the property. In turn, one might expect in such cases that the landholder would be more likely to take account of the full range of user cost involved in utilizing or husbanding this group of wild animals, if granted commercial property rights in these.

In most cases, it is technically possible to enclose wildlife species within a property. But frequently, this is too costly to be economic. Moreover, for some species their enclosure would result in their available range being too small to enable them to survive on the available habitat. Cooperation between several adjoining landholders could overcome the externality problem involving inter-property mobility of animals. How well such cooperation will work requires further consideration.

Patterns of movement of biota vary considerably. Geographical variation in this movement and differences in its time patterns can have different economic consequences for the private utilization of wildlife, and for efforts by private entities, NGOs and government bodies to conserve wildlife species.

Some species follow established migratory paths in response to seasonal changes and variations in weather conditions. The characteristics of the population of a migratory species normally alter with its migratory movements. For example, its age distribution, the size and weight of its individual members, the total weight of its population, and so on, often change. When migratory species pass through several jurisdictions or there is open access to them at some points in their migration and life-cycle, harvesting of them usually does not maximize their sustainable economic yield. This has proven to be a continuing problem for the harvesting of tuna, for example (Tisdell, 2003b, Ch. 1).

Some migratory species have an attribute which can be conducive to their conservation on land managed by private entities, NGOs or government bodies: the adults of these species return, for breeding purposes, to the locality where they were born. This is true, for example, of some turtle species and some species of birds such as the yellow-eyed penguin and the royal albatross. The rebound or boomerang effect from protecting these species at their breeding sites can be substantial. Protecting these species at their natal sites can increase the numbers returning to these sites, if this is not offset by increased mortality rates off-site. Examples and analyses of several cases of this type can be found in Tisdell and Wilson (2012).

The cases considered in Tisdell and Wilson (2012) include the conservation of some species of penguins, the royal albatross and species of marine turtles. An interesting case considered in Tisdell and Wilson (2012) is the conservation of the yellow-eyed penguin in New Zealand (Chapter 13). Some private landholders engage in protecting nesting sites of these penguins for economic gain. In addition, the Yellow-eyed Penguin Society, an NGO, undertakes conservation activities to conserve these penguins at their breeding sites but this NGO is not commercially oriented. Another example, in Tisdell and Wilson (2012, Chapter 12) involves the conservation of little (fairy) penguins in Victoria, Australia, at a breeding site on Port Phillip Bay. The conservation activities supporting this species at this site are managed by an NGO which obtains substantial revenue from payments by tourists to watch the Penguin Parade. The conservation of the royal albatross at its rookery on Taiaroa Head in New Zealand is financed in a similar manner by an NGO and benefits from the rebound effect from conservation of this species (Tisdell, 1990, Ch. 7; Tisdell and Wilson, 2012, Ch. 13). Examples involving marine turtles and the rebound effect are also discussed in Tisdell and Wilson (2012). They are the conservation of sea turtles by a government body, the Queensland Parks and Wildlife Service at the Mon Repos Rookery in Queensland (Chapter 9) and their conservation at breeding sites at Sri Lanka by small NGOs and private enterprises (Chapter 10). All of these cases involve a payment for the non-consumptive use of wildlife for tourism, except in the case of the Yellow-eyed Penguin Society. This society relies basically on donations/grants to conserve yellow-eyed penguins for their non-use value.

Off-site, these species face varied risks to their population. For example, marine turtles are harvested by some subsistence communities for food, may die from ingesting plastic bags and can be a part of the accidental catch in marine fishing. Albatross may be baited (hooked) incidentally by attempting to eat fish caught on fishing long lines. Penguin species when offshore may be deprived of fish caught by fishing vessels and suffer a reduction in their populations.

Note also that some species are territorial and keep to established territories in their travels. The size of these territories and patterns can vary. Sometimes landholders' individual properties can encompass the movements of groups of these species, but not always.

13.4 MORE ON NON-CONSUMPTIVE USE RIGHTS IN WILDLIFE

Many landholders are able to obtain de facto non-consumptive use rights to non-captive wildlife. In some cases, this is used to support tourism and recreational enterprises. Examples from Australia include the use by O'Reilly's of wild birds as a tourist attraction. O'Reilly's is a private resort located within Lamington National Park in Queensland on freehold land. Birds visit from the surrounding park and are viewed and fed by visitors with food purchased usually from an outlet in O'Reilly's resort centre. If feeding wild birds attracts more tourists to the site, commercial benefits to O'Reilly's are not so much a consequence of sales of grain to feed the birds but extra business generated in their nearby tourist shops as well as sales of extra accommodation.

Whether this results in greater conservation of wild birds is not known. It probably favours the populations of parrots that come for feeding. The process most likely adds to political support for conservation of this national park.

In an urban setting, on the Gold Coast of Queensland, Currumbin Wildlife Sanctuary feeds rainbow lorikeets. They are attracted to the site at feeding time in considerable numbers. A similar feeding programme has started at Lone Pine Koala Sanctuary in Brisbane. This tends to favour rainbow lorikeet populations relative to competing species of other birds in the nearby areas.

At Lone Pine Koala Sanctuary, catfish in the nearby Brisbane River are also fed on bread at regular times, and on Moreton Island near Brisbane, hand-feeding of porpoises is a tourist attraction, as it is also at Monkey Mia in Western Australia. While these activities are non-consumptive of the wildlife involved, some environmentalists object to those practices on the grounds that the animals become dependent on humans, and in some cases the local composition of populations of species may alter. Furthermore, hand-fed animals may become aggressive towards humans.

The farm-holiday sector may also benefit from the presence of non-captive wildlife. Depending upon how important seeing wildlife is as an attraction for farm visitors, this can have positive consequences for wildlife conservation. It may, for example, encourage landholders to leave habitat intact that will attract wildlife of interest to those visiting for farm holidays. But, of course, landholders will take no account of non-use values in their conservation decisions.

Another type of non-consumptive use of wildlife is tourism based on 'headstart' programmes designed to increase the stock of targeted wildlife. In Sri Lanka, 'private' sea turtle nurseries are operated for this purpose. These nurseries have de facto rights in their collected turtle eggs and the resulting hatchlings, before the hatchlings are released to the ocean. Their financial viability depends on the willingness of tourists to pay to visit their hatcheries, on donations by tourists and sales of souvenirs at those hatcheries. How effective such nurseries are likely to be in increasing or sustaining populations of sea turtles in the wild is discussed by Tisdell and Wilson (2005). Several other case studies of the consequences for biodiversity conservation are presented and discussed (as was mentioned above) in Tisdell and Wilson (2012). In the cases considered there, all the landholders adopt measures to conserve focal species and rely mostly on tourism for their economic viability.

13.5 CONSUMPTIVE USE RIGHTS IN NON-CAPTIVE FUGITIVE WILDLIFE

Most economic theory dealing with the utilization of open access resources has been developed on the assumption that consumptive use is the sole purpose of management and utilization of resources (Tisdell, 1972). This is particularly evident in the early literature on fisheries economics

(Gordon, 1954; Smith, 1969). Although for some species, consumptive use value may constitute most or all of their economic value, for many this is not so. For example, respondents to interviews in Sri Lanka indicated that almost 70 per cent of the amount they said they were willing to contribute towards the conservation of the Asian elephant was based on the non-use value of these elephants (Tisdell, 2014, p. 72). Furthermore, many other species have little or no use value but have a high non-use value, for example Australian tree kangaroos (Tisdell, 2014, p. 93).

It should not be forgotten, however, that the economic value attributed to different species has shown considerable alteration with the passage of time. In many cases, significant changes have occurred in their economic use value as well as their non-use values as a result of economic development and changed social attitudes. The nature and reasons for changes in the ascribed non-use values of wild species are discussed, for example, in Tisdell (2014, see especially Ch. 10). In addition, major changes have occurred historically in the use value of some wild species. For instance, the use value of the Asian elephant (*Elephas maximus*) has altered dramatically in the last few centuries.

The Asian elephant was once mainly prized in Asia (when domesticated) for its contribution to warfare, as a beast of burden, its ability to clear land of vegetation and so on. Consequently, rulers on the Indian subcontinent adopted measures to conserve its wild populations (Kautilya, 1961; Tisdell, 2005b). Today, the Asian elephant is not used in warfare because its functions have been replaced by machines. Similarly, mechanical substitutes have replaced its value for most of its productive purposes, even though it still finds some use for tasks such as snigging logs and lifting these onto trucks. Its main use value now is for tourism but the largest contributor to its total economic value appears to be its non-use value (Tisdell, 2014, p. 72).

As economic development has occurred, the consumptive economic value of many, but not all, wildlife species has declined. Take the case of tree kangaroos. Their harvest by Australian Aborigines once contributed to the threat to Australian species of tree kangaroos but is no longer considered to be a significant threat because Australian Aborigines now have higher incomes and preferred substitutes are available for the meat and fur of these animals, meaning that Aborigines rarely harvest them now. However, in Papua New Guinea, a low income country, the hunting of species of tree kangaroos for subsistence purposes continues to be a major threat to their survival (Martin, 2005, Ch. 10). On the other hand, with economic development, the demand for the consumptive use of wild species has, it seems, increased in some societies. As a result of increasing incomes, many Chinese, for example, appear to have increased their demand for the consumption of some wildlife species, especially for use in traditional medicine.

Product substitution has had a major influence on the declining consumptive use value of many wild species. For example, whales, seals and penguins no longer have much (if any) economic value for their oil because several low-cost economic substitutes are available, many of which have preferable attributes. At one time in the past, for example, penguins were captured and boiled down on Macquarie Island in sub-Antarctica for their oil.

The economic value of wild species varies considerably in the extent to which non-consumptive and consumptive values contribute to their total value. However, let us consider the case in which consumptive use value is the only relevant economic value of a non-captive wildlife species. Then let us consider the extent to which the economics of exploitation of a fugitive wildlife 'visiting' private land differs from the open access case.

Suppose that private (exclusive) use rights in a wildlife species are only held by the landholder while it is on his/her property. Consequently, there is no open access to this species. If harvest of the wildlife is profitable, the landholder will harvest it (if a single or composite factor is used to harvest it) by employing the factor up to the point where the marginal factor cost of its harvest equals the value (to him or her) of the marginal product of the wildlife harvested. This condition differs from an open access one because in this case employment (effort) for harvesting occurs up to the level where the marginal factor cost equals the value of the factor's average product. Rent dissipation occurs in the latter case and no consideration is given to user cost.

However, in the fugitive wildlife case, landholders may give some consideration to user costs even though this is likely to be less than is socially desirable. The two situations can differ in the extent to which an account is taken of user costs. Under open access, no consideration is given to user costs, whereas in the fugitive wildlife case, consideration may occur.

In the fugitive wildlife case, the future population of a species visiting a property, and consequently, future benefits of the species to the property holder, may be influenced by the landholder's current levels of harvesting of the wildlife and by his/her other land-use practices. Whether or not an influence is present on future economic benefits to the landholder as a consequence of his/her present actions, depends on the mobility pattern of the wildlife concerned and their ecological requirements. Thus a second difference emerges between the economic theory of exploitation of open access resources and that for the exploitation of fugitive resources.

This can be illustrated by Figure 13.2, where x signifies the quantity of the focal species, X, harvested by the landholder. The line ABC represents

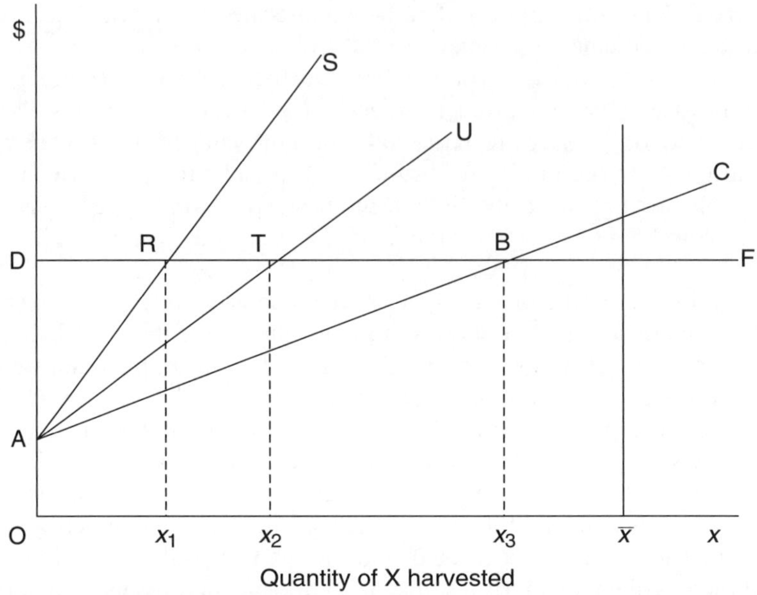

Figure 13.2 Diagram to illustrate that even when no account is taken of the rebound effect by a private property holder commercially harvesting a wild species (or if it is absent), a lower level of harvest occurs than if open access is the case: this diagram also shows that as the rebound coefficient increases, the economic incentive of the property holder to conserve the harvested species increases

the landholder's average costs of harvesting X when the user costs of doing so are ignored and line ATU is the corresponding level of marginal cost. The price per unit received for the harvested species is assumed to equal OD (and constant) and consequently, marginal revenue equals average revenue, as shown by line DF. Let the total population of the focal species on the property be designated by $\overline{x}$. This is a boundary constraint, namely, the maximum possible level of the harvest.

Several conclusions can be drawn from Figure 13.2. First, if the cost of harvesting the focal species is the same under conditions of open access as those experienced by the private property holder, then when exclusive rights are given to the private property holder to harvest a wildlife species on his/her property and market it, his/her harvest will be lower than under open access, even if the species is extremely fugitive. In the extreme case, there is a zero rebound coefficient for the harvested species. In this case, the quantity of the species harvested under open access is x_3. However, if the private property rights scenario applies, it is only x_2. If, on the other hand, the landholder experiences a positive rebound effect from limiting his or her harvest, then user costs should rationally be taken into account by the landholder in making harvesting decisions. Given a positive rebound coefficient, user costs are positive, and therefore the marginal costs of the current harvest inclusive of marginal user costs might be as shown by line ARS. Consequently, the landholder's optimal harvest of X falls from x_2 to x_1. The landholder's economic incentive to conserve the focal species increases with a rise in the rebound effect.

Even if a species is highly fugitive (mobile), the private property solution and the open access one differ significantly. Market failure is less pronounced in the former case, a point which is not clear from the perspective provided by Ciriacy-Wantrup (1968). However, there is no guarantee that the private property solution will result in the survival of the focal harvested species. For example, x_1 and x_2 can be so high that most or all the population of X on a property is harvested. If this is so on most or all of the properties where X occurs, consumptive use of the species *can result in the total population falling below a minimum viable level.* Consequently, permitting consumptive use of the species on private land without any limitations on the harvest can result in its demise in such cases.

Where the rebound effect is sufficiently large or the species is sufficiently sessile relative to the size of the property, it can pay the landholder to undertake investments favourable to the survival of the focal species. For example, this may take the form of changing vegetation cover so that the change favours the focal species and adopting measures to control the population of predators of this species. There is evidence that this is practised by some private property holders, NGOs and government bodies managing land to conserve wild species for conservation and tourism purposes (Tisdell and Wilson, 2012).

It is also important to consider the size of a landholding on which a landholder has exclusive rights to harvest a wild species. In many cases, the larger is the landholding relative to the movement pattern of commercially valuable wild species, the larger tends to be the amount of use value of a wild species able to be appropriated by the landholder. In other words, in these cases, externalities are much reduced. This, for example, is likely to apply to some large properties, such as large grazing properties in Australia, and to large 'protected' areas in Africa in which some harvesting of elephants is undertaken by the state, for instance in South Africa's Kruger National Park.

13.6 A DISCUSSION OF SOME RELEVANT ARTICLES ON COMMERCIAL RIGHTS TO WILDLIFE ON PRIVATE LAND

Articles which are particularly relevant to the above analysis are those by Swanson (1994b) and by Skonhoft (1999). Both emphasize that open access models developed mainly for the captive fisheries require modification in the case of the use and conservation of terrestrial wildlife. Swanson (1994b) emphasizes that land conversion for alternative commercial uses in preference to wildlife maintenance is always an important option in terrestrial land use. It results in destruction of habitat and loss of wildlife species. Skonhoft (1999) comes to a similar conclusion. Both authors suggest that such possibilities affecting conservation are uncommon in marine areas. However, with growth in mariculture, it should also be observed that conversion of marine habitats is also becoming more common.

Moreover, it is important to qualify the views of Swanson (1994b) and of Skonhoft (1999). Actually human use of land merely changes the range (type) of available habitats for wildlife species and these changes may favour some species. For example, the extension of cattle grazing in Australia favoured red kangaroos by creating marsupial lawns and providing them with extra watering points. Hence, the population of red kangaroos in Australia now exceeds that at European settlement, but some other species of kangaroos disappeared as a result of loss of suitable habitat. See for example, relevant chapters about kangaroos in Tisdell (2002). Nonetheless, it is probable that human use of land mainly reduces diversity of ecological niches or environments (Tisdell, 1999) and that this is the main reason why, overall, wild biodiversity declines.

As in the open access case, Swanson (1994b) treats wildlife as a private good and considers only its consumptive value. In any case, his analysis leads him to the conclusion that it is only 'the high-value, high-growth resources that will ultimately survive' and that policies that restrict the economic gains of landholders from such species can only ultimately accelerate their extinction (1994b). However, in light of the above analysis, a private property solution to conserving such species is likely to be ineffective if a large proportion of their population moves off each individual private landholding and there is a small rebound effect. The economic motivation to harvest a high proportion of these species while on individual properties is high in such cases if the market value of the harvest is high. Furthermore, it is likely in such cases that the market price for each unit of harvest of such a species will rise as its population declines and eventually will drive the species to extinction. The dynamics are similar to those in the open access case as discussed, for example, by Tisdell

(2005a, pp. 137–8). In this private property case, each landholder is guided only by his/her immediate profit in making harvesting decisions. On the other hand, if the harvested species are sufficiently sessile or the rebound coefficient is high enough, allowing landholders to appropriate economic gains from high-value high growth species can be an effective policy for promoting their conservations.

Although Skonhoft extends Swanson's model, both models formally treat wildlife as being non-fugitive. As can be seen from Equation (3) of Skonhoft (1999), the non-consumptive benefit (use value) of wildlife from tourism appropriated by the landholder is included in his analysis as well as the nuisance value of the focal species. However, the nuisance value recorded is only that internal to a landholding, not external damages caused by movement of animals.

Basically, Skonhoft models this aspect in this way because he does not give specific attention in his modelling to the movement of animals between properties. This he does despite his observation that 'the motivation for including those nuisance costs can be found from African wildlife where large mammals frequently destroy agricultural production of agro-pastoralists living in the proximity of wildlife habitats' (Skonhoft, 1999). However, this involves an externality as a result of movements of wildlife and is not captured by Skonhoft's analysis. Nevertheless, an alternative possible interpretation of Skonhoft (1999) is that landholdings in his model are combined in such a way that all wildlife movements are internal to the combined property, and cooperation occurs between landholders to maximize joint benefits from the utilization of wildlife. But this is a special rather than a general case.

Much of the current state of economic analysis of this subject is unsatisfactory because it fails to explicitly model and allow for movements of wildlife between properties. Current analysis seems only useful for species with zero mobility or very low mobility between landholdings, or for those that can be enclosed economically on a private property. It may also be useful if a few landholders are willing to cooperate fully to manage mobile wildlife.

13.7 GLOBAL RIGHTS TO TRADE IN WILDLIFE

Various aspects of global property rights in wild and heritage germplasm were discussed in Chapters 10 and 11. It was concluded that the granting of global property rights in such germplasm is unlikely to provide much economic incentive for promoting biodiversity conservation. This is likely to be so even if states (national governments) are the main beneficiaries from these rights. They are in a strong position to gain because they can draw on biodiversity on their own (large) landholdings, such as national parks. The shortcomings of this mechanism are likely to be magnified when the biodiversity to be conserved is located on private property and fees for the use of the germplasm involved are collected by the state and distributed to landholders to provide an economic incentive to private landholders to conserve germplasm that may have future use and commercial use value. This is because the sum of money available for distribution is likely to be small and the transaction costs (agency and administrative costs) relative to this sum are likely to be high. Similar problems arise in making payments to private landholders for conserving wild germplasm to those that occur for paying these landholders to conserve heritage germplasm. These problems were identified in Chapter 10 of this book when the International Treaty on Plant and Genetic Resources for Food and Agriculture (ITPGRFA) was discussed.

Apart from matters involving property rights in wild germplasm, rights to trade internationally in wildlife species, or in parts of these species or products derived from them, can have important implications for the conservation of wild biodiversity. Whether or not allowing (international) trade favours the conservation of wild species having substantial commercial consumptive value depends upon whether or not there is open access to them or if they are fugitive resources. It varies with the nature of their mobility and rebound effects in relation to properties able to gain from their exploitation. As is well known, in the open access case, market extension increases the likelihood of such species being driven to extinction by over-harvesting (see for example,Tisdell, 2005a, Ch. 6), and this may happen in the case of some fugitive species.

However, restricting pre-existing trade in such species is not necessarily enough to ensure their survival. It is only likely to be effective if the trade restriction does not result in landholders converting the habitat of these species to alternative uses or does not reduce the willingness of landholders to protect these species from poachers because of the reduced economic value of the species to landholders. If the costs of land conversion are so high relative to the economic returns from it that they make this conversion uneconomic, landholders will have no economic incentive to alter habitats if they are no longer able to benefit from trade in wildlife species when the market is restricted. In the latter case, restricting trade in an endangered species (and its products) that possess significant market value can be an effective policy for conserving the species, but in the former case this policy is unfavourable to a species' survival. In the latter case, it results in a reduction in the quantity of species harvested and no change in its available habitat. In the former case, while the policy may result in a reduction in the quantity of the harvest, it triggers habitat conversion, which is unfavourable to the species' survival.

When assessing policies denying market access for harvested wild species or their products as a means for nature conservation, one must consider not only:

- the impact of these policies on the quantity harvested of the species, but also
- their effect on habitat conversion.

The latter effect often seems to be overlooked in formulating conservation policies. While in the case of many ocean-going species (such as whales), restricting trade in their products is likely to be effective because this reduces the economic incentive to harvest them *and* there is no economic incentive to convert their habitats to alternative uses. This seems less likely to be so in the case of terrestrial wild species.

At the global level, international agreements about how to protect wildlife species have adopted potentially conflicting approaches to market access. The Convention on International Trade in Endangered Species (CITES) basically adopts the view that restricting trade in endangered species is the most effective way to conserve them. On the other hand, the Convention on Biological Diversity (CBD) is based on the view that when the use value of wild species (and germplasm) can be appropriated by landholders, this is usually an effective means to encourage the conservation of wild species and wild biodiversity. However, the effectiveness of these different approaches to nature conservation vary with individual cases, and in many instances, neither provides adequate means for nature conservation. The CITES approach is based on market denial for endangered species or their products and the CBD relies on granting market access as an instrument for nature conservation. Some highly valued species and biodiverse systems have a high economic value but little commercial value, and are unlikely to be conserved other than by the state or by collective action by NGOs. Neither the rights to access markets nor its denial result in this case in the conservation of these species.

There are interesting cases in which rights to trade in endangered species or in their products are able to make a positive contribution to their conservation. In some cases, access to international markets is important in supporting the consumptive economic use value of wildlife species. Let us briefly consider a couple of cases.

The hunting of the African elephant (*Loxodonta africana*) has been a controversial subject. This elephant is a mixed economic good. Its economic value consists of positive and negative components. It has non-use values (for example, for existence value), non-consumptive use value (for tourism) and consumptive use value (for ivory, hides and meat). At the same time, it is an agricultural pest. Moreover, in the territories that it now occupies, its population growth tends to outstrip the availability of its food resources and, if not culled, this can lead to destructive changes in vegetation cover, with negative impact on the populations of other wildlife species, and the control of elephant populations by starvation (Webb, 2014, p. 183). Deciding on a socially optimal population of African elephants requires all of its economic attributes to be taken into account. The 1989, CITES ban on elephant ivory exports was a particularly contentious issue because it undermined the economics of sustainable management of elephant populations in African countries adopting sustainable use programmes, such as South Africa and Botswana.

In the case of Kruger National Park in South Africa, apart from the presence of elephants contributing to its revenue from tourism, culled elephants are processed for ivory, hides, meat and other by-products. Whether or not they have been over-harvested for consumptive use relative to their value for other purposes is not clear. What is clear, however, is from the point of view of the managers of Kruger National Park and the South African Government, the elephant population in this park is a valuable economic resource. Its use value for tourism and for consumptive purposes provides a strong economic incentive to conserve it sustainably. Because Kruger National Park covers a large area and exclusion of poachers is relatively effective, a high degree of state property rights is maintained in species located within it.

There is considerable evidence that a blanket ban on all ivory exports can be expected to undermine the conservation of African elephants (Barbier et al., 1990). Moreover, lack of managed culling of elephants in national parks (to which they are now mostly confined) can result in serious damage to the vegetation cover of these parks due to increased elephant populations and this could potentially have negative impacts on wildlife tourism in Africa. Overall, land use has been so much altered by human beings that adopting a laissez-faire approach to elephant conservation is an ineffective conservation strategy. This is also so for some other wild species.

The managed use of resources provided by populations of wild saltwater crocodiles (*Crocodylus porosus*) in Australia has contributed to the sustainable conservation of this species for the reasons discussed in Tisdell (2014, Ch. 9). This case differs from that of the utilization of the African elephant because it relies heavily on the collection of the eggs and hatchlings of those crocodiles which are utilized by commercial crocodile farms in Australia to rear crocodiles mainly for their skins, with their meat being a by-product. The difference is that the consumptive use value of the

African elephant is based primarily on the harvesting of adults from the wild, even though whole family groups are usually culled.

The commercial viability of the Australian crocodile industry is heavily dependent upon its having access to international markets (Tisdell, 2014, Ch. 9). Prior to 1984, Australia was prevented by CITES (for a significant period of time) from exporting saltwater crocodile skins and products based on the use of this wild resource. However, in the mid-1980s, CITES agreed to allow exports of these products based on an acceptable ranching programme being submitted by the Northern Territory Government (Webb, 2014, Ch. 22). In this case, 'ranching meant restricting the wild harvest of eggs or hatchlings, with low survival rates in nature, to supply crocodile farms that would raise them for skin production' (Webb, 2014, p. 149).

By contrast, efforts to allow international trade in consumptive use of marine turtles have not been politically successful. This is so despite the potential of the farming and ranching of some turtle species for the consumptive use to be sustainable. Political and other reasons why the consumptive use of marine turtles has been opposed by conservation groups are outlined and are discussed by Webb (2014, Ch. 23). Tisdell (2014, Ch. 8) considers how public value, attitudes and policies have affected the conservation of the hawksbill turtle and how efforts to develop the farming of it have been stymied by the CITES ban on the exports of hawksbill shells (see also Webb, 2014, Ch. 23).

Even the farming of green turtles on the Cayman Islands (based on a closed breeding cycle rather than ranching) has been restricted in its commercial value by the inability of the farm there to gain access to international markets for the meat of these animals due to the opposition of some conservationists to this activity. This case is analysed in Tisdell (2005a, Section 6.5).

Whether or not policies for the sustainable consumptive uses of wild species are adopted depends heavily on biopolitics, as is evident from Webb's coverage of this aspect (Webb, 2014) and also on public attitudes, as is evident in the findings in Tisdell (2014, Chs 15 and 16). However, public attitudes are subject to influence by a variety of pressure groups, some of which are likely to exploit or foster these influences for their own economic advantage.

13.8 FURTHER DISCUSSION AND CONCLUSION

It has been shown that economic conditions that would theoretically be satisfied in exploiting fugitive wildlife resources (and wildlife generally) can be expected to differ from those for open access if landholders are given exclusive commercial property rights in wildlife on their property. Those property rights prevent rent dissipation from entry, which would otherwise occur with open access. Furthermore, while private user cost is zero under open access, it can be positive in the case of fugitive wildlife resources. This depends upon the mobility patterns/amount of mobility of the wildlife concerned and their general ecological requirements.

By granting commercial property rights in wildlife, some species that would disappear in the absence of such rights (given existing land rights,) may be saved from extinction. However, this is not so for all species. Some will still disappear, including some species that it would be socially optimal to save if account is taken of their non-use value. Some will disappear that have considerable use value but are highly mobile and that possess unfavourable movement patterns from a landholder's point of view. This is because in such cases the landholder's user cost approaches zero. Finally, even if landholders can appropriate all economic benefits from a wildlife species for its use value, it may be uneconomic to conserve it because competing alternatives provide a higher economic return.

Hence, we can conclude that while the provision of commercial property rights in non-captive wildlife species can lead to the conservation of some species that would otherwise disappear, this policy is incapable of saving from extinction some species that, from a strictly economic perspective, should be saved from extinction. It can even hasten the extirpation of some wild species that have a high consumptive use value, as was demonstrated. Consequently, it seems that this policy can only be effective in conserving some wildlife (in some regions, probably only a few species) that otherwise would disappear. In relation to the conservation of biodiversity, this policy of granting commercial property rights to landholders is best considered to be a supplement to state provision of protected areas and their supply by NGOs, rather than an alternative. It is also necessary to be aware that while some wildlife species using private lands stand a better chance of being conserved if they are commercial private property, others will stand a better chance (if not a perfect one) of survival if they remain protected as state or crown property. While errors have been made in the past by retaining most wildlife on private land as protected state or crown property, it would also be an error to believe that wildlife on private property everywhere will stand a better chance of being conserved, if it is made private commercial property.

Furthermore, even if this were not so, the question still remains unresolved of whether the preservation of biodiversity ought to be determined to maximize the total *economic value* of wildlife species. Concerns have been raised that such an approach (based on economics) is unethical as well as being flawed because of empirical limitations, such as the impossibility of accounting fully for the preference of future generations (Gowdy, 1997; Hampicke, 1999a; 1999b).

From the above analysis, it can be seen that the granting of property rights to landholders to wildlife present on their land is of limited value for conserving wild biodiversity. It is limited in (*at least*) three ways:

- The financial growth rate of a wildlife resource may be too low compared with the return on capital elsewhere, for the landholder to have an economic incentive to conserve wild species, even if it or its products are marketable.
- Not all economic values, especially non-use economic values, of some species are marketable and this leads to economically inefficient decisions.
- Migration (movements) of wildlife can result in ineffective private property rights in wildlife species even though rights *de jure* are granted. However, the larger the size of properties, the less likely this is to be a problem, keeping other things equal.

Further limitations are considered in the next chapter.

To some extent, the last problem just mentioned could be overcome by several adjoining property owners cooperating in the management and utilization of migratory wildlife. This is being experimented within South Africa but there are still problems in evolving communal rules acceptable to all participants, as was once pointed out to me by one South African farmer involved in such an arrangement. The prospects for efficient communal management of common resources have been discussed, for example, by Bromley and Cernea (1989), Ostrom (1990) and Ostrom et al., (1994). There is, however, no guarantee that communal governance will be socially optimal (Tisdell, 1995) even though it may be superior to some centrally imposed alternatives intended to promote nature conservation.

The effectiveness for nature conservation of the presence (or denial) of the right to market wild germplasm and to harvest wild species for their consumptive commercial use was assessed. It was noted that CITES and CBD adopt different views on the latter matter. It was pointed out that the effectiveness of these different mechanisms for nature conservation varies according to the species involved and other conditions. In some cases, denying market access can help to foster nature conservation but in others the opposite result follows. However, even when the consumptive use and marketing of some endangered wildlife species seems to be able to support the conservation of particular wildlife species, such as African elephants and some species of marine turtles, their sustainable consumptive use may be opposed by most of the general public (Tisdell, 2014, Ch. 15) and by NGOs (Webb, 2014). This objection may be partly based on ethical objections to killing members of these species, doubts about the effectiveness of this approach for conserving the species involved, and emotional factors that might be 'exploited' by NGOs to raise funds.[4]

NOTES

1. In Australia native animals are the property of the state (the crown) and apart from a few exceptions, it is illegal to hunt them. Exceptions include the commercial hunting of large kangaroos (for their meat and skins), regulated by a permit system, and restricted harvesting of saltwater crocodiles in the Northern Territory for similar purposes (Tisdell, 2014, Ch. 9). Also Indigenous Australians can hunt native animals for their own use (Tisdell, 2014, Ch. 16). Recreational hunting of most native animals is illegal. On the other hand, no restrictions apply to the hunting of feral animals such as feral pigs (Tisdell, 1982), goats and buffalo.
2. It cannot, however, be concluded on the basis of partial microeconomic analysis that a low rate of interest necessarily favours the preservation of wild biodiversity and natural capital. One has to take into account the macroeconomic determinants of the interest rate, as is shown in Chapter 15. For example, a low interest rate can be associated with a high level of aggregate investment resulting in the accelerated loss of natural ecosystems.
3. This interdependence problem is not amenable to analysis by considering a single landholding in isolation. More attention is given to the problem in section 13.5 following Figure 13.2.
4. There is very limited overlap between the material in this chapter and one of my previously published articles, namely C.A. Tisdell (2004), 'Property rights in non-captive wildlife and biodiversity conservation', *International Journal of Global Environmental Issues*, **4** (4), 195–208. This article was published by *Inderscience*.

REFERENCES

Albani, M. and D. Romano (1998), 'Total economic value and evaluation techniques', in R.C. Bishop and D. Romano (eds), *Environmental Resource Valuation: Applications of the Contingent Valuation Method in Italy*, Boston: Kluwer Academic, pp. 47–71.

Bandara, R. and C.A. Tisdell (2002), 'Conserving Asian elephants: Economic issues illustrated by Sri Lankan concerns', in C.A. Tisdell (ed.), *The Economics of Conserving Wildlife and Natural Areas*, Cheltenham, UK and Northampton, MA, USA: Edward Elgar Publishing, pp. 193–211.

Bandara, R. and C.A. Tisdell (2003), 'Use and non-use values of wild Asian elephants', *Sri Lankan Economic Journal*, **4** (2), 3–30.

Barbier, E.B., J.C. Burgess, T. Swanson and D.W. Pearce (1990), *Elephants, Economics and Ivory*, London: Earthscan.

Benson, D.E., R. Shelton and D.W. Steinbach (1999), *Wildlife Stewardship and Recreation on Private Lands*, College Station, TX: A&M University Press.

Bromley, D.W. and M.M. Cernea (1989), 'The management of common property natural resources: Some conceptual and operational failures', *World Bank Discussion Papers*, No. 57. Washington, DC: World Bank.

Ciriacy-Wantrup, S.V. (1968), *Resource Conservation: Economics and Policies*, 3rd edn, Berkeley, CA: Division of Agricultural Science, University of California.

Clark, C. (1973), 'Profit maximization and the extinction of animal species', *Journal of Political Economy*, **81**, 950–61.

Clark, C. (1976), *Mathematical Bioeconomics: The Optimal Management of Renewable Resources*, New York: John Wiley.

Ehrlich, P.R. (1995), 'The scale of human enterprise and biodiversity loss', in J.H. Lawton and R.M. May (eds), *Extinction Rates*, New York: Oxford University Press, pp. 214–26.

Erickson, J.D. (2000), 'Endangering the economics of extinction', *Wildlife Society Bulletin*, **28** (1), 34–41.

Freese, C.H. and D.L. Trauger (2000), 'Wildlife markets and biodiversity conservation in North America', *Wildlife Society Bulletin*, **28** (1), 42–51.

Gordon, H.S. (1954), 'The economic theory of a common property resource: The fishery', *Journal of Political Economy*, **62**, 124–42.

Gowdy, J. (1997), 'The value of biodiversity: Markets, society, and ecosystems', *Land Economics*, **73**, 24–41.

Hall, C.A.S., P.W. Jones, T.M. Donovan and J.P. Gibbs (2000), 'The implications of mainstream economics for wildlife conservation', *Wildlife Society Bulletin*, **28** (1), 16–25.

Hampicke, U. (1999a), 'The limits to economic valuation of biodiversity', *International Journal of Social Economics*, **26**, 158–73.

Hampicke, U. (1999b), 'Conservation in Germany's agrarian countryside and the world economy', in A.K. Dragun and C.A. Tisdell (eds), *Sustainable Agriculture and Environment*, Cheltenham, UK and Northampton, MA, USA: Edward Elgar Publishing, pp. 135–52.

Isaacs, J.C. (2000), 'The limited potential of ecotourism to contribute to wildlife conservation', *Wildlife Society Bulletin*, **28** (1), 61–9.

Kautilya (1961), *Arthasastra (Science of Polity)*, 7th edn, R. Shamasastry [trans.], Mysore: Mysore Printing and Publishing House.

Martin, R. (2005), *Tree-kangaroos of Australia and New Guinea*, Collingwood, Vic: CSIRO.

May, R.H., J.H. Lawton and N.E. Stork (1995), 'Assessing extinction rates', in J.H. Lawton and R.M. May (eds), *Extinction Rates*, New York: Oxford University Press, pp. 1–24.

Muir-Leresche, K. and R.H. Nelson (2000), 'Private property rights to wildlife: The Southern African experiment', Washington, DC: Competitive Enterprise Institute [Online]. Available at: https://cei.org/studies-issue-analysis/private-property-rights-wildlife-southern-africa-experiment. (Accessed 3 February, 2015).

Ostrom, E. (1990), *Governing the Commons: The Evolution of Institutions for Collective Action*, Cambridge, UK: Cambridge University Press.

Ostrom, E., R. Gardner and J. Walker (1994), *Rules Games and Common-Pool Resources*, Ann Arbor, MI: University of Michigan Press.

Seasholes, B. (2002), 'Private conservation of wildlife in Africa', *Property Rights Foundation of America Inc.* [Online] Albany, NY. Available at: http://prfamerica.org/speeches/6th/PrivateConservationOfWildlifeInAfrica.html. (Accessed 3 February, 2015).

Skonhoft, A. (1999), 'On the optimal exploitation of terrestrial animal species', *Environmental and Resource Economics*, **13**, 45–77.
Smith, V. (1969), 'On models of commercial fishing', *Journal of Political Economy*, **77**, 181–98.
Swanson, T.M. (1994a), *The International Regulation of Extinction*, New York: New York University Press.
Swanson, T.M. (1994b), 'The economics of extinction revisited and revised: A generalised framework for the analysis of the problems of endangered species and biodiversity losses', *Oxford Economic Papers*, **46**, 800–821.
Swanson, T.M. (1995), *The Economics and Ecology of Biodiversity Decline: The Forces Driving Global Change*, Cambridge, UK: Cambridge University Press.
Tisdell, C.A. (1972), 'The economic conservation and utilisation of wildlife species', *The South African Journal of Economics*, **40** (3), 235–48.
Tisdell, C.A. (1982), *Wild Pigs: Environmental Pest or Economic Resource?*, Sydney, Oxford and New York: Pergamon Press.
Tisdell, C.A. (1990), *Natural Resources, Growth and Development*, New York, Westport, CT and London: Praeger.
Tisdell, C.A. (1995), 'Issues in biodiversity conservation including the role of local communities', *Environmental Conservation*, **22** (3), 216–22.
Tisdell, C.A. (1999), *Biodiversity, Conservation and Sustainable Development*, Cheltenham, UK and Northampton, MA, USA: Edward Elgar Publishing.
Tisdell, C.A. (2002), *The Economics of Conserving Wildlife and Natural Areas*, Cheltenham, UK and Northampton, MA, USA: Edward Elgar Publishing.
Tisdell, C.A. (2003a), 'Socio-economic causes of loss of animal diversity: Analysis and assessment', *Ecological Economics*, **45**, 365–76.
Tisdell, C.A. (2003b), *Economics and Ecology in Agriculture and Marine Production: Bioeconomics and Resource Use*, Cheltenham, UK and Northampton, MA, USA: Edward Elgar Publishing.
Tisdell, C.A. (2005a), *Economics of Environmental Conservation*, 2nd edn, Cheltenham, UK and Northampton, MA, USA: Edward Elgar Publishing.
Tisdell, C.A. (2005b), 'Elephants and polity in Ancient India as exemplified by Kautilya's *Arthasastra* (Science of Polity)', *Economics, Ecology and The Environment*, Working Paper no. 120. Brisbane: School of Economics, The University of Queensland.
Tisdell, C.A. (2014), *Human Values and Biodiversity Conservation: The Survival of Wild Species*, Cheltenham, UK and Northampton, MA, USA: Edward Elgar Publishing.
Tisdell, C.A. and C. Wilson (2004), 'The public's knowledge of and support for conservation of Australia's tree-kangaroos and other animals', *Biodiversity and Conservation*, **13** (12), 2339–59.
Tisdell, C.A. and C. Wilson (2005), 'Do open-cycle hatcheries relying on tourism conserve sea turtles? Sri Lankan developments and economic–ecological considerations', *Environmental Management*, **35** (4), 441–52.
Tisdell, C.A. and C. Wilson (2012), *Nature-based Tourism and Conservation: New Economic Insights and Case Studies*, Cheltenham, UK and Northampton, MA, USA: Edward Elgar Publishing.
Webb, G.J.W. (2014), *Wildlife Conservation: In the Belly of the Beast*, Darwin: Darwin University Press.

14. Economic incentives to conserve wildlife on private lands: analysis, policy and examples

14.1 INTRODUCTION

This chapter extends the coverage of the previous one. First it does this by further considering the ability and desirability for conservation purposes of granting property rights to landholders to utilize wildlife when it is present on their land. Secondly, it also does this for other policy measures that can induce landholders to conserve wildlife when it is present on their land. These policy instruments include those that are commonly classified as involving payment for the provision of ecosystem services (PES). Several examples illustrate the analysis.

It is important to consider such matters because some free-market advocates believe that the provision of private property rights in wildlife on private land provides a powerful (and virtually sufficient) economic incentive for nature conservation because it enables property owners to market such wildlife or its attributes. If such marketing is the most profitable land use, private landholders will conserve the wildlife concerned and its required habitat. But land is not always most profitably used for exploitation of wildlife, and many economic values of wildlife (such as non-use economic values) cannot be marketed. Moreover, as analysed in the previous chapter, the mobility of some wildlife adds to the limitations of the private-property approach. While some species may be conserved by this approach, it is suboptimal as a single policy approach to nature conservation. Nevertheless, it is being experimented with in the Northern Territory of Australia. There have been both successes and failures. For example, landholders in the Northern Territory had the opportunity to harvest a quota of eggs and chicks of redtailed black cockatoos for commercial sale on their properties. This scheme was expected to provide an incentive to private landholders to retain trees with hollows essential for the nesting of these birds, but failed. This case and others (some successful) are analysed. Despite private-property failures, the long-term survival of some wildlife species depends on their ability to use private lands without severe harassment, either for their migration or to maintain their population. This is true, for example, of the Asian elephant in Sri Lanka. Nature conservation on private land is often a useful, if not essential, supplement to its conservation on public lands. Community and public incentives for such conservation are outlined.

With the resurgence of economic liberalism in the last few decades extolling the virtues of private property rights and market-based solutions for directing economic and social choices about resource use, there has also been increased advocacy of such approaches as means to sustain biodiversity (including biodiversity of wildlife) and some see these policies as the key to wildlife conservation (for example, see Smith, 1981; Brubaker, 1997). For reasons presented in this chapter, I am sceptical of claims that strengthening of private property rights and free markets is *the* answer to conserving wildlife and sustaining biodiversity. It may help in some cases, but the policy has opposite effects in other cases. The situation is in fact quite complex and 'one size does not fit all'. Progress in conservation management requires these complexities to be taken into account.

Several questions need to be raised when considering the desirability of conservation of wildlife on private land and the provision of economic incentives for such conservation. First, why is such conservation desirable and why are economic incentives required for it and justified? Secondly, what types of economic incentives are available, how effective are they likely to be, and which ones are desirable on economic grounds? The answers to these questions depend to some extent on one's perspective. Consider the first matter.

Those with strong ecocentric values argue that more effort is needed to conserve wildlife on private land because a large proportion of Earth's land is in private hands and the survival of many species requires the use of private land by wildlife as well as the preservation of critical habitats needed by wildlife on this land. Without appropriate economic incentives, private landholders are likely to destroy these critical habitats, directly exterminate some species as pests, and in most cases, disregard wildlife in their land use decisions. In the absence of such incentives, continuing biodiversity loss is inevitable. Many conservationists believe that such loss is undesirable because biodiversity has intrinsic merit, or it is a stewardship imperative of mankind to protect biodiversity, or the sustainability of ecosystem services will be threatened by continuing biodiversity loss with potentially negative consequences for economic welfare. Those with ecocentric values basically believe that the conservation should be regarded as supplying a merit 'good'; that is, as a desirable goal in itself.

Assessment of the situation by economists is from a different viewpoint. Economics is based upon the view that social systems should be organized in a manner that enables human wants or desires to be satisfied to the maximum possible extent given the limited availability of resources, including wildlife resources. Some economists argue that perfectly operating market systems are theoretically capable of achieving the goal of satisfying human desires to the maximum possible extent, subject to the limited availability of resources. They often favour liberal economic policies in order to promote perfect market systems. However, in practice, perfect market systems are not used to manage all of society's resources. This is partly because perfect systems are unattainable. Even when markets are developed to the maximum extent attainable, they are not able to allocate or conserve all resources efficiently. For the supply of some commodities and the conservation of some natural resources, such as wildlife, irreducible market failures remain. In such cases, government intervention may be required to ensure that human wants are more fully satisfied than otherwise, as for example, has been argued by Bishop (1981) in relation specifically to wildlife on private land.

Economic arguments in favour of greater economic incentives for wildlife conservation on private land are not normally based on the satisfaction of special economic interests. Nevertheless, special interest groups do often put forward arguments for greater economic incentives for conservation of wildlife on private land. Special interest groups include landholders who have based, or would like to base, much of their business on the economic utilization of wildlife. Also the customers or potential customers of these landholders, such as recreational hunters or farm-stay visitors interested in wildlife, may form a special interest group. Moreover, some nature conservation NGOs may also advocate such measures. In addition, some landholders themselves obtain pleasure from wildlife and nature conservation and are prepared to forgo monetary economic gains for greater conservation of nature. They would, therefore, appreciate economic incentives for wildlife conservation. Such special interest groups usually seek their own self-interest rather than the collective or public interest, although they may incidentally also promote the latter.

First this chapter examines, primarily from a collective perspective, the role that private property rights in wildlife and the complementary facilitation of markets in wildlife can play in conserving wildlife on private land. Limitations of this approach are then highlighted. In doing so, it provides additional perspectives about this subject to those presented in the previous chapter. Specific economic incentives to address failures in or to promote wildlife conservation on private land are subsequently outlined, and policy approaches based on the granting of private property rights in wildlife in Africa and the Northern Territory of Australia, and elsewhere, are assessed.

14.2 PRIVATE PROPERTY RIGHTS AND FREE MARKETS AS A MEANS FOR WILDLIFE CONSERVATION

It is sometimes argued that the best way to encourage private landholders to conserve wildlife is for governments to give them private property rights in wildlife, strengthen these rights where they exist, and promote the operation of free markets in the exchange and use of wildlife on private lands. According to this view, deficiencies in the conservation of wildlife on private lands and lack of economic incentives for such conservation arise from institutional inadequacies, and these can be overcome by adopting the policies just mentioned.

In most nation states, wildlife has historically been the property of the crown or state, not private property. However, the state has not always protected every species of wildlife. Therefore, on occasions, some wildlife species have, in fact, been private property or common property. In some countries, some species of wildlife still do not have legal protection.

Proponents of private property rights in wildlife species usually argue that in the absence of such rights, private landholders have no economic incentive to conserve wildlife because they can make little or no money out of its presence. They will in such conditions develop commercial activities on their land which destroy wildlife habitat and reduce wild biodiversity. In these circumstances, protection of critical wildlife habitat on private land will require direct intervention by governments. Furthermore, even if the wildlife concerned is legally protected and regarded by landholders as a pest, many landholders may kill it illegally.

Proponents of private property rights contend that if private landholders are given full property rights to wildlife, these consequences would not arise, or they would be less serious than now. They believe that the wildlife concerned will as a result, have economic value for landholders and they will be able to make money or obtain more income from wildlife on their land than otherwise. If this is combined with free market trading in wildlife or their products, this should provide private landholders with even greater economic incentive to conserve wildlife because it will enhance the level of income of landholders from wildlife-related economic activities.

Assessment of such a claim is quite complex. While this policy should result in the conservation of those species of wildlife from which adequate economic returns can be obtained by landholders, it may accelerate the loss of those species yielding inadequate economic returns. The composition of wildlife species can be expected to move in favour of those species that have greatest use value or more specifically, in favour of those wildlife species having the greatest net economic use value able to be appropriated by private landholders.

The likely impact of this policy on the pattern of wildlife conservation on private lands is unclear. Compared to the current situation, it may result in a greater biomass of wildlife on private land but the species conserved may be mostly those of value for hunting, for game meat and other by-products, and for non-consumptive recreational use on properties. Management practices are likely to be adopted on private properties to encourage the presence and survival of such species and discourage other species of lesser commercial value or those that are in competition with the economically favoured species. Publications exist (Burger, 1973) suggesting how landholders can change their land management to encourage particular types of wildlife. The resulting regime of wildlife biodiversity on private land is likely to be distorted in favour of wildlife species with significant commercial value.

Nevertheless, this policy does extend the range of species possessing commercial value. Hence, it can (but need not) result in more wildlife species being conserved on private land than with current institutional arrangements. On the other hand, in some localities, husbandry of existing domesticated or cultivated species will continue to be preferred by private landholders to the commercial use of wild species.

If commercial use of a particular set of wildlife species becomes economically more valuable than other wildlife species, private land management practices are likely to be altered in favour of the commercially more valuable wildlife species. This could be at the expense of the commercially less valuable wildlife species, those that lack commercial value, and species regarded by private landholders as pests, that is, those having negative commercial value.

Wildlife species that are in ecological competition with species commercially favoured by this policy will be disadvantaged. However, those in a complementary ecological relationship should be favoured. Thus, the pattern of wildlife conservation is 'distorted' in a manner that is unlikely to be acceptable to supporters of the 'land ethic'. For example, Thirgood et al. (2000) and Thirgood and Redpath (2008) report that persecution of raptors such as hen harriers (*Circus cyaneus*), peregrine falcons (*Falco peregrinus*) and golden eagles (*Aquila chrysaetos*) is widespread in Britain and is mostly attributable to their predation on red grouse (*Lagopus scotica*). Landowners possessing large areas of moorland privately managed for shooting grouse are hostile to these raptors, as are hunters of grouse, because they reduce their grouse populations. Furthermore, they point out that it is unclear whether the management of heather moorland for grouse benefits other upland birds.

Again, the re-introduction of the grey wolf (*Canis lupus*) to parts of North America is controversial because of its potential negative effect on marketable commodity production, including on populations of some game in which landholders have on-property rights for hunting, such as some deer. Private landholders may destroy the wolf whenever they can, and so may deer hunters (Lohr et al., 1996).

Granting private property rights to landholders in all wildlife on their land is not a recipe for solving social conflict about ecosystem management and wildlife conservation (Guynn and Yarrow, 1995). This is because it can in particular cases result in the extermination of species not wanted by landholders but the conservation of which is highly valued by other members of society. However, in particular cases, granting such rights does help to reduce social conflict. For example, the granting of limited property rights to landholders in the Northern Territory of Australia to market saltwater crocodile eggs on their land has reduced social conflict about the government's programme to conserve saltwater crocodiles and appears to have had little effect on the functioning of natural ecosystems according to research results (Tisdell, 2014, Ch. 9).

The provision of fishing rights to private property holders appears in some cases (in the UK and the USA) to have had positive impacts on the environmental ecosystem functioning (Anderson, 1991). However, this does not offset the potential loss in biodiversity in other cases. Consequently, fewer wildlife species are likely to be conserved globally than desired by those strongly favouring biodiversity conservation as a goal in itself. Also, the private property policy will not give ideal economic results because commercially oriented landholders will not take account of unmarketable economic values, such as the existence value which some members of the general public attribute to predators. In such cases, private property rights in wildlife can interfere with natural ecosystem functioning.

Note that this critical view does not imply that use of private property rights and free markets as a wildlife conservation technique have no place. Conservation of some wildlife species can be effectively promoted by such policies, but not all. As a result of private property rights and free markets for economic use of wildlife, nature conservation on private land may be better than in the absence of such policies but not ideal. However, to rely on these policies alone to conserve wildlife would be socially unsatisfactory. Better social results can be attained by using mixed systems of various kinds of which state protected land areas are a part. To appreciate this position, it is necessary to be aware of market failures that result intrinsically in the lack of marketability of important economic values of some wildlife species.

14.3 MARKET FAILURES LIMIT THE EFFECTIVENESS OF PRIVATE PROPERTY RIGHTS AND THE USEFULNESS OF MARKETS AS MEANS FOR CONSERVING WILDLIFE

Markets and private property rights for commodities work best in promoting a social economic optimum when non-owners can be economically excluded from the benefits from those commodities. In such circumstances, non-owners wishing to obtain benefit from the commodities can be forced by their owners to pay for their use, thereby providing an incentive to their owners to supply the commodities and in the case of wildlife, conserve those species that can provide their owners with sufficient economic gains. However, it can be too costly or impossible to effectively make all wildlife private property. In such cases, landholders cannot appropriate economic benefits from the species concerned or this benefit is limited. Three types of problems arise.

First, some species of wildlife are fugitive or mobile and cannot be economically confined to a private property (Bishop, 1981; Ciriacy-Wantrup, 1968). Depending upon how mobile wildlife species are, other landholders can benefit from efforts by an individual landholder to conserve these if the species are economically valuable. This (as was shown in the previous chapter) limits the economic incentive of an individual landholder to conserve a highly mobile wildlife species if the landholder only has private property rights to it while it is on his or her land.

Second, the total economic value of some species depends primarily on their non-use economic values rather than their use values. Many ecological economists divide the total economic value of commodities into economic use value plus non-use economic value. Non-use values cannot, as a rule, be marketed. This is true, for example, of existence value. Some individuals value the continuing existence of wildlife species and this valuation does not depend upon their viewing these species or using them in other ways. For example, almost all the economic value of Australian tree kangaroos appears to be accounted for by their existence value (Tisdell and Wilson, 2004) and more than half the economic value of the Asian elephant in Sri Lanka seems to be accounted for by its non-use values (Bandara and Tisdell, 2003).

Non-use economic values involve public goods and usually cannot be made private property and be marketed. Therefore, government intervention is required to conserve species that have high non-use economic values and species that are highly mobile. For example, if a wildlife species is mobile and an agricultural pest, the government may need to compensate farmers for agricultural damages caused by it in order to increase their tolerance of it and promote collective welfare. This can be appropriate from an economic viewpoint if society *as a whole* values the species highly. For example, despite the agricultural damage done by the Asian elephant in Sri Lanka, its collective economic value is positive because the collective economic value placed on it, especially by urban dwellers, exceeds the cost of the agricultural damage caused by it (Bandara and Tisdell, 2004).

The third issue of relevance is the cost of excluding would-be users of the wildlife resource from utilizing it on a private property. If this is high, it can be uneconomic for a property-holder to manage wildlife on his/her property for economic gain. High exclusion costs can, therefore, result in some species not being conserved which it would be economic for a landholder to conserve on his/her property if exclusion costs were lower. This is true both of species that either have commercial consumptive or non-consumptive use value. Some specific examples of the importance of exclusion for the economics of ecotourism enterprises are given in Tisdell and Wilson (2012).

While private property rights will save some wildlife species from extinction, they are not an effective means for preserving all species. They are ineffective when owners of wildlife find that the proportion of total economic value appropriated by them is too low to provide an economic incentive to conserve the wildlife concerned. This can happen when a species is highly mobile, exclusion is a problem or when it has a low economic use value.

Put differently, private property rights in wildlife species can save some species only from extinction. These are wildlife species that ensure landholders greater economic returns than from any other alternative form of land use. These will usually be wildlife species possessing high economic use value, low mobility or low cost of confinement to a private property, and not requiring a very large geographical range for their survival. Differentiation of private property rights in the UK in fish appear to reflect this aspect – private property rights exist for species that are not very mobile but do not exist for highly mobile wild fish (Anderson, 1991).

Note that even in cases where a private landholder is able to appropriate all, or a substantial part of, the economic value of a wildlife species, the landholder may not find it profitable to conserve the species. First, an alternative land use incompatible with survival of the species may be more profitable. Or, secondly, it may be most profitable to liquidate the wildlife asset and invest the monetary sum obtained in the capital market (Clark, 1973; 1976). Maximizing net economic value (even net total economic value) does not ensure the survival of all species of wildlife. Thus, economic considerations lead to some species being extinguished, even if there are no market failures. Additional species are likely to be extinguished because of the presence of market failures. Some of these might be saved by institutional change that provides for private property rights in wildlife and for liberal market regimes. Other wildlife species will not be conserved in the absence of government intervention to support their conservation. Such support may come, for example, from the setting aside of state-protected areas or by the government providing economic incentives, such as subsidies, to private landholders, for critical habitat retention.

The above, however, does not take account of the fact that organizations, apart from private enterprises and governments, can contribute directly to nature conservation. The role of not-for-profit conservation organizations should be taken into account as well as cooperative arrangements between landholders. Because of free-riding, NGOs are only likely to be able to fill part of the gap in the social demand for conserving wild species and there are also limits to the extent to which different landholders are likely to agree to cooperate in order to conserve a species and contribute to nature conservation. Even taking into account these additional contributions to nature conservation, a case exists for government's intervention to provide economic incentives for nature conservation on some landholdings managed by private enterprises, communal groups (tribal groups in some countries) and by NGOs directly supplying nature conservation services. The range of policies available is much wider than simply paying for or subsidizing the ecosystem services provided, that is, policies involving PES schemes. It is worthwhile considering the range of policies that may be used for this purpose.

14.4 GOVERNMENT ECONOMIC INCENTIVES AND OTHER POLICIES FOR CONSERVING WILDLIFE ON PRIVATE LAND AND ELSEWHERE

It is useful to identify three different categories of public policies which may be adopted by government to conserve wild nature on lands other than those directly managed by governments. These are:

1. The provision of economic incentives to landholders (such as payments for ecosystem services, PES) for conservation of nature with landholders being left to make a *free choice* about their land use decisions.
2. The above policy, but landholders compelled by government laws and regulations to adopt specified conservation practices.

3. Cases where landholders are compelled by regulations or law to adopt particular conservation practices and are subject to penalties for non-compliance.

Consider each of these types of policies in turn.

Government Economic Incentives for Nature Conservation with Landholders Having Free Choice about Land Use

A number of lists have been drawn up of the types of public economic incentives that may be used to conserve wildlife on private land assuming that landowners have free choice about their land use. For example, Benson et al. (1999, p. 41) provide a relatively comprehensive list of public policies (many of an economic nature) that may encourage wildlife conservation on private lands. A considerably modified form of this list, with additions, is shown in Table 14.1. Crosthwaite (1995) also provides a useful overview

Table 14.1 A list of public policies that might be used to provide economic incentives to conserve wildlife on private lands

1. Subsidies for the cost of habitat restoration and protection.
2. Annual subsidies for conserving areas of habitat, as in the EU.
3. Subsidies on marketing of species with use (market) value but a value too low to make their conservation by a landholder economic.
4. Tax concessions for wildlife conservation activities on private land at least as generous as those allowed to agriculture and similar industries, for example, full tax deductability of expenses incurred in wildlife conservation. Also taxation concessions on local government taxes or rates can be important.
5. Adequate public compensation/insurance for damages caused by selected wildlife species on farms.
6. Harness community and NGO voluntary contributions for conservation efforts on private land. This will increase the financial gearing of any public contribution of funds for conservation. This has been done successfully in Australia, for example, with Landcare.
7. Benson et al. (1999) recommend stricter trespass law as and reduced legal liability of landholders.
8. Increased public leasing of private land needed for wildlife conservation.
9. More partnerships between state conservation bodies, private landholders and non-government bodies with all contributing some resources to wildlife conservation efforts.
10. For some species (but not all), the granting of private property rights in these (or in some cases, communal property rights) can provide an effective economic incentive for their conservation.

of financial incentives that could be used to encourage nature conservation on private and leasehold land.

While drawing up lists of this kind is of value for rational decision-making about public policy because it helps to ensure that no possible strategies are overlooked, this is likely to be the most straightforward part of the exercise. Before rushing into policy changes, several more fundamental issues need to be considered. These include the following:

1. Why are economic incentives for wildlife conservation on private land needed?
2. What type/nature of wildlife conservation on private land is being aimed for?
3. Why is the aim (socially?) justified?
4. How effective are the proposed economic incentives likely to be in promoting the desired wildlife conservation goals?
5. How cost-effective are the various alternative economic incentives, or how efficient are they in achieving the desired wildlife goals in relation to the costs involved?
6. How practical or workable are the various policy proposals, taking into account transaction costs and public agency costs in their implementation?
7. How equitable are the alternative proposals for encouraging wildlife conservation on private land?

The essential starting point for rational consideration of these matters is to clarify what is or should be the objective(s) to be achieved in conserving wildlife on private land. Until that is decided, it is impossible to evaluate the alternative available (economic) instruments for achieving increased wildlife conservation on private land. For example, is the purpose to increase the population of the wildlife species with the greatest economic use value? To what extent should non-use economic values of wildlife be part of the objective? Is the goal to maximize the total economic value of wildlife? Is the goal primarily one of sustaining biodiversity or saving certain rare species? Maybe, it is only a particular wildlife species or a small set of wildlife species that is being targeted for wildlife conservation on private land in a particular region.

Where wildlife species have little economic non-use value that can be appropriated by private landholders, strengthening private property rights to wildlife will be, in itself, an ineffective policy for conserving the species concerned (see previous chapter). In such cases, subsidies for required habitat preservation on private land may be necessary. In addition, where the species have very limited commercial value, some subsidy on their marketing may be needed to stimulate their conservation on private land.

Not all landholders are profit-driven and opposed to the conservation of nature. Nevertheless, conserving wildlife on private land can often involve landholders in financial outlays. For instance, habitat restoration and protection is needed on some private properties to protect particular wildlife species, but this can usually only be achieved with a financial outlay. For example, trees may have to be planted and watered when young, weeded and so on and fenced in so they are not damaged by livestock. While an individual landholder may have some willingness or demand to do this, the cost of carrying out the project may exceed his/her willingness to do so. However, with some financial subsidy for habitat restoration and protection, a landholder may undertake it to some extent.

This can be illustrated by Figure 14.1. Suppose that the cost per unit area of restoring and protecting targeted species of wildlife on a private property is OC, for example, \$500 per hectare. The property owner's marginal willingness to pay for it is assumed to be indicated by the line AB. Because OA is less than OC, the property owner is not prepared to restore any habitat. But if a subsidy of AG for each hectare restored is given, x_2 units of the owner's land would be rehabilitated. Only a portion of the cost in value is met by the government and a part is met by the landholder. In

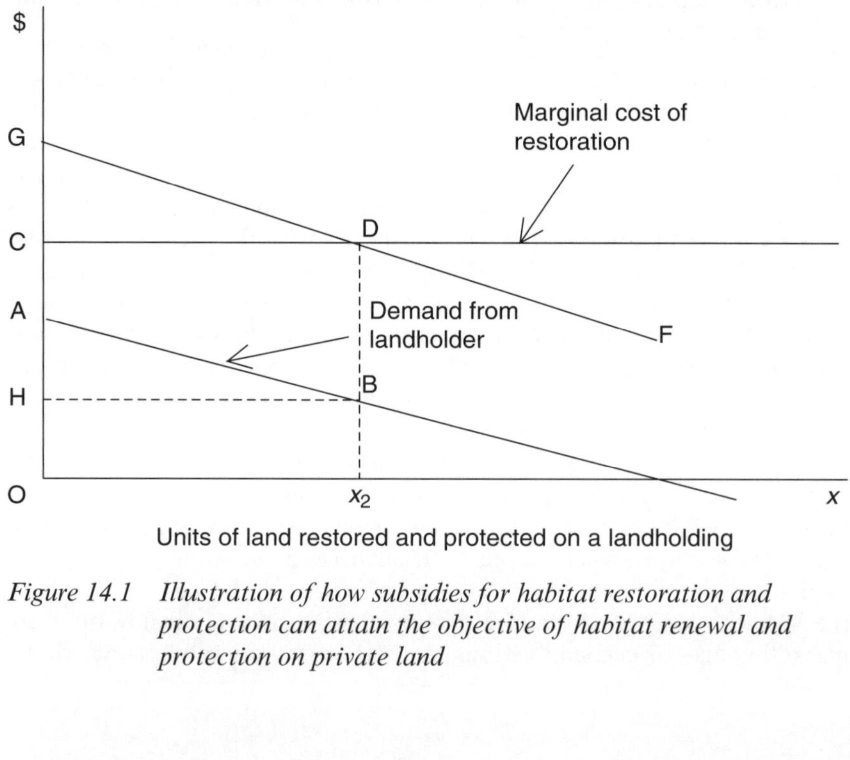

Figure 14.1 Illustration of how subsidies for habitat restoration and protection can attain the objective of habitat renewal and protection on private land

the particular case shown, OHx_2 of the total cost is met by the landholder and HCx_2 by the government.

The actual financial outlay by the government can be considerably reduced if non-governmental agencies are prepared to contribute some funds and volunteers provide labour. Individuals are often prepared to donate their time and effort in helping with habitat restoration projects. Such cooperative efforts are well established in Australia and many Western countries. An example of this is the effort to provide tree corridors on private land for wildlife conservation on the Atherton Tablelands of northern Queensland. One possible beneficiary of this is the Lumholtz tree kangaroo (*Dendrolagus lumholtzi*).

If landholders can be convinced that they will obtain greater benefits from habitat restoration and preservation than they initially believe, this will increase the prospects of habitat conservation on private land and can reduce the necessary public outlay for it. It would have the effect in Figure 14.1 of shifting the landholder's demand curve for habitat restoration, AB, upwards, so increasing the likelihood that the landholder will engage in more restoration than otherwise.

Subsidies of this nature can be justified on economic grounds when external benefits arise from wildlife species preservation made possible by habitat restoration or protection on private lands. Nearly all of the economic benefits from the conservation of some species of wildlife are external to a private landholder because the economic benefit from them consists primarily of non-use economic value. This is true, for example, of Australian tree kangaroos (Tisdell, 2014, Ch. 5; Tisdell and Wilson, 2004).

It is however, appropriate to point out limitations of adopting some of the policies listed in Table 14.1. One problem with payments or subsidies given by governments for the supply of ecosystem services (such as nature conservation) is that landholders may be uncertain about their continuation. Governments and government policies can change and result in such subsidies being withdrawn. If a subsidy for habitat restoration is withdrawn, a landholder can be left with a financial burden. The restored habitat may reduce the value of his/her land for supplying marketable goods, and converting it back to a form in which it is more productive in supplying marketable goods can be costly. In any case, once the subsidy is removed, there is unlikely to be an incentive for landholders to continue with the once subsidized conservation activity.

A major problem also is that landscapes on different landholdings are usually very heterogeneous. Consequently, areas of suitable habitat conserved by a general subsidy for habitat conservation may only result in patches of habitat being conserved or restored. This may not be a problem for conserving some wild species, but others often require connected habitats or large areas of continuous habitat for their survival. Therefore, variations in the size and spatial allocation of subsidies can be required on different properties to achieve acceptable conservation outcomes. In effect, grants with conditions rather than uniform subsidies may be needed. This can make such schemes costly to administer. Furthermore, the costs of checking on compliance can be considerable. The question also needs to be considered of who should pay for the subsidy and how the funds should be collected.

Taxation concessions for protecting or restoring habitat for wildlife conservation raise similar issues. There is, however, the further complication that the value of such concessions to a landholder will depend on the level of the landholder's tax liability in the absence of the concession. It may also be the case that public administrators are likely to show little enthusiasm for tax concessions to landholders for conserving nature if this results in a reduction in the tax base and a fall in public revenue. Institutional biases seem to exist in favour of the provision of marketable goods (which potentially add to tax revenue) rather than those that are not marketable or that are insignificantly marketable, and consequently, shrink the tax base. Aspects of these issues are discussed, for example, in Tisdell and Pearson (2001).

Prohibition on Consumptive Wildlife Use and/or on Habitat Destruction

In some cases, but not always, prohibiting wildlife use or habitat destruction can be an effective conservation measure, as was pointed out earlier in this chapter (also see previous chapter). Often, however, public policies focus on just prohibiting consumptive uses of species or on just forbidding the destruction of their habitat, but not both. Consequently, in many cases, they fail to achieve their conservation objectives. For example, although prohibiting the harvesting of wildlife may prevent the consumptive use of a species, its habitat may disappear in the absence of restrictions on the destruction of its habitat and consequently drive the species to extinction.

Prohibition on the consumptive use of wildlife and habitat change on landholdings may be accompanied by public compensation or, in some cases, no compensation may be paid. Whether or not compensation is paid and the amount of it raises questions of justice. Such intervention frequently results in the state taking away pre-existing perceived or actual property rights of landholders and involves the state's exercise of eminent domain (Tisdell, 2009, pp. 126–7; Seidl et al., 2002).

14.5 OBSERVATIONS ON SOME ACTUAL SCHEMES TO PROVIDE ECONOMIC INCENTIVES TO CONSERVE WILDLIFE ON PRIVATE LAND

In southern Africa, property rights in wildlife on private land and communal land have been used extensively to encourage the conservation of wildlife. This strategy seems to have been successful as far as the conservation of some species are concerned (Seasholes, 2002; Child, 1996). Its comparative success is probably due to the relatively high economic net value of southern African mammals for viewing, trophy hunting, game meat and other products. While a considerable range of mammals is conserved under these institutional arrangements, the pattern of species conservation nevertheless displays selectivity in favour of wildlife species with the greatest commercial value on private properties.

If a fully balanced pattern of wildlife conservation on private land were to emerge as a result of private rights, there would be little or no economic justification for government provision of protected areas and national parks as a means of wildlife conservation. Furthermore, it is unlikely that this 'success' in southern African can be repeated everywhere because in some countries and regions, the cost of enclosing the relevant species is too high or their economic use value is quite low compared to their economic non-use value.

In some countries, public compensation is paid to private landholders for agricultural damages caused by some species of wildlife. For example, such schemes exist in Sri Lanka and in Yunnan, China, to compensate farmers for damages caused by elephants. Their purpose may be to provide some equity to affected farmers and make them more tolerant of the species. However, in developing countries, the amount of compensation paid in relation to damages sustained by agriculturalists and delays in its payment usually make this mechanism ineffective as a conservation device (Tisdell and Zhu, 1998; Bandara and Tisdell, 2002). However, if compensation is of a sufficient magnitude and is paid promptly, it can be of positive assistance in helping to conserve species that need a considerable range for their survival, such as the Asian elephant. The economic benefits that urban dwellers obtain from the continued existence of a species, such as the Asian elephant in Sri Lanka, seems to provide an economic justification for more generous compensation schemes for private landholders (Bandara and Tisdell, 2003).

Australian Examples

In Australia, policies involving the management of wildlife on private property vary according to the state. The Northern Territory of Australia has displayed the greatest interest in policies for conserving wildlife through their sustainable use (Parks and Wildlife Commission of the Northern Territory, 1996). In line with this strategy, it has implemented several different wildlife management programmes. Those for the management of crocodiles (*Crocodylus porosus* and *Crocodylus johnstoni*), and for the sub-species of the red-tailed black cockatoo (*Calyptorhynchus banksii macrorhynchus*) are worth commenting on here (Parks and Wildlife Commission of the Northern Territory, 1997a; 1997b).

The crocodile management programme of the Northern Territory (NT) is intended to ensure the long-term conservation of crocodiles in the NT. The aim is 'to establish a commercial value for crocodiles and use it to make crocodiles and their habitats assets to owners, rather than liabilities' (Parks and Wildlife Commission of the Northern Territory, 1997a, p. 2). The crocodile management programme of the Northern Territory has continued with modifications. Its continuing development is discussed in Webb (2014, Ch. 22) and its sustainability consequences are given particular consideration in Tisdell (2014, Ch. 9). By this means and regulation of sustainable commercial utilization of crocodiles, it is intended to 'maintain viable wild populations of crocodiles and conserve the wetland habitats upon which they depend' (Parks and Wildlife Commission of the Northern Territory, 1997a, p. 2).

The overall programme has been successful in increasing wild saltwater crocodile populations in NT (Webb, 2014, Ch. 22; Tisdell, 2014, Ch. 9), but this may be due more to quotas on the harvest (mainly of eggs and hatchlings) and other policies than the economic incentives it provides to landholders to conserve wetland habitat favourable to the nesting of saltwater crocodiles, or to the encouragement it gives a landholder to create a more favourable habitat for the nesting of saltwater crocodiles. Responses to a survey of pastoralists in the NT indicated that virtually none changed their land use practices as a result of the crocodile management programme (Tisdell et al., 2005). The following extract from this working paper summarizes the relevant responses:

> With the exception of one, most pastoralists stated that income from allowing saltwater crocodile harvesting did not alter their land use practices or farm management nor did most pastoralists conserve any habitat suitable for saltwater crocodiles as a result of income from saltwater crocodile harvesting. The only pastoralist who said that she altered land use practice as a result of receiving income from crocodile harvesting explained that she (i) cleared infestations of an introduced weed, mimosa (*Mimosa pigra*), off the river banks that are potential crocodile nesting areas to increase their available nest areas, and (ii) aerial-sprayed mimosa weeds to maintain other crocodile nesting sites. She also said that if crocodiles nest where you know then it does not cost you helicopter time to search for them (for egg harvesting, for example) (Tisdell et al., 2005, p. 16)

Nevertheless, the fact that landholders were able to receive some income from the programme to protect saltwater crocodiles (albeit insufficient to compensate them for their livestock losses) contributed to their support for the conservation programme (Tisdell, 2014, Ch. 9).

The sustainable use programme for conservation of the red-tailed black cockatoo (*Calyptorhynchus banksii macrorhynchus*) in the NT has not been successful. The key aim of this management programme was stated to be to conserve this species outside protected areas in NT and 'to further this aim, the commercial use of the species will be permitted on private lands, under strict controls' (Parks and Wildlife Commission of the Northern Territory, 1997b, p. 1). The aim of sustainable commercial use of this species on private land was to 'promote retention and management of habitats on private lands and establish with landowners the concept that wildlife, wildlife habitats and biodiversity in general can be valuable economic assets worth considering' (Parks and Wildlife Commission of the Northern Territory, 1997b, p. 3).

Quotas for the harvesting of eggs and nestlings of black cockatoos were available to approved landholders satisfying particular requirements. For these quotas, approved landholders needed to pay the Commission royalties (fees). Other costs would, in addition, have to be borne by landholders to ensure compliance and arrange for the marking of birds reared under this programme, for example, microchip identification of live birds. Quite a few records were also to be kept by landholders participating in the scheme. Furthermore, landowners were required to enter into a formal agreement with the Parks and Wildlife Commission to manage and protect some wildlife habitat of *C. banksii macrorhynchus* such as trees having hollows in which these birds nest (Parks and Wildlife Commission of the Northern Territory, 1997b, p. 4).

In fact, no landholder was prepared to enter into such an agreement. The obvious reason why this scheme failed is that the cost to landholders of participating in it exceeded any economic benefit they could hope to gain from it. The transaction costs were too high. For the scheme to work, it would have been necessary to subsidize eligible landholders for eggs and hatchlings made available under the programme and sold. Whether such a subsidy would have been justified economically is not clear because the species does not seem to be endangered.

Given that the marketable value of the caged red-tailed black cockatoo was (is) high, it may be difficult to appreciate why the private property solution to conserving these species did not work in the Northern Territory of Australia. In the Australian market (Australian Petlinks Classified ads) the price of a red-tailed black cockatoo as a caged bird was AU$2000 (US$1400 approx. as at 17 September 2004). By comparison, the price of a caged eclectus parrot (*Eclectus roratus*) was AUS$800–900, a sulphur-crested white cockatoo (*Cacatua galarita*) was AU$50–60, and a rainbow lorikeet (*Trichoglossus haematodus*) was AU$35. So the red-tailed black cockatoo was highly priced. Furthermore, the market could be thin, and, therefore, any significant increase in supply could drive the market price down substantially. Apart from the high transaction costs imposed on landholders wanting permits for rights to harvest red-tailed black cockatoos in the NT, landholders may have had limited knowledge about how to undertake such harvesting successfully and market these birds. It was a new venture and considerable learning is involved in the early stages of a new business venture.

One can learn both from the successes and failures of policies designed to provide economic incentives to conserve wildlife on private land. The above review of some cases relying on allowing landholders to obtain commercial value from species on their land indicates that some have succeeded and that others have failed for economic reasons.

14.6 CONCLUDING COMMENTS

In this chapter, additional consequences were identified to those considered in the previous one of giving property rights to landholders for wildlife on their land. It was pointed out that in some cases this favours the conservation of species having substantial market value to the detriment of those that are in ecological competition with these favoured species. Various alternative public policy measures for conserving wildlife on private land were assessed. These measures included payments for conserving wildlife on private and communal properties. Although many scholars have, in recent times, expressed support for payments to landholders for supplying desired ecosystem services, including the conservation of socially desired populations of wild species, this policy is not always an efficient means of achieving these aims. The reasons for this were identified; for example, doubts which landholders may have about the length of time such payments will continue, high agency costs and the often heterogeneous nature of landscapes needed to conserve particular species.

Furthermore, financial incentives for nature conservation on private lands can take various forms. Apart from direct payments (subsidies and grants), other possibilities include low interest loans and taxation concessions. It was, however, suggested that taxation concessions are unlikely to be favoured by public bureaucrats. In addition, the value of taxation concession to the landholder varies with the landholder's income or asset values (generated by marketed produce) and the agency costs involved in monitoring compliance with the conditions required for the concessions can be high.

It was also pointed out that financial penalties, prohibitions and legal regulations can also be used to influence the conservation of nature on private land. The latter may in some cases be combined with financial payments or other concessions to landholders, such as allowing landholders limited rights to market wildlife species or their products.

The alternative policies need to be assessed, taking into account how equitable they are, the costs of administering these, and by considering their effectiveness for achieving the goals being sought. The policy problems involved in government intervention designed to manage the conservation of wildlife resources on private land are much more complex than those involved in regulating the supply of private goods. Policy measures that have merit in the latter case can have severe limitations in the former case. Therefore, considerable caution is needed in the latter cases in recommending policies that are appropriate and effective as means to conserve wildlife resources in private lands.

The above does not imply that public policies should not be adopted to help conserve wildlife resources on private land. Such policies can increase social economic welfare, and are important given that the management of most land is now in the hands of private persons or bodies. Furthermore, this situation is unlikely to change in the foreseeable future.

It may also be noted that private land rights do not seem to have evolved in a socially optimal manner. Land ownership has been mainly determined by its economic use value; primarily the ability of owners to extract economic benefit from land. Some land that is now privately owned would have been best left as land to be protected (as a natural or near natural ecosystem) by the state. Consequently, the current allocation of land rights does not maximize the total social economic value of land taking into account both use and non-use values. Put differently, it does not maximize the level of human welfare obtained from land use or ecosystem use. However, any major re-allocation of private property rights seems unlikely given 'the sanctity' associated with the possession of private property rights. Furthermore, the scope is often limited for reversing the ecological nature of land that has been transformed for commercial use. Therefore, policies to conserve nature and ecosystem services will continue to have to contend with an imperfect distribution of private rights to land use. Given institutional constraints, it will be necessary to be satisfied with 'second best' solutions to promoting nature conservation.[1]

NOTE

1. This chapter contains extracts from a previously published article of mine, namely C.A. Tisdell (2004), 'Economic incentives to conserve wildlife on private lands: analysis and policy', *The Environmentalist*, **24**, 153–63. I am grateful to *Springer* for permission to re-use material from this article.

REFERENCES

Anderson, T.L. (1991), 'The market process and environmental amenities', in W. Block (ed.), *Economics and the Environment: A Reconciliation*, Vancouver: The Fraser Institute, pp. 137–57.

Bandara, R. and C.A. Tisdell (2002), 'Asian elephants as agricultural pests: Economics of control and compensation in Sri Lanka', *Natural Resources Journal*, **42** (3), 491–519.

Bandara, R. and C.A. Tisdell (2003), 'Use and non-use values of wild Asian elephants', *Sri Lankan Economic Journal*, **4** (2), 3–30.

Bandara, R. and C.A. Tisdell (2004), 'The net benefit of saving the Asian elephant: A policy and contingent valuation study', *Ecological Economics*, **48** (1), 93–107.

Benson, D.E., R. Shelton and D.W. Steinbach (1999), *Wildlife Stewardship and Recreation on Private Lands*, College Station, TX: A&M University Press.

Bishop, R.C. (1981), 'Economic considerations affecting land owner behaviour', in R. Dumke, G.Y. Burger and J.R. March (eds), *Wildlife Management on Private Lands*, Proceedings of a Symposium: Wildlife Management on Private Lands, 3–6 May 1981, Milwaukee, WI,. Printed by La Crosse, WI: La Crosse Printing Co., pp. 73–86.

Brubaker, E. (1997), 'Beyond quotas: Private property solutions to overfishing', in R. Amason, E. Brubaker, K. De Alessi, K. Erikson, P. Macgillivray, T. McClurg, B. Runolfsson, P. Sprout and M. Walker (eds), *Fish or Cut Bait: The Case in Individual Transferable Quotas in the Salmon Fishery British Columbia*, Vancouver: Fraser Institute, pp. 151–78.

Burger, G.Y. (1973), *Practical Wildlife Management*, New York: Winchester Press.

Child, B. (1996), 'The practices and principles of community-based wildlife management in Zimbabwe: The CAMPFIRE programme', *Biodiversity and Conservation*, **5**, 369–98.

Ciriacy-Wantrup, S.V. (1968), *Resource Conservation: Economics and Policies*, 3rd edn, Berkeley, CA: Division of Agricultural Science, University of California.

Clark, C. (1973), 'Profit maximization and the extinction of animal species', *Journal of Political Economy*, **81**, 950–61.

Clark, C. (1976), *Mathematical Bioeconomics: The Optimal Management of Renewable Resources*, New York: John Wiley.

Crosthwaite, J. (1995), 'A review of potential financial incentives to achieve nature conservation goals on private and leasehold land', in A. Bennett, G. Backhouse and T. Clark (eds), *People and Nature Conservation: Perspectives on Private Land Use and Endangered Species Recovery*, Mosman, NSW: The Royal Zoological Society of New South Wales, pp. 191–200.

Guynn, D.C. and G.K. Yarrow (1995), 'Ecosystems management and wildlife management: Compatible or conflicting?', *Seventh Eastern Wildlife Damage Management Conference 1995* Paper 13 [Online] Lincoln, NE, USA: University of Nebraska. Available at: http://digitalcommons.unl.edu/ewdcc7/13. (Accessed 16 February, 2015).

Lohr, C., W.B. Ballard and A. Beth (1996), 'Attitudes toward gray wolf reintroduction to New Brunswick', *Wildlife Society Bulletin*, **24**, 414–20.

Parks and Wildlife Commission of the Northern Territory (1996), 'A management program for *Crocodylus porosus* and *Crocodylus johnstoni* in the Northern Territory of Australia', Palmerston, NT: Parks and Wildlife Commission of the Northern Territory.

Parks and Wildlife Commission of the Northern Territory (1997a), 'A strategy in conservation through the sustainable use of wildlife in the Northern Territory of Australia', Palmerson, NT: Parks and Wildlife Commission of the Northern Territory.

Parks and Wildlife Commission of the Northern Territory (1997b), 'A management programme for the red-tailed black cockatoo *Calyptorhynchus banksii* in the Northern Territory of Australia', Palmerston, NT: Parks and Wildlife Commission of the Northern Territory.

Seasholes, B. (2002), 'Private conservation of wildlife in Africa', *Property Rights Foundation of America Inc*. [Online] Albany, NY. Available at: http://prfamerica.org/speeches/6th/PrivateConservationOfWildlifeInAfrica.html. (Accessed 3 February, 2015).

Seidl, I., C. Tisdell and S. Harrison (2002), 'Environmental regulation of land use and public compensation: Principles, and Swiss and Australian examples', *Environment and Planning C: Government and Policy*, **20** (5), 699–716.

Smith, R.J. (1981), 'Resolving the tragedy of the commons by creating private property rights in wildlife', *CATO Journal*, **1**, 438–68.

Thirgood, S. and S. Redpath (2008), 'Hen harriers and red grouse: Science, politics and human–wildlife conflict', *Applied Ecology*, **45**, 1550–54.

Thirgood, S., S. Redpath, I. Newton and P. Hudson (2000), 'Review raptors and red grouse: Conservation conflicts and management solutions', *Conservation Biology*, **14**, 95–104.

Tisdell, C.A. (2009), *Resource and Environmental Economics: Modern Issues and Applications*, Singapore, Hackensack (NJ), London and Beijing: World Scientific.

Tisdell, C.A. (2014), *Human Values and Biodiversity Conservation: The Survival of Wild Species*, Cheltenham, UK and Northampton, MA, USA: Edward Elgar Publishing.

Tisdell, C.A. and L. Pearson (2001), 'Analysis of property values, local government finances and reservation of land for national parks and similar purposes', *Economic Analysis and Policy*, **3** (2), 175–85.

Tisdell, C.A. and C. Wilson (2004), 'The public's knowledge of and support for conservation of Australia's tree-kangaroos and other animals', *Biodiversity and Conservation*, **13** (12), 2339–59.

Tisdell, C.A. and C. Wilson (2012), *Nature-based Tourism and Conservation: New Economic Insights and Case Studies*, Cheltenham, UK and Northampton, MA, USA: Edward Elgar Publishing.

Tisdell, C.A. and X. Zhu (1998), 'Protected areas, agricultural pests and economic damage: Conflicts with elephants and pests in Yunnan, China', *The Environmentalist*, **18** (2), 109–18.

Tisdell, C.A., C. Wilson and H. Swarna Nantha (2005), 'A report on the management of saltwater crocodiles (*Crocodylus porosus*) in the Northern Territory: Results of a survey of pastoralists', *Economics, Ecology and the Environment*, Working Paper no. 130. Brisbane: The School of Economics, The University of Queensland.

Webb, G.J.W. (2014), *Wildlife Conservation: In the Belly of the Beast*, Darwin: Darwin University Press.

15. Biodiversity conservation, loss of natural capital and interest rates: the relationships are irregular

15.1 INTRODUCTION

The view is widely accepted in natural resource economics that a rise in the market rate of interest usually increases the rate of exploitation of natural resources (reduces their conservation) because it lowers the net present value of deferring their use, that is their user cost. For example, Tietenberg (2003, pp. 92–3) states in his widely adopted textbook that 'The general conclusions which hold for all models we consider, is that higher discount rates tend to skew resource extraction toward the present because they give the future less weight in balancing the relative value of present and future resource use'. Similarly, Colin Clark (1973; 1976) has argued that a rise in the market rate of interest tends to increase the rate of exploitation of privately owned natural biological resources and increases the likelihood of their extirpation, other things being held constant.

The purpose of this chapter is to emphasize that there is no regular general relationship between the level of the market rate of interest and the rate of depletion of natural capital, particularly biological capital. This is most clearly seen when macroeconomic relationships are taken into account. The results depend on the hypothesis that the rate of exploitation of natural resources and the extent of transformation of natural environments tend to rise with the level of aggregate economic activity, for example, as measured by the level of GDP. Although all growing economic activity tends to have this effect (Georgescu-Roegen, 1971), it is possible that the major impact on the depletion of natural resources, including biological resources, arises from increasing levels of aggregate investment, that is, from the positive rate of accumulation of man-made capital.

In developing the above theme, I shall focus mainly on biodiversity loss but the analysis applies to natural resource loss generally. The analysis is developed by first considering the relationship between loss of natural capital (especially natural biodiversity) and the general level of economic activity. The role of man-made capital in that process is given particular attention. Natural resource economists have mostly relied on microeconomic models to claim that a rise in the market rate of interest tends to increase the rate of natural resource extraction. This is true within the assumptions of some microeconomic models but it is not true for others. However, the view that a rise in the market rate of interest usually results in accelerated extraction of natural resources and growing biodiversity loss, is shown to be dubious when macroeconomic factors determining the market rate of interest are taken into account. It is shown that there is no regular relationship between the level of aggregate investment, the level of aggregate economic activity and the market rate of interest. In some circumstances, it is found that the market rate of interest increases as the levels of aggregate investment and economic activity rise and vice versa. Therefore, when macroeconomic considerations are taken into account, accelerated natural resources extraction and biodiversity loss can easily occur when market rates of interest are rising or falling. After this is established, a discussion of the results and the conclusions follow.

15.2 THE RATE OF NATURAL RESOURCE DEPLETION AND BIODIVERSITY LOSS DEPENDS ON THE LEVEL OF AGGREGATE ECONOMIC ACTIVITY AND INVESTMENT

Most neo-Malthusian economists hypothesize that natural resource depletion and deterioration of natural environments tend to increase with the level of economic activity, for example, the growth of GDP (see, for example, Daly, 1980). At the same time, it is recognized that this is not an exact relationship. As pointed out by Ehrlich (1989), for instance, it is possible in special circumstances to have economic growth without increasing environmental deterioration. This can occur, for example, if new technologies are developed and applied that are less damaging to natural environments than earlier ones and if everything else remains unchanged. However, a problem is that other things do not always remain the same. A reduction in natural resource utilization per unit of aggregate output is often more than offset by a large increase in output so that consequently, the total use of natural resources rises (Tisdell, 2001). Neo-Malthusians, such as Paul Ehrlich, claim that the rate of environmental deterioration depends basically on the level of human population, consumption (or economic production) per head of population and the environmental impact of the technology used in economic production. Despite technological improvements, it is clear that globally the environmental and natural resource footprint resulting from economic growth globally has accelerated with the passage of time. One of the consequences of this has been a loss of natural biodiversity.

For example, the Millennium Ecosystem Assessment (2005, p. 2) concluded that: 'Virtually all Earth's ecosystems have been dramatically transformed through human actions'. Its main finding was as follows:

> Human actions are fundamentally changing the diversity of life on Earth, and most of these changes represent a loss of biodiversity. Changes in important components of biological diversity were more rapid in the last 50 years than at any time in human history. Projections and scenarios indicate that these rates will continue or accelerate in the future (Millennium Ecosystem Assessment, 2005).

Most of this loss in biodiversity can be attributed directly or indirectly to continuing global economic growth.

As indicated by early neoclassical theories of economic growth (for example Solow, 1956; Swan, 1956), investment in man-made physical capital (capital accumulation) has been a significant contributor to economic growth, but increases in the quantity of labour, improved technology, greater investment in education and other factors (such as economies resulting from greater market size) have also played a very important role in promoting economic growth. Denison (1962) found that the latter qualitative contributors to economic growth have risen in significance with the passing of time to become more important than the quantitative factors (physical capital and labour) as contributors to economic growth. Prior to this, Deane (1955) argued that the prime mover in Britain's economic growth resulting in the Industrial Revolution was technological progress rather than a high savings and investment ratio.

In any case, it seems that a combination of factors contribute to the continuing transformation of natural ecosystems by humans and consequent biodiversity loss. Both the availabilities of labour and man-made physical capital provide means to transform natural ecosystems. In addition, technological change (for example, growing knowledge about more effective methods to extract economic production from natural resources and ecosystems), greater education (by transmitting skills and knowledge to facilitate natural resource extraction), and increases in market size (for example, as a result of the processes involved in economic globalization) tend also to contribute to the growing transformation of natural ecosystems and to biodiversity loss. They facilitate extended as well as intensified economic utilization of natural resources, including natural environments. Myers (1981) provides an example of how expanding international trade (market extension) results in the loss of natural ecosystems and Tisdell (2003, see also Chapter 6 in this book) has identified several ways in which globalization (market extension) and technological change have contributed to loss of genetic diversity in livestock. Similar processes also contribute to biodiversity loss in crops.

Swanson (1994) was one of the first economists in modern times to emphasize the impact of the economic transformation of natural ecosystems on the loss of natural genetic diversity. Habitat loss is the primary cause of biodiversity loss in the wild (Millennium Ecosystem Assessment, 2005). However, the significance of such changes for loss of species was noted much earlier. For example, Harting (1880, p. 209), writing about the extinction of wild animals in Britain observed:

> Lake and moor have become fields of yellow grain; forest has been changed into morass, morass into moor, and moor again into forest, until finding nowhere to rest in peace, the deer, the beaver, the reindeer, the wild boar, and the wolf, have become in Britain amongst the things that were.

While loss of habitat due to land and water conversion by humans (mainly for economic reasons) is not the only source of biodiversity loss in the wild, it is probably the most important factor contributing to such loss (see Millennium Ecosystem Assessment, 2005, p. 8). Other factors include climate change, the introduction of invasive species, overexploitation of wild populations of species or their destruction as pests, and pollution. Virtually all these sources of biodiversity loss are due to a combination of factors, but economic factors seem to be important in all cases. The Millennium Ecosystem Assessment (2005, p. 42) states that indirect drivers of biodiversity loss include demographic, economic, socio-political, cultural and technological factors.

Aggregate economic investment plays a major role in biodiversity loss in the wild. Usually, land and water conversion involves an initial investment (for example, for land clearing and the establishment of agriculture) and is facilitated by the use of physical capital and new technologies. A similar situation prevails for the increased exploitation of wild species or their destruction as pests. Climate change and increased pollution are influenced by the nature of economic growth which, in turn, is related to the level and nature of economic investment to a significant extent. Because much new technology is embodied in new physical capital and this new capital is often required in order for societies to take advantage of improved education and skills, the level of investment still appears to be of prime significance for economic growth and biodiversity loss. Thus, because of interdependence of the growth variables, the level of aggregate net investment remains a major determinant of economic growth even if improved education and technological advances have become more important as contributors to economic growth as, for example, postulated by Denison (1962).

It is observed in the Millennium Ecosystem Assessment (2005, p. 4) that the distribution of species on Earth is becoming more homogeneous and at the same time the number of species and their varieties is declining (cf. Tisdell, 1999, Ch. 4). This is true not only for wild species, but is also so for varieties of cultured species. Continuing investment, increased knowledge and technological change have all contributed to this process of genetic loss (Tisdell, 2003).

In the light of the above considerations, I argue in this chapter that the level of the market rate of interest has neither a regular relationship with the rate of biodiversity loss nor with the rate of exploitation of natural resources. This is so when factors determining the level of the market rate of interest are taken into account in macroeconomic models. The established view that higher rates of interest usually accelerate biodiversity loss and the rate of extraction of natural resources has evolved from the consideration of particular microeconomic models specified using partial analysis. This dominant hypothesis needs to be reconsidered in view of the implications of macroeconomic models, and when some modifications are made to the microeconomic models which are normally used to illustrate this hypothesis. This will now be shown.

15.3 MICROMODELS IN WHICH THE RATE OF INTEREST IS AN EXOGENOUS VARIABLE

Colin Clark (1976) observed that an increase in the rate of interest (other things held constant) is likely to increase the probability of extinction of commercially exploited species if the cost of realizing (liquidating) their stocks is zero or very low and they are privately owned. As the rate of interest rises, those who own a biological asset are likely to increase their monetary gain by harvesting this asset to extinction and investing the net revenue obtained at the going rate of interest. Thus, for example, if a 5 per cent annual net return is earned on the sustainable harvest of a species, it will be profitable to retain it if the rate of interest is less than 5 per cent but not if it is in excess of 5 per cent, say 8 per cent. Note that it will also pay not to retain the species if it can be replaced at little upfront cost by another that gives a higher rate of private return. The lesson drawn from this example is that markets (free of economic failures) do not necessarily ensure the conservation of biodiversity.

If (on the other hand) the biological resource is an open access one, harvesters of it take no account of user costs and their conservation decisions are not influenced by the rate of interest. Despite this, the rate of exploitation of the targeted resource (species) can be sensitive to the rate of interest. Other things held constant, a higher rate of interest can be expected to reduce investment in the exploitation of the species; for example, less investment in capital equipment is likely. Thus, in the case of open access resources that are commercially exploited, a higher rate of interest is likely to be favourable to the conservation of biodiversity.

Even in the private ownership case considered by Colin Clark (1976), biodiversity conservation can be favoured by a high rate of interest if the upfront costs of conversion or realization of a biological asset are high and the benefits obtained by conversion or realization are not immediate. Usually, conversion or realization involves significant upfront costs because conversion, substitution or liquidation of stocks of species does not take place instantaneously. Therefore, decisions to engage in land or water conversion that lead to the extinction of some species or to liquidate species as assets may, at first, result in costs exceeding benefits, with net benefits only becoming positive at a later time.

Consequently, the flow of net benefits from conversion of habitat resulting in species extinction as a consequence of the substitution of species, or the liquidation of biological assets, may typically display the type of pattern shown in Figure 15.1 by curve ABC. This implies that, as the rate of interest rises, the net present value of these strategies falls. This is because, other

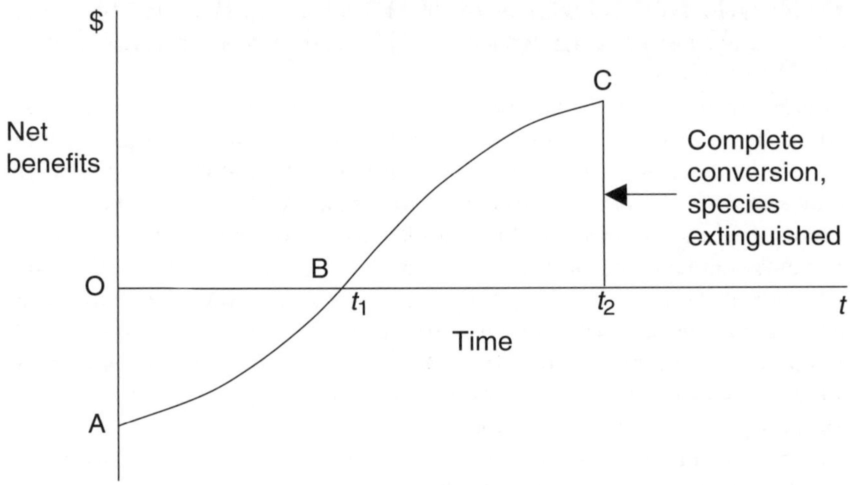

Figure 15.1 The 'typical' flow of possible net benefits from liquidating a species as an asset, substituting it by another or by converting habitat which results in the loss of species

things being held constant, an increase in the market rate of interest raises the discount rate applied by businesses to determine their most profitable economic strategy (Hicks, 1946, pp. 195–6). This results in businesses giving a higher relative weight to present or near returns compared to those further into the future (Tietenberg, 2003, p. 3). Therefore, the profitability of adopting these strategies declines. In these circumstances, a rise in the rate of interest is favourable to biodiversity conservation.

These microeconomic examples indicate that a rising rate of real interest can be favourable to biodiversity conservation if the relevant resources are private property. This is because the activities described usually require considerable upfront costs and their benefits are delayed (see also Tisdell, 2011 for further analysis). It seems only in the special case described in the opening paragraph of this section that an increase in the rate of interest is unfavourable for biodiversity conservation.

Furthermore, even when there is open access to a natural resource, a higher rate of market interest may deter use of this natural resource if capital is a significant input required for its exploitation. This is because a rise in the market interest rate tends to increase the cost of capital and reduce the profitability of new investment, everything else remaining unchanged.

15.4 MACROECONOMIC CONSIDERATIONS

In order to obtain a wider perspective on the relationship between the rate of interest and the extent to which biodiversity conservation occurs, it is necessary to take into account factors that determine the rate of interest as specified by macroeconomic models. Macroeconomic models of the determination of the rate of interest are complex and varied. However, I shall use two simple models to make the point that there is no regular association between the market rate of interest as such, extent of biodiversity conservation, and the rate of exploitation of natural resources.

This is shown by using the loanable funds theory of interest (Wicksell, 1936) in the first instance, and then by applying the IS–LM macroeconomic model (Mankiw, 2006), which is a more recent theory used to determine simultaneously the market rate of interest and the aggregate level of economic activity.

The implications of the loanable funds theory can be illustrated by Figure 15.2. In it the curve marked D_1D_1 represents the initial demand for loanable funds and that marked S_1S_1 indicates the supply of loanable funds. The dependent variable, r, is the market rate of interest and

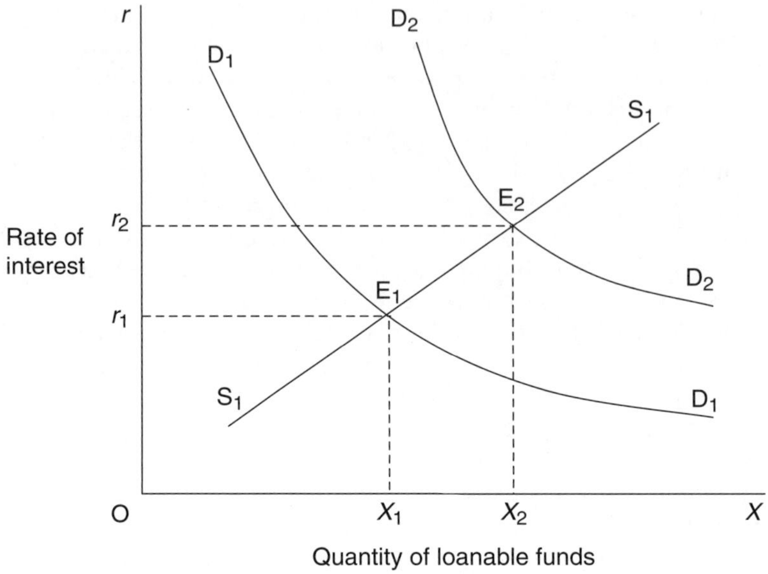

Figure 15.2 A case in which an increase in the market rate of interest is associated with a rise in the level of aggregate investment: in this case, biodiversity loss and natural resource extraction are expected to accelerate

the independent variable, X, is the quantity of loanable funds. Initially equilibrium is established at E_1 with a corresponding market interest rate of r_1 and the quantity of funds X_1 is lent. Now suppose that the demand for loanable funds shifts upward so that the above curve alters from D_1D_1 to D_2D_2. Assuming that everything else remains unaltered, the market equilibrium shifts from E_1 to E_2; the market rate of interest rises from r_1 to r_2 and the quantity of loanable funds supplied and borrowed increases from X_1 to X_2. Consequently, aggregate investment goes up and biodiversity loss and the rate of natural resource extraction tend to increase. In this comparative analysis in which the supply and demand curves have normal slopes, this will happen if the magnitude of the movement of the demand curve to the right exceeds the shift of the supply curve to the right. This association between changes in the market interest rate and biodiversity loss and the rate of natural resource extraction is consistent with that inferred from some microeconomic models (namely that a rise in the rate of interest accelerates biodiversity loss and natural resource extraction) but the mechanisms involved are different.

In the case illustrated in Figure 15.2, the rise in the demand for loanable

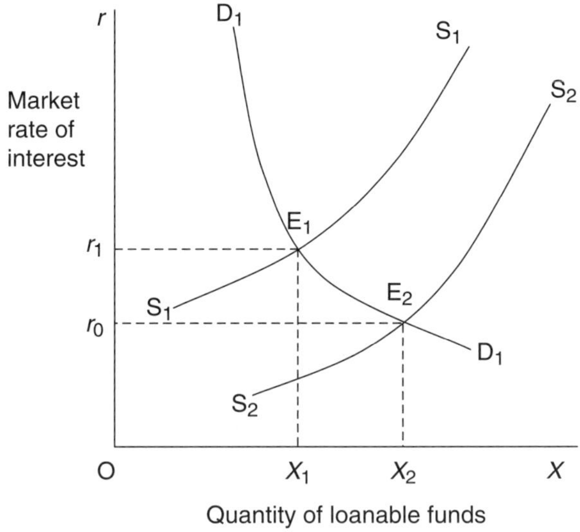

Figure 15.3 *A case in which a fall in the rate of interest is associated with an increase in investment and lending and consequently with a likely acceleration in biodiversity loss and natural resource extraction*

funds for investment purposes can occur for several reasons. For example, technological advances may increase the profitability of extracting natural resources and new sources of natural resources may be discovered, such as new oil and gas fields and new ocean-dwelling species which prove profitable to exploit.

However, it is also possible for biodiversity loss and natural resource extraction to accelerate when the market rate of interest falls. This is illustrated by Figure 15.3. In this case, the demand curve for loanable funds, D_1D_1, is assumed to be stationary and the supply curve of loanable funds is supposed to shift from S_1S_1 to S_2S_2. The supply curve of loanable funds may shift in this way because the quantity of money is increased as a consequence of government policy, or liquidity preference is reduced as a result, for example, of greater economic optimism. Market equilibrium alters from E_1 to E_2, the market rate of interest falls from r_1 to r_0 but lending increases from X_1 to X_2. Thus, the amount of investment rises and biodiversity loss and natural resource extraction accelerate. A similar result follows if the demand for loanable funds shifts to the right but by a smaller amount than a similar shift in the supply curve. This underlines the point that in a macroeconomic setting, there is no regular association between

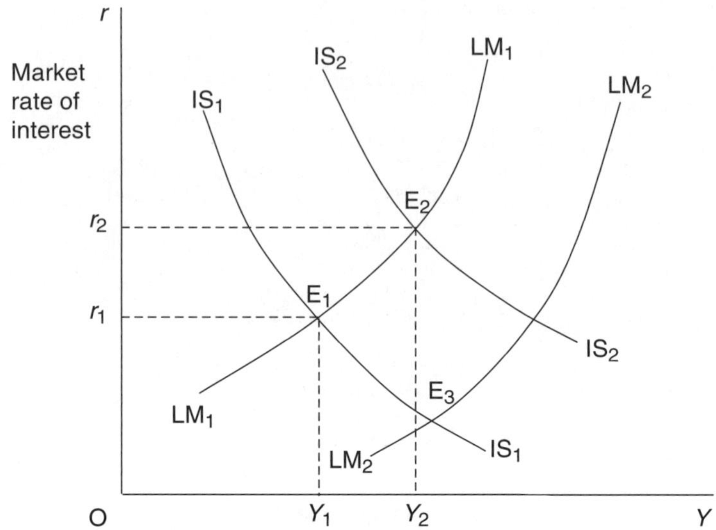

Figure 15.4 Application of the IS–LM macroeconomic model indicates that there is no regular connection between the level of the market rate of interest and the rate of biodiversity loss: this is so for the rate of extraction of natural resources generally

the market rate of interest and changes in the rate of biodiversity loss and variations in the rate of natural resource extraction.

Loanable funds theory has been used to illustrate the above-mentioned point because of its simplicity. Snippe (1985) has argued that if appropriately interpreted, there is not a substantial difference between loanable funds theory and liquidity preference theory. However, it is usually accepted that IS–LM theory (which relies on an investment savings schedule and a liquidity preference supply schedule) is more general. The proposition mentioned above is readily seen to hold when IS–LM theory is applied. This is illustrated in Figure 15.4. There Y represents the level of aggregate demand (level economic activity) and r continues to indicate the market rate of interest. Suppose that initially the investment savings schedule is similar to that identified by IS_1 and that the liquidity preference money supply schedule is as identified by LM_1. The equilibrium rate of interest is then r_1 and the level of aggregate economic activity is Y_1. Should the demand to invest (for example, due to technological progress or increased optimism) shift upwards, the IS curve may shift up to IS_2. If all else remains constant, the equilibrium rate of interest rises from r_1 to r_2

and the level of economic activity goes up from Y_1 to Y_2. Assuming that the rate of biodiversity loss and the rate of natural resource extraction are positively related to the level of aggregate economic activity, both these dependent variables increase. Biodiversity loss and natural resource extraction accelerate. However, this is not due to the alteration in the market rate of interest. This is underlined by the fact that these consequences (that is, increased biodiversity loss and natural resource extraction) can occur when the rate of interest falls. For example, if the LM curve shifts to the right by a greater amount than the IS curve, the market rate of interest declines and the level of economic activity increases. For example, in Figure 15.4, if the LM curve shifts from LM_1 to LM_2 and the IS curve remains at IS_1, market equilibrium alters from E_1 to E_3. Thus, in a macroeconomic setting, there is neither regular association between changes in the market rate of interest and alterations in the rate of biodiversity loss, nor in the rate of natural resource extraction. This is so, despite some microeconomic models suggesting otherwise.

15.5 DISCUSSION AND CONCLUSION

From the above analysis, it follows that there is no regular relationship between changes in the level of the market rate of interest and variations in the rate of biodiversity loss. More generally, there is no regular relationship between these changes and alterations in the rate of extraction (or utilization) of natural resources. Macroeconomic models highlight the fact that the market rate of interest is determined by macroeconomic forces, many of which are the primary sources of biodiversity loss and drivers of resource extraction. The rate of interest is not the driver of these effects; it is merely the result of macroeconomic relationships, which are the primary drivers.

While the lack of a regular relationship between changes in the level of the market rate of interest and the extent of biodiversity loss (as well as variations in the rate of natural resource extraction) is clear from macroeconomic models, the lack of general relationship is also apparent from microeconomic models. It was demonstrated using microeconomic models that businesses can react in quite different ways to a rise in the market rate of interest depending on the economic circumstances faced by them. Depending on these circumstances, the analysis indicates that they can have economic incentives to engage in behaviours that increase or decrease the rate of natural resource extraction or to adopt behaviours that increase the rate of biodiversity loss or moderate it.

Note that the theory posited here does not suppose that there is an exact relationship between the rate of loss of biodiversity (or more generally between reductions in natural capital) and either aggregate investment, economic growth or changes in the level of economic activity. Furthermore, it is quite possible that this relationship, although altering in direction, could change in magnitude for several reasons. For example, once biodiversity loss proceeds far enough, the scope for further loss tends to decline. The Millennium Ecosystem Assessment (2005, p. 2) points out that 'globally, the net rate of conversion of ecosystems has begun to slow, although in some instances this is because little habitat remains for further conversion'.

Keynes thought it possible that capital could accumulate in modern times to such an extent that the marginal efficiency of capital would become zero or close to zero in the long term (Keynes, 1936, Chs 16 and 24). But of course, he only had in mind man-made capital. This would result, in Keynes' view, in the rate of interest falling to zero or close to it in the long run. Yet it can be hypothesized that in order to reach this stationary state, there would have been a tremendous conversion of natural resources into man-made capital, resulting in great biodiversity loss. Consequently, in this case, a falling rate of interest would be associated with a major cumulative loss of biodiversity and natural capital.

The above analysis suggests that the level of market rate of interest and its variations are not very relevant for the sustainability of economic development. This sustainability depends fundamentally on the nature and magnitude of macroeconomic variables and on alterations in these. These appear to be real drivers of the loss of natural capital, and the level of the market rate of interest is merely a consequence of the interaction of these macroeconomic forces. As demonstrated, in circumstances where variations in macroeconomic variables accelerate the loss of natural capital, the market rate of interest can either be increasing or decreasing. The converse also holds. Apart from indicating that the level of the rate of interest has no regular influence on the rate of biodiversity loss, this chapter makes it clear that the rate of interest is not, in itself, of central significance in determining the conservation of natural capital, particularly when macroeconomic factors are considered. This is because (as shown in this chapter) the rate of interest can rise or fall when the level of aggregate investment and economic activity increases or decreases. It is the level of aggregate economic activity (particularly, the aggregate level of investment) that is of greatest significance for the depletion of natural resources.[1]

NOTE

1. This chapter draws on a previously published paper of mine, namely C.A. Tisdell (2011), 'Biodiversity conservation, loss of natural capital and interest rates', *Ecological Economics*, **70**, 2511–15. I am grateful to *Elsevier* for permission to use extracts from this publication.

REFERENCES

Clark, C. (1973), 'Profit maximization and the extinction of animal species', *Journal of Political Economy*, **81**, 950–61.
Clark, C. (1976), *Mathematical Bioeconomics: The Optimal Management of Renewable Resources*, New York: John Wiley.
Daly, H.E. (1980), 'Introduction to the steady-state economy', in H.E. Daly (ed.), *Economics, Ecology and Ethics: Essays Toward a Steady-State Economy*, San Francisco: Freeman and Company, pp. 1–31.
Deane, P. (1955), 'The implications of early national income estimates for the management of long-term growth in the United Kingdom', *Economic Development and Cultural Change*, **4**, 3–38.
Denison, E.F. (1962), *Source of Economic Growth and the Alternatives Before Us*, New York: Committee for Economic Development.
Ehrlich, P.R. (1989), 'Facing the habitability crisis', *BioScience*, **39**, 480–82.
Georgescu-Roegen, N. (1971), *The Entropy Law and The Economic Process*, Cambridge, MA: Harvard University Press.
Harting, J.E. (1880), *British Animals Extinct within Historic Times*, London: Trubner & Co.
Hicks, J.R. (1946), *Value and Capital*, 2nd edn, Oxford: Oxford University Press.
Keynes, J.M. (1936), *The General Theory of Employment, Interest and Money*, London: Macmillan.
Mankiw, N.G. (2006), *Macroeconomics*, 5th edn, New York: Worth Publishers.
Millennium Ecosystem Assessment (2005), *Ecosystems and Human Well-being: A Synthesis*, Washington, DC: World Resources Institute.
Myers, N. (1981), 'The hamburger connection: How Central America's forests became North America's hamburgers', *Ambio*, **18**, 3–8.
Snippe, J. (1985), 'Loanable funds theory versus liquidity preference theory', *De Economist*, **133**, 129–49.
Solow, R.M. (1956), 'A contribution to the theory of economic growth', *Quarterly Journal of Economics*, **70**, 65–74.
Swan, T.W. (1956), 'Economic growth and capital accumulation', *Economic Record*, **32**, 334–61.
Swanson, T.M. (1994), *The International Regulation of Extinction*, New York: New York University Press.
Tietenberg, T. (2003), *Environmental and Natural Resource Economics*, 6th edn, Boston, MA: Addison-Wesley.
Tisdell, C.A. (1999), *Biodiversity, Conservation and Sustainable Development*, Cheltenham, UK and Northampton, MA, USA: Edward Elgar Publishing.
Tisdell, C.A. (2001), 'Globalisation and sustainability: Environmental Kuznets curve and the WTO', *Ecological Economics*, **39** (2), 185–96.

Tisdell, C.A. (2003), 'Socio-economic causes of loss of animal diversity: Analysis and assessment', *Ecological Economics*, **45**, 365–76.

Tisdell, C.A. (2011), 'Core issues in the economics of biodiversity conservation', in R. Costanza, Karin Limburger and Ida Kubiszewski (eds), *Ecological Economics Reviews*, *Annals of the New York Academy of Sciences*, **1219** (1), 99–112.

Wicksell, K. (1936), *Interest and Prices (*Geldzins und Güterpreise*): A Study of the Causes Regulating the Value of Money* (trans. from German by R.F. Kahn), London: Macmillan.

16. Valuing and sustaining natural ecosystem services: assumptions, estimates and public policies

16.1 INTRODUCTION

The valuation of the supply of ecosystem services, particularly on a global or a macro-scale, is an extremely challenging but important subject requiring a multidisciplinary approach. In recent decades, there has been increased interest in using economic techniques to value these services. Costanza et al. (2014) describes this as a huge surge in interest and points out that 'hundreds of projects and groups are currently working toward better understanding, modelling, valuation and management of ecosystem services and natural capital' (Costanza et al., 2014, pp. 152–8). Although the goal of such effort (namely to improve human well-being or avoid avoidable reductions in it) is commendable, it is also important to scrutinize scholarly and other efforts intended to achieve this goal. This chapter extends earlier discussions in this book of ecosystem valuation and the role of natural capital in sustaining human well-being.

It does this first by adding to the previous discussion about the contribution of natural capital to human well-being and by further considering the classification of the Millennium Ecosystem Assessment (2005) of ecosystem services. Attention is then given, focusing mainly on Costanza et al. (2014), to assessing the significance of recent economic estimates of the value of ecosystem services. Subsequently, some basic limitations of economic techniques commonly used to value ecosystem services are identified. The important role which economics can play in valuing changes in ecosystems is then highlighted. However, valuing such changes does not depend on economics alone. Both the biophysical and the economic consequences of such changes need to be predicted. Faulty assessment may arise from either source or both. Examples of some faulty assessments or lack of such assessments and their consequences are given. It is also pointed out that in many cases, only limited information is needed to decide whether an ecosystem change is socially desirable, and examples are suggested. Subsequently, before concluding this chapter, issues are raised (additional to those covered in Chapter 14) about the use of payments for the supply of desired ecosystem services.

16.2 THE SUPPLY OF NATURAL CAPITAL, HUMAN WELL-BEING AND TYPES OF ECOSYSTEM SERVICES

The supply of natural capital is a significant contributor to human well-being. It includes biotic as well as abiotic forms of this capital. In the case of biotic types of natural capital, it includes not only genetic material of value to humans, but also associations between living things as exemplified by their interdependence in ecosystems. But it is not only biotic resources that constitute natural capital: the stock of all natural resources can be considered to constitute natural capital. This includes mineral resources, soils of varying qualities, and even the presence of harbours suitable for ports. Furthermore, not all genetic material and ecosystems can be classified as natural. Much of this material and many ecosystems (such as agroecosystems) are a result of human selection and effort and have significant influences on human well-being. This is an important point because these forms of human developed capital produce valuable ecosystem services often at the expense of ecosystem services supplied by natural capital. While there are some ecologists who wish to confine ecosystem analysis purely to the functioning of natural ecosystems (see Chapters 2 and 8 of this book), this view is too narrow, especially when it comes to analysing the implications of the functioning of ecosystems for human welfare. Some economists, however, appear to have adopted the narrower rather than the wider view of ecosystems by focusing exclusively (or almost so) on the value of natural ecosystems. This is evident in the TEEB study of the economics of ecosystems and biodiversity (see, for example, Kumar, 2010), as I have pointed out elsewhere (Tisdell, 2014; see also Chapter 2 of this book).

More recently, Costanza et al. (2014) presented a simplified model in which natural capital is shown as being the exclusive generator of ecosystem services (see their Figure 1, p. 153). They suppose that human well-being depends on four types of capital:

1. built capital;
2. social capital;
3. human capital; and
4. natural capital, which is represented as the sole source of ecosystem services.

In my view, this representation overlooks several other important contributors to human well-being. These include:

1. the stock of knowledge;
2. the stock of human developed ecosystems and germplasm (both heritage and non-heritage types);
3. labour as part of the welfare equation; and
4. cultural heritage.

In this valuation context, the lack of adequate consideration of the items listed in point 2 above seems to be a serious oversight since the supply of services associated with these forms of human developed capital are often at the expense of ecosystem services provided by natural ecosystems. Although Costanza et al. (2014) do identify human built capital as a contributor to human well-being, there is no indication that they include human developed ecosystems and germplasm in this category, or elsewhere in their model. Furthermore, their approach to analysis and measurement of the value of ecosystem services for human well-being is more limited in the types of services considered than are those identified in the Millennium Ecosystem Assessment (2005) (see Table 16.1). For example, Costanza et al. do not pay enough attention to the supply of food and fibre. The Millennium Ecosystem Assessment identifies four broad types

Table 16.1 Types of services provided by ecosystems according to the Millennium Ecosystem Assessment (2005) classification

Category of Service	Examples
1. Provisioning services	Food/fibre; genetic resources; biochemical, natural medicines and pharmaceuticals; ornamental resources; fresh water.
2. Regulating services	Air quality regulation; climate regulation; water regulation; erosion regulation; water purification and waste treatment; disease regulation; pollination; natural hazard regulation.
3. Cultural services	Cultural diversity; spiritual and religious values; knowledge systems; educational values; inspiration; aesthetic values; social relations; sense of place; cultural heritage values; recreation and ecotourism.
4. Supporting services	Soil formation; photosynthesis; primary production; nutrient cycling; water cycling.

Source: Based on Millennium Ecosystem Assessment (2005, Table 2.2, pp. 33–7).

of ecosystem services of value to human beings (Millennium Ecosystem Assessment, 2005, pp. 33–7). These are:

1. provisioning services;
2. regulating services;
3. cultural services; and
4. supporting services.

Provisioning services mainly comprise the supply of commodities that provide material benefits to humans (such as food and fibre), regulating services include environmental services of value to humans, cultural services include religious and recreational services, and supporting services cover phenomena such as nutrient and water recycling. Table 16.1 provides a more detailed coverage of the types of ecosystem services included in each of these categories.

It can be seen that Table 16.1 includes types of services provided by both natural ecosystems (a part of natural capital) and by ecosystems altered by humans. For example, food and fibre are mainly provided today by human altered ecosystems, with a small proportion of total supply coming from natural or near-natural ecosystems; a situation very different from that when humankind depended mainly on hunting and gathering for the supply of its food and clothing.

There is no doubt that living things and their assemblages play a key role (in many cases, an irreplaceable role) in providing the type of services listed in Table 16.1 and consequently in maintaining human well-being. However, determining their exact role (and especially their exact economic value) is a formidable task, particularly when their global valuation is attempted. In fact, it is impossible to measure in monetary units the absolute value of commodities or services (for example, the availability of water) essential for human life and for which no substitutes exist. The economic value of such commodities can, at best, be measured at the margin of their supply. This is so if willingness to pay or to accept monetary compensation is used as the basis of economic valuation. The larger the scale on which economic valuation of ecosystems is attempted, the greater is the likelihood that the absolute economic value of some ecological services can't be determined.

In many cases, for example, the supply of water for drinking and other purposes is marketed. But its market prices do not indicate its absolute value and the absolute economic value of having access to water cannot be measured, unlike in the case of diamonds, which are inessential to human life and for which the absolute willingness to pay for their supply could be measured. I shall return to this subject and to the diamond–water paradox in Section 16.4.

16.3 OBSERVATIONS ON RECENT ESTIMATES OF THE GLOBAL VALUE OF ECOSYSTEM SERVICES

Particularly since 1997, considerable effort and resources have been allocated to estimating the global economic value of natural capital as a supplier of ecosystem services. Major milestones in this quest were the estimates of Costanza et al. (1997) for 1997, and the comparative estimates of Costanza et al. (2014) for 2011, taking into account the findings of De Groot et al. (2014). Costanza et al. (2014) estimated the mean aggregate global flow of ecosystem values for 2011 at 2007 prices to be US$124.8 trillion. This was 1.66 times global GDP for 2011,which was estimated to be US$75.2 trillion at 2007 prices. These authors also found that compared to its value in 1997, the aggregate global value of the flow of (natural) ecosystem services in 2011 decreased by US$20.2 trillion, that is, by 14 per cent. This involved a reduction in the value of the flow of marine (natural) ecosystem services of 18 per cent and a fall of 11 per cent in the flow of economic value from terrestrial biomes. Their basic method relied on estimating the global size of each biome in hectares and multiplying this by the estimated average economic value of flows of (natural) ecosystem services produced by each biome per hectare in each year. Then these values were added for all the biomes to provide an aggregate global estimate. Per hectare estimates of economic value were based on an array of estimates available in the economic literature, with considerable use being made of the benefit or transfer value method. Limitations of this procedure were noted in Tisdell (2014). Here I want to raise some different considerations.

First, the extent to which these estimates measure only the value of natural ecosystem services supplied by natural capital is unclear. Valuations may not be independent of the supply of other forms of capital and may include some of the value due to the presence of services of human modified ecosystems. This will tend to inflate these values if they are purported to represent only the value of natural ecosystem services due to the presence of natural capital. Second, the dimension of economic value that is being measured is unclear but it is definitely the case that the absolute global economic value of natural ecosystem services is not being measured because it is impossible to express this value in meaningful monetary terms because many of these services are essential to human existence. There are many essential commodities and services that cannot be assigned an absolute value but only a value at the margin of their positive supply. As discussed below, water is one such commodity.

One interpretation of the global values obtained by Costanza et al. (2014) is that they are rough estimates of the exchange value generated by natural ecosystem services (at the time estimated), *if* these services could all be marketed at the marginal economic values estimated for them. Therefore, they might be viewed as representing an attempt to assign to these services similar monetary units to those used to determine the level of GDP. This is one way of bringing public attention to how important these services are for human well-being.

This seems to be the interpretation that Costanza et al. (2014) favour. In relation to their global estimates of the global value of ecosystem services for 2011, they state:

> What we have estimated is the relative contribution of natural capital now, with the current balance of asset types. Some of this contribution is already included in GDP, embedded in the contribution of natural capital to marketed goods and services. But much of it is not captured in GDP because it is embedded in services that are not marketed or are not fully captured in market products and services. Our estimates show these services (i.e. storm protection, climate regulation, etc.) are much larger in relative magnitude right now than the sum of marketed goods and services (GDP) (Costanza et al., 2014, p. 157).

This approach helps to drive home the point that natural capital and natural ecosystem services are very important contributors to human welfare. The main aim of proponents of global estimates appears to be to increase the awareness of members of the public about the economic value of natural capital, and therefore presumably to make them more receptive to conserving natural capital on economic grounds.

What about the direct policy applications of such an approach? These are not so apparent. Given that Costanza et al. (2014) have estimated a total decrease in the flow of economic value supplied mainly by natural ecosystems (due primarily to the reduced stock of natural capital) of US$20.2 trillion in 2011 compared to 1997, this might be expected to activate alarm bells. However, according to their estimates of global GDP (Costanza et al., 2014, p. 156), this was approximately US$46.3 trillion in 1997 and US$75.2 trillion in 2011, an increase of US$28.9 trillion (all valued at 2007 prices). *Consequently*, the increased value of global GDP exceeded the estimated loss in value of natural ecosystem services in this period by US$8.2 trillion. Therefore, the devil's advocate could argue that the trade-off was justified on economic grounds. The problem is that making such trade-offs may result in an unsustainable increase in GDP.

In any case, we probably do not need global economic valuations of ecosystem services to increase the public's awareness of the type of problems that are resulting in natural ecosystem services and natural capital being undervalued in economic decision-making. Studies on more limited spatial states can be used to drive home the point, and may be more suited to the array of valuation techniques available to economists. These techniques seem to be most 'robust' when applied to marginal changes in supplies of resources using partial analysis. However, these techniques also have their basic limitations.

16.4 WILLINGNESS TO PAY AS A MEANS OF VALUING ECOSYSTEM SERVICES – LIMITATIONS

It is not intended here to consider the limitations of *techniques* to elicit willingness-to-pay values. These limitations have been well covered in the scholarly literature. The main purpose is to point out that willingness-to-pay techniques are unable to provide an indication of the absolute value of services or commodities essential to human life. However, absolute values can be estimated, based on willingness to pay (or willingness to accept compensation), for commodities that are not essential to life and for which substitutes exist.[1] Furthermore, variations in willingness to pay for small (marginal) changes in the supply of essential services (the flow of which exceeds the threshold for the existence of human life) are relevant to economic valuation and for policy purposes.

Pearce (1998) expressed concerns about the economic values attributed to essential ecosystem services by those undertaking valuation of global ecosystem services, fearing that they might be considered by some to be an indicator of their absolute value. Costanza et al. (2014, p. 157) strongly reject the view that their estimates of the global value of the flow of ecosystem services are measures of their absolute values and state the following:

> Some have argued that estimating the global value of ecosystem services is meaningless, because if we lost all ecosystem services, human life would end, so their value must be infinite (Chaisson, 2002). While this is certainly true, as we clearly point out in the 1997 paper (Costanza et al., 1997), it is a simple misinterpretation of what our estimates refer to.[2]

As I have pointed out above, they suggest that their estimates of the economic value of global ecosystem services correspond to values similar to those used in accounting for GDP. However, different types of accounting procedures are usually followed to obtain these values. This can be clarified using a diagram similar to that presented by Howarth and Farber (2002, p. 423) for heuristic purposes.

Consider Figure 16.1. There line ABC represents the marginal valuation curve (or demand curve) for the supply of a particular type of ecosystem service, X. Suppose, for simplicity, that it is a pure public good and

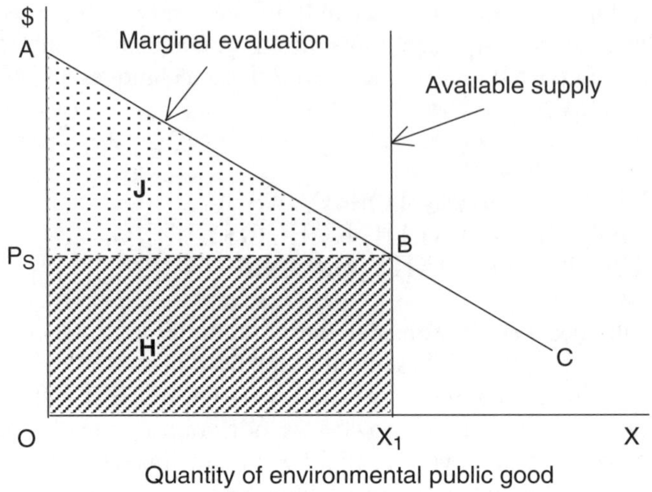

Figure 16.1 Macroeconomic accounting for the value of ecosystem services versus microeconomic welfare accounting for their value: an illustration

represents the demand of a single individual for that good. Also assume that the available quantity supplied of that good per unit of time is X_1. Consequently, the individual's shadow price for the supply of this commodity is P_S. If the individual had to pay a uniform price for each unit of commodity, the individual's payment for it would amount to the hatched area identified by H. It might be argued that this is the type of hypothetical shadow payment corresponding to items included in GDP. However, this does not represent the full willingness of the individual to pay for the supply, X_1, of the environmental service, that is, the absolute value placed on this supply by the individual. That is equal to the hatched area plus the dotted area in Figure 16.1.

Several observations are prompted by this heuristic model. First, this absolute valuation measure is only relevant for non-essential ecosystem services. Furthermore, the amount which an individual is estimated to be willing to pay, based on the shadow price of such an environmental service, is not a very reliable indicator of its absolute economic value to the individual. For example, if the line ABC is steeper or less steep in slope but still passes through point B, the area of the dotted triangle will be correspondingly larger or smaller but the area of the hatched rectangle will remain unchanged. Economic activity measures of the value of ecosystem services can differ greatly from those adopted in traditional welfare economics, as I have discussed elsewhere (Tisdell, 2012).[3] There is also the further complication in relation to the supply of public goods that different individuals will most likely have different shadow prices for the available supply. For the price of marketed goods this does not occur. It means that the research costs involved in determining shadow prices are likely to be quite high for such commodities.

A problem with many estimates of the economic value of ecosystem services is that they utilize heterogeneous measures of their value and extend partial values obtained at a limited geographical scale to a global or macro-scale. This makes the estimates contentious. Nevertheless, economic methods are useful for evaluating the changes in economic value occasioned by alterations to ecosystems of limited size. In addition, attempts to do so on a broader scale can increase awareness about important environmental problems facing humankind and can help in identifying solutions to them. At the same time, there is an urgent need to provide more clarity about what is being measured when the economic valuation of ecosystems is done for the globe, or even sometimes on a smaller scale. At present, a hodgepodge of measures are often combined in a single analysis, depending on what data is available.

16.5 COST-SIDE ALTERNATIVES TO DEMAND-SIDE METHODS OF ECOSYSTEM VALUATION

Concerns have been raised about the subjectivity of willingness to pay and willingness to accept compensation as measures of the value of environmental services based on the perceptions of individuals. That raises the question of what the alternatives are. Costanza et al. (2014, p. 153) state that some 'existing valuation methods like avoided and replacement cost estimates are not dependent on individual perceptions of value'. Up to a point this is true but it is not entirely so. While one can, for example, take into account the objective costs escaped by avoiding illness as a result of stopping a deterioration in some environmental services, the subjective costs of the pain and trauma of ill-health are not (for instance) captured by such an approach. Furthermore, individual perceptions are likely to have to be taken into account if social costs, rather than private costs, of the loss of natural ecosystem services are an important consideration in ecosystem valuation.

Although these difficulties exist, it is pertinent to observe that in many cases fine-tuning of measurement is not necessary to choose between the economic values of alternative changes to ecosystems, including the possibility of no change. This is because even given some uncertainties about the estimated values for one ecosystem, A, compared to another changed one, B, the economic values for one may dominate those for the other. Furthermore, even when uncertainty exists, some possible choices can usually be ruled out as irrational (see, for example, Tisdell, 2009, pp. 143–5). This narrows the set of possible rational choices. Moreover, optimal choice is possible in many cases when bounded rationality exists (Tisdell, 1996). Consequently, the presence of uncertainty does not rule out the exercise of rationality.

Return now to specific assessment of the cost of replacement method: while it can provide a reliable measure of economic value of ecosystem services in some circumstances, in many cases it overstates this value. This seems to be so in several studies of ecosystem services which utilize it as part of the total valuation (for example, Xue and Tisdell, 2001; Ninan and Inoue, 2013). However, many of these studies were designed to raise awareness about the value of ecosystem services rather than to provide precise measures of their value (Costanza, 2014).[4]

Situations in which use of the cost of replacement methods results in biases in economic valuation, and where they do not, are illustrated by the following examples. In Figure 16.2, x_n represents the annual volume of water provided by an ecosystem, and line DEF indicates the extra value that human beings put on the supply of this water. In aggregate, they value this total supply by an amount equivalent to the hatched area. The cost of

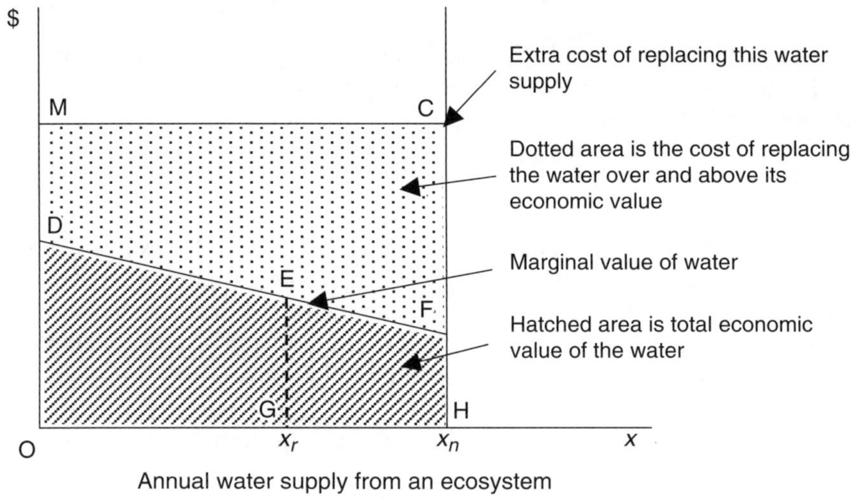

Figure 16.2 An illustration that the cost of replacement of an ecological service can overstate its economic value: in this case, the economic cost of replacing the water supply exceeds the economic value of the water supply

supplying this water by alternative means (for example, by a dam or pipeline from elsewhere) is shown by the line MC and its supply is supposed to cost \$OM per unit. Therefore, the total cost of supplying x_n of water will be equal to the dotted area plus the hatched area. Thus, the economic value of this water will be overstated by an amount equal to the dotted area because the loss in economic value if all water supplies disappear due to a change in the ecosystem will be equivalent to the area of the hatched quadrilateral. A further problem is that when the ecosystem is changed, not all water supplies may be lost. For example, supplies may be reduced to x_r annually. In that case, the total loss in the economic value of this water service will be equal to the area of quadrilateral GHFE.

In some instances an ecological service is not fully demanded. In the case illustrated in Figure 16.3, the extra value placed on water is shown by line DNF. Even if the water supply is free, a maximum quantity of water of x_2 is demanded. The supply $x_n - x_2$ of water from this ecosystem has no economic value. If line MNC represents the per unit cost of replacing this water, it will be optimal to only replace x_1 of the water if all the water supply from the area is lost. When the ecosystem disappears, this will result in an economic loss of an amount equal to the area of rectangle OGNM

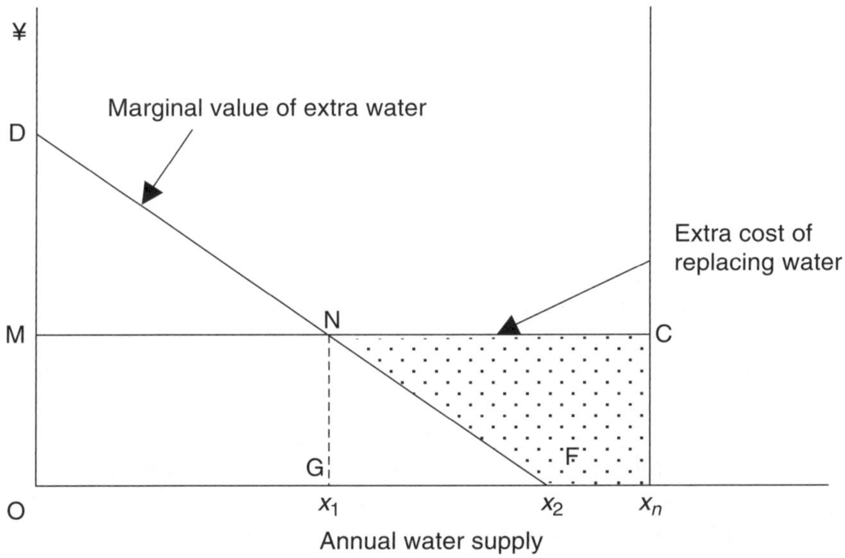

Figure 16.3 An illustration of another case in which the cost of replacement method can overstate the economic value of an ecological service: the dotted area represents the excess of the cost of replacing water compared to its economic value

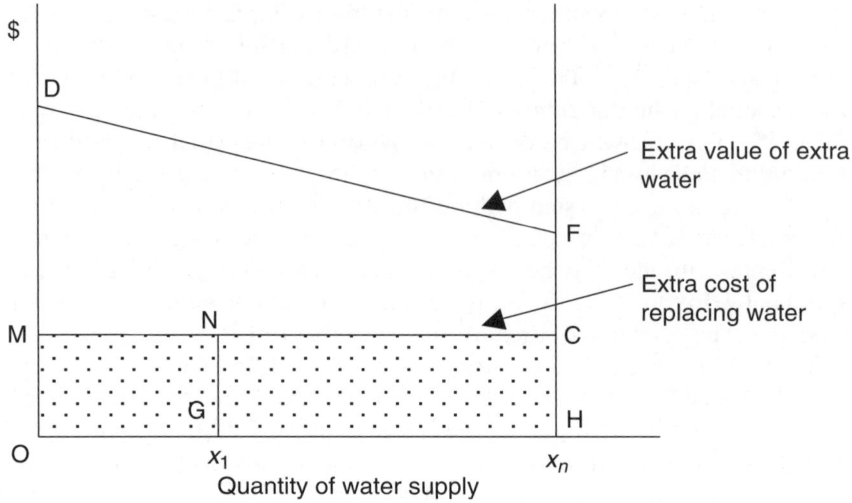

Figure 16.4 *An illustration that the cost of replacement method can sometimes provide an accurate measure of economic value of an ecosystem. However, whether it does so depends on what services are provided by the changed ecosystem: in the case shown, the cost of replacing lost water in an ecosystem is less than its economic value*

(the cost of supplying water by alternative means) plus a stemming loss equivalent to the area of triangle GFN. This is a lower economic loss than would be incurred by replacing the whole of the water supply, x_n. This type of estimate of replacement costs will overestimate the economic burden of losing all the water supplies (provided by the ecosystem) by an amount equivalent to the dotted area in Figure 16.3.

Of course, there are some situations in which the cost of replacement would be an accurate measure of an ecological service. Such a case is illustrated in Figure 16.4. There line DF represents the extra value of water from the ecosystem and MC is the per unit cost of supplying it by other means. A cost equivalent to the dotted area will be incurred if the ecosystem disappears and *all* water supply is lost. However, this may not be so. If x_1 of water continues to be supplied by the changed ecosystem, then the additional cost incurred, as a result of the changed ecosystem, is equivalent to the area of rectangle GHCN, *not* the area of rectangle OHCM. This shows the complexity of the problem, and the need to do comparative analysis.

Despite these problems, it is important to complete economic valuations of ecosystems because some ecosystems which appear at first sight to have little economic value in fact have a high economic value when their unpriced or under-priced services are taken into account. These valuations have significant implications for the optimal use of terrestrial and marine areas. However, methods for economic valuation of alternative ecosystems need further assessment and development. This was illustrated by showing that the application of the cost of replacement method of valuing ecological services associated with an ecosystem can provide misleading estimates of the economic value of these services.

16.6 FURTHER OBSERVATIONS ON THE VALUATION OF CHANGES IN ECOSYSTEMS: EXAMPLES AND THE ROLE OF ECONOMICS

The above critique of economic methods which have been used for valuing natural ecosystems, particularly at a global scale, does not imply that economic methods are unable to play a useful role in valuing changes in ecosystems. These methods can be very useful for assessing (from an anthropocentric point of view) the desirability or otherwise of *changes* to ecosystems for which the geographical scales are not excessively large. For example, the biophysical changes due to economic development or activity in a river system may be studied and these changes provide opportunities for economic valuation. Or again, the biophysical consequences of alterations to a natural area undergoing desertification can be studied and the benefits and costs associated with these are amenable to economic valuation. Similarly, attempts to 'restore' ecosystems can often be valued from an economics point of view, such as measures in China to counteract advancing desertification in some of its western regions. Considerable scope also exists for economic valuation of the consequences of the introduction of exotic species to new regions or countries. If such species prove to be a pest, economic valuation has an important role to play in determining the net benefit of their control or their elimination. It is also relevant to assessing the effort which should be put into border control to prevent the introduction of potential pests.

Valuing changes to an ecosystem is basically a two-stage process:

- The biophysical impacts of a disturbance to an ecosystem and its relationship to the flow of ecosystem services need to be specified. This is generally the role of natural scientists and engineers, for most services. This is akin to specifying a production function.
- It is these relationships that provide the basis for economic valuation.

The reliability of the economic valuation depends on the accuracy of both these components. Shortcomings in specifying biophysical relationships frequently result in flawed evaluation of environmental changes, as can failures in the purely economic component of the valuation of biophysical changes.

Sometimes, it is easy (at least, in retrospect) to assess changes in economic value brought about by alterations to ecosystems, but in most instances, this is not so because there are losers and gainers from such changes, or some beneficial and detrimental effects occur. Consider some different cases involving the introduction of exotic species to Australia. The first of these involves clear overall economic loss because of the negative effects on Australian ecosystems, the second is definitely of positive economic value and the third is of an uncertain overall economic value.

The cane toad (*Rhinella marina*) was introduced to Australia in 1935 from Hawaii (but came originally from South America) by the Bureau of Sugar Experiment Stations in order to control the grey-backed cane beetle (*Dermolepida albohirtum*) and the Frenchi beetle (*Lepidiota frenchi*), both of which damage sugar cane. Unfortunately, the introduction of the toads has not only caused large environmental damage but there is also no evidence that they have had an impact in reducing populations of the cane beetles they were introduced to predate (Anon, 2015). Most native predator species in Australia are susceptible to fatal poisoning as a result of eating cane toads and this has seriously diminished the populations of several native species. Also the populations of some native insectivores and skinks have declined due to cane toad predation. The cane toad in Australia has virtually no economic benefits, although one enterprise produces curios for tourists from them. It is, nevertheless, quite clear that their introduction to Australia has resulted in the economic value of Australian ecosystems being reduced.

Prickly pears (*Opuntia* spp.) were introduced to Australia in the eighteenth and nineteenth centuries for varied purposes, for example to develop a cochineal industry and for providing hedges as forms of enclosures. They soon became a major pest. They invaded considerable areas of agricultural land and made them unusable, and in addition, they colonized many natural habitats. The environmental and agricultural cost was enormous but areas of infestation were greatly reduced by biological controls, mainly the introduction of the cactoblastis moth (*Cactoblastis cactorum*) in the 1920s (Seabrook and McAlpine, 2015). There is no evidence that these biological controls have had any adverse effects on natural and human modified ecosystems in this case. Their introduction into Australian ecosystems resulted in a clear net economic benefit.

In the above two cases, detailed economic evaluation is unnecessary. The first case, the introduction to Australia of the cane toad, involved a Paretian economic loss, whereas in the second case, the introduction of the cactoblastis moth essentially resulted in a Paretian economic gain. However, many other cases involving the introduction of exotic species are more complicated because there are significant gainers and losers. This appears to be so in the case of buffel grass, which is an exotic species introduced to Australia. It has caused natural ecosystem degradation, habitat loss and species decline in arid and semi-arid Australia (Australian Government, 2015). Some graziers see it as a beneficial source of fodder but others regard it as a pest. So it is not possible a priori to assess the net economic benefit or detriment of its presence. Even if it were found (using the Kaldor–Hicks criterion) that the presence of buffel grass in Australia resulted in a substantial overall economic loss, it may only be economic to control it to a limited extent, and consequently its presence will still continue to have a negative economic effect after controlling it to the most economic extent.

Once an invasive species is introduced to a new region, it is often uneconomic to eradicate it, even though its presence results in a substantial economic burden. Frequently, changes to the affected ecosystems are economically and ecologically irreversible. This is why considerable caution needs to be exercised before introducing exotic species to a new region.

In the case of the introduction of the cane toad to Australia, some (but not enough) experimental precaution (Anon, 2015) was taken before its release in Queensland. No precaution was taken in first introducing various species of prickly pear to Australia in the early years following European settlement.[5] It seems that buffel grass was first introduced to Australia in the stuffing of saddle-packs – its introduction was accidental not deliberate. However, later there were deliberate introductions of its seed. Some graziers sowed the seed and still sow it and the seed is commercially available. One Australian state, the State of South Australia, has declared buffel grass to be a noxious weed, but this is unlikely to happen in the northern states of Australia because a significant group of graziers (but not all) believe that the presence of buffel grass is economically advantageous.

There are many other introduced species that adversely affect the functioning of ecosystems in Australia, but Australia is not unique in that regard.[6] A similar situation exists, for example, in the United States and sometimes the same species are involved, and the same types of economic issues arise.

A recent Australian case is the accidental introduction of fire ants (*Solenopsis invicta*). It is believed that it was introduced in dirt on second-hand farm machinery imported from the United States entering Australia through the port of Brisbane. It has invaded the Brisbane area and its population is spreading despite efforts to stop this and eradicate this ant. The species is also a pest in the United States to which it was also accidentally introduced. So far, efforts to eradicate it in the United States have not succeeded. The presence of the fire ant when introduced to ecosystems where it is not naturally present causes marked ecological damage and has no economic benefit.

Economics has an important role to play in the valuation of changes in ecosystems which can occur as a result of the introduction of exotic species. This includes policies for preventing the introduction of economically deleterious species, and policies for their control or eradication if introduced. However, the accuracy of the economic assessments depends on the quality of the available information about the biophysical relationships involved. There are also many other types of disturbances to existing ecosystems which are amenable to economic analysis.

Faulty assessment of biophysical relationships frequently results in faulty policies being adopted to alter ecosystem functioning. The release of the cane toad in Australia is one such example. A more recent example appears to be China's efforts to halt desertification in some of its western regions. China has engaged in tree planting in some of these regions to control desertification, but in many instances the trees have failed to survive and might have accelerated sand erosion in some cases. Cao et al. (2011) argue that planting grasses would be a more effective policy and they claim that the planting of thirsty trees in dry regions is questionable strategy for reducing desertification (Cao et al., 2010). Furthermore, Cao et al. (2009) argue that bans on grazing have been more important in reducing desertification in several parts of China rather than tree planting. In any case, since major projects for the restoration of ecosystem functions in China are undertaken by the State Forest Administration, this authority may be biased in favour of tree planting rather than other forms of restoring the vegetation cover of degraded ecosystems (Tisdell and Xue, 2013). If the Chinese critics of the biophysical impacts of China's greening projects are correct, then the State Forest Administration has failed to correctly predict the biophysical consequences of its policies in some areas, or it is hampered by its mission being focused on forest-based solutions to the provision of ecosystem services.

In concluding this section, I would like to express the view that economic valuation seems more relevant to assessing the desirability of the following types of changes rather than in accounting for the global value of ecosystems. These are:

- assessing the value of changes in ecosystems;
- determining the (net) economic value of preventing undesirable changes to ecosystems;
- determining the (net) economic benefit of reversing changes in ecosystems to restore one or more of their functions.

16.7 MORE ON PAYMENTS FOR THE PROVISION OF ECOSYSTEM SERVICES

In recent years, there has been strong support among scholars for governments or NGOs paying for the supply of wanted ecosystem services which would not be provided or which would be insufficiently supplied in the absence of targeted intervention payments for the provision of ecosystem services. However, these payments have their drawbacks and there are also alternative types of intervention which should be assessed, as was pointed out in Chapter 14. One problem is that payments for ecosystem services can be short-lived, and once they cease, the services targeted for provision can cease to be supplied, sometimes permanently. This is because governments frequently change their policies and the financial fortunes of NGOs supporting such schemes vary.

The lack of permanency aspect can be illustrated by China's Grain for Green Project. This is a part of China's tree planting programme (the largest such contemporary programme undertaken by any country), which is intended to restore particular ecosystem services, such as to reduce desertification, soil erosion and flooding, depending on the location of this tree planting. For example, Cao et al. (2009, p. 1182) found from a survey of those participating in the Grain for Green Project in northern Shanxi Province that around 40 per cent of participants 'planned to return to cultivating forested areas and grassland once the project's subsidies end in 2018'. Consequently, the long-term environmental benefits of the project will be greatly reduced. A long-term solution is likely to depend on alternative means of livelihood becoming available to those farm families that now depend on fragile marginal agricultural areas for their subsistence, for example, the possibility of being employed in urban areas.

The problem is that farm families in areas subject to severe environmental degradation due to agricultural use are poor and have few alternatives to using this land for agriculture, if they are to survive. In such cases, imposing financial penalties on them for using the land for agriculture (despite the loss of natural ecosystem service which results) is not a policy option. In this respect, Uchida et al. (2007) found that China's Grain for Green Project had been very successful overall in reducing soil erosion but has been only moderately successful in alleviating the poverty experienced by its participants. They also expressed doubts about its long-term benefits in relieving rural poverty because few beneficiaries had shifted to off-farm work (Uchida et al., 2007, p. 617). This was probably because of the limited availability of such work.[7]

A second issue involves principal–agent problems. It is often very difficult or costly for the principal or donor of funds paying for ecosystem services to monitor how the funds are actually used. This has been evident in China's Grain for Green Projects. The funding for these projects is from the central government (the principal) but their implementation relies on local governments (the agents). Zheng et al. (2011, p. 3) found that 'on average only about 49 per cent of the grain subsidy and 23 per cent of the cash subsidy has been received by program recipients participants' in the Grain for Green programme because of 'local government reallocation of subsidies for other uses', and other factors. The other factors could include a large amount of these funds being used 'to support' administrators. The 'loss of funds' before reaching participants is really quite staggering. If Zheng et al. (2011) are correct, more than half the grain intended for farmers for this programme and more than three-quarters of the cash was not received by farmers for the purpose intended by the central government.

Also there is the question of compliance at the farm level. It seems that in some cases the subsidy for greening was received but lack of compliance occurred. For example, Uchida et al. (2007, p. 217) report that some farmers received the subsidy but used it to increase their livestock numbers. Presumably, a few privileged members of the local communities were able to take advantage of the scheme in this way. Increased livestock numbers accelerate desertification in areas subject to this phenomenon.

It should not be thought that these types of problems are peculiar to China, or that they only occur in developing countries. They can arise in a variety of different contexts and they are not confined to payments made by governments. NGOs may experience the same type of problems in paying for the supply of ecosystem services as well as international agencies, such as UN organizations. For example, payments for restoring forest cover and for avoiding its depletion in order to maintain or increase their level of carbon sequestration under international REDD schemes, are presumably subject to problems similar to those identified for China's Grain for Green Project.

As was pointed out in the previous section, it has been claimed that the most effective techniques for countering desertification in China have not been adopted by its State Forest Administration. Once again, such problems are not peculiar to China or to payments for ecosystem services made by governments. Well-meaning NGOs may make similar mistakes. For example, if the funding body were merely to support tree planting on farms, species of trees may be planted by farmers that do not maximize social benefits (see Jindal and Kerr, 2014, especially p. 392 for the results of a case study in Kenya).[8]

Another aspect of payment for the supply of ecosystem services worthy of study is consideration of the factors influencing the willingness of landholders to participate voluntarily in such schemes. Kramer et al. (2014) have investigated this in North Carolina and have identified several factors which seem to be important there. They found, for example, from their sample, that there was a preference for shorter-term contracts and for programmes provided by government rather than those of NGOs or private companies. It is possible, however, that the factors influencing the willingness of landholders to participate in such schemes will vary between geographical areas and in different communities. However, presumably in most cases there will be greatest willingness to allocate land for conservation purposes which has a low use value to the landholder in return for payment for doing this.

16.8 CONCLUDING COMMENTS

It has been stressed that although natural capital is an important contributor to human welfare, and its importance has frequently been overlooked, some ecological economists in their eagerness to rectify this neglect have paid insufficient attention to other types of resources that contribute to human well-being, such as human modified ecosystems and the stock of human developed germplasm. Similarly, a number of limitations of estimates of the value of the global flow of (natural) ecosystem services were noted. The imprecision of these estimates has been defended by their proponents on the basis that they are mainly intended to raise awareness of the economic value of such services in comparison to values accounted for by GDP estimates. Although in principle it would be possible to value the flow of ecosystem services on a similar basis to GDP (as was shown), the available estimates do not consistently incorporate this approach. In the main, they rely on a variety of methods to provide estimates for different types of ecosystem services, the values of which are aggregated. The methods often include estimates of willingness to pay, the cost of replacement and the costs avoided. Significant limitations of such methods for valuing ecosystem services were noted, such as their inability to determine the absolute values of essential ecosystem services, and in the case of the cost of replacement method, its proneness to overstating economic values, even though it can provide an accurate estimate of economic value in some cases.

It has been contended that economic methods are most helpful in valuing changes in ecosystems. However, these assessments depend on adequate specification of biophysical changes as well as appropriate economic valuation of the changes involved. It is not uncommon for biophysical changes in ecosystems to be incorrectly predicted, or not to be predicted at all, and consequently, failures occur in ecosystem management, sometimes resulting in large (and irreversible) economic losses. A few of the many possible examples were given.

The policy of paying for the supply of wanted ecosystems was given consideration additional to that in Chapter 14. Matters given extra consideration included the permanency of such payments, whether they are likely to provide a long-term solution to ensuring the supply of such services, and principal–agent problems often associated with these policies. Although the illustrations of these difficulties were mainly drawn from China's Grain for Green programme, it was stressed that such problems are not peculiar to China or to government conducted programmes of this type.

This scrutiny of economic methods which have been used to value ecosystem services is not intended to convey the message that such valuation is futile or that it is of no value. In fact, there are many areas where economics has made and can make a worthwhile contribution to determining changes in economic values associated with ecosystem change. Undoubtedly, further progress will be needed, but we also need greater appreciation of what economic valuation methods can and cannot do.

NOTES

1. This problem has similarities with the diamond–water paradox (Anon, 2014) but is somewhat different in its nature to this paradox, as presented, for example, by Adam Smith (1910, Vol. I, p. 25). This is because there is no difference in absolute value of commodities essential to life and it is meaningless to speak of their relative absolute value. Nevertheless, relative marginal values of commodities essential for life are relevant when their supply exceeds minimum essential levels.
2. However, it would not be correct to contend that these values are infinite in terms, for example, of willingness to pay. It is more appropriate to say that they are not amenable to economic monetary valuation.
3. In fact, existing global and regional estimates of the value of ecosystem services are based on microeconomic welfare accounting and usually include a selection of demand-side and cost-side values. Certainly, the estimates of Costanza et al. (2014) do not use macroeconomic type values, that is economic activity measures. Therefore, their suggestion is dubious that their estimates of the economic value of global ecosystem services are similar to those used in accounting for GDP.

4. In Table F.1 in a Foreword, Costanza (2014) outlines a range of uses for ecosystem valuation and suggests the degree of precision required in their economic valuation at different spatial scales. He considers that the prime use of economic valuation at the regional to global scale is to raise awareness and interest in the value of ecosystem services and that a low degree of precision is required in economic valuation at these levels for these purposes.
5. The same is true of introduced rabbits, which became a very serious Australian agricultural pest, as well as a menace to Australia's natural ecosystems. Control of their populations has, for now, been achieved by the introduction of pathogens.
6. These include several types of feral animals. The presence of all of these has had negative impacts on native ecosystems and most also have negative consequences for the functioning of human developed ecosystems. These feral animals include cats, dogs, donkeys, pigs, goats, buffalo, camels, horses and cattle. Tisdell (1982) considers various impacts of feral pigs on ecosystems in Australia.
7. One possible strategy would be to provide subsidies for relocation of younger family members for training and movement to urban areas.
8. The scheme which they studied was not funded by an NGO but was part of the Western Kenyan Integrated Management project and was funded by the Global Environment Facility of the United Nations.

REFERENCES

Anon (2014), 'Paradox of value', *Wikipedia.* Available at: http://en.wikipedia org/w/ index.php?title=Paradox_of_value. (Accessed 17 February, 2015).

Anon (2015), 'Cane toads in Australia', *Wikipedia.* Available at: http://en.wikipedia.org/wiki/cane_toads_in_Australia. (Accessed 24 February, 2015).

Australian Government, Department of Environment (2015), 'Ecosystem degradation, habitat loss and species decline in arid and semi-arid Australia due to the invasion of buffel grass (*Cenchrus ciliaris* and *C. pennisetiformis*) – threat abatement actions', Commonwealth of Australia. Available at: http://www.environment.gov.au/biodiversity/threatened/threat-abatement-advices/buffel-grass-threat-abatement-actions. (Accessed 24 January, 2015).

Cao, S., L. Chen and X. Yu (2009), 'Impact of China's Grain for Green Project on the landscape of vulnerable arid and semi-arid agricultural regions: A case study in northern Shaanxi Province', *Journal of Applied Ecology*, **46**, 536–43.

Cao, S., G. Wang and L. Chen (2010), 'Questionable value of planting thirsty trees in dry regions', *Nature*, **465**, 31.

Cao, S., G. Sun, Z. Zhang, L. Chen, Q. Feng, B. Fu, S. McNulty, D. Shankman, J. Tang, Y. Wang and X. Wei (2011), 'Greening China naturally', *Ambio*, **40**, 828–31.

Chaisson, E.J. (2002), *Cosmic Evolution: The Rise of Complexity in Nature*, Cambridge, MA: Harvard University Press.

Costanza, R. (2014), 'Foreword: The importance of valuing ecosystem services', in K. Ninan (ed.), *Valuing Ecosystem Services: Methodological Issues and Case Studies*, Cheltenham, UK and Northampton, MA, USA: Edward Elgar Publishing, pp. xvii–xxiii.

Costanza, R., R. De Groot, P. Sutton, S. Van der Ploeg, S.J. Anderson, I. Kubiszewski, S. Farber and R.K. Turner (2014), 'Changes in the global value of ecosystem services', *Global Environmental Change*, **26**, 152–8.

Costanza, R., R. d'Arge, R. De Groot, S. Farber, M. Grasso, B. Hannan, K. Limburg, S. Naeem, R.V. O'Neill, J. Paruelo, R.G. Rasking, P. Sutton and M. Van den Belt (1997), 'The value of the world's ecosystem services and natural capital', *Nature*, **387**, 253–60.
De Groot, R., et al. (2014), 'Global estimates of the value of ecosystems and their services in monetary units', *Ecosystem Services*, **1**, 50–61.
Howarth, R. and S. Farber (2002), 'Accounting for the value of ecosystem services', *Ecological Economics*, **41**, 421–9.
Jindal, R. and J. Kerr (2014), 'Explaining demand for tree planting in a payment for environmental service project in Lake Victoria Basin, Kenya', in K.N. Ninan (ed.), *Valuing Ecosystem Services: Methodological Issues and Case Studies*, Cheltenham, UK and Northampton, MA, USA: Edward Elgar Publishing, pp. 378–94.
Kramer, R., A. Jenkins and A. Lesser (2014), 'The role of ecosystem payments in achieving conservation goals: Attitudes among farm operators', in K.N. Ninan (ed.), *Valuing Ecosystem Services: Methodological Issues and Case Studies*, Cheltenham, UK and Northampton, MA, USA: Edward Elgar Publishing, pp. 395–411.
Kumar, P. (ed.) (2010), *The Economics of Ecosystems and Biodiversity: Ecological and Economic Foundations*, London and Washington, DC: Earthscan.
Millennium Ecosystem Assessment (2005), *Ecosystems and Human Well-being: A Synthesis*, Washington, DC: World Resources Institute.
Ninan, K.N. and M. Inoue (2013), 'Valuing forest ecosystem services: What we know and what we don't', *Ecological Economics*, **93**, 137–49.
Pearce, D.W. (1998), 'Auditing the earth', *Environment*, **40** (2), 23–8.
Seabrook, L. and C. McAlpine (2015), 'Prickly pear', *Queensland Historical Atlas* [Online]. Available at: http://www.qhatlas.com.au/content/prickly-pear. (Accessed 24 February, 2015).
Smith, A. (1910), *The Wealth of Nations*, London: Dent and Sons.
Tisdell, C.A. (1982), *Wild Pigs: Environmental Pest or Economic Resource?*, Sydney, Oxford and New York: Pergamon Press.
Tisdell, C.A. (1996), *Bounded Rationality and Economic Evolution*, Cheltenham, UK and Brookfield, VT, USA: Edward Elgar Publishing.
Tisdell, C.A. (2009), *Resource and Environmental Economics: Modern Issues and Applications*, Singapore, Hackensack (NJ), London and Beijing: World Scientific.
Tisdell, C.A. (2012), 'Economic benefits, conservation and wildlife tourism', *Acta Turistica*, **24**, 127–48.
Tisdell, C.A. (2014), 'Ecosystem functions and genetic diversity: TEEB raises challenges for the economics discipline', *Economic Analysis and Policy*, **44** (1), 14–20.
Tisdell, C.A. and D. Xue (2013), 'Managing ecosystem services for human benefit: Economic and environmental policy challenges', in E. Creighton and P. Davovich (eds), *Environmental Policy: Management, Legal Issues and Health Aspects*, New York: Nova Science Publishers, pp. 87–106.
Uchida, E., J. Xu and S. Rozelle (2007), 'Are the poor benefiting from China's land conservation program?', *Environment and Development Economics*, **12**, 593–620.
Xue, D. and C. Tisdell (2001), 'Valuing ecological functions of biodiversity in

Changbaishan Mountain Biosphere Reserve in Northeast China', *Biodiversity and Conservation*, **10** (3), 467–81.

Zheng, H., P. Glenwe, S. Polasky and J. Xu (2011), 'Reputation, policy risk and land use: A study of China's "Grain for Green" programme', *UNU-WIDER Working Paper*, WP 2011/39, Helsinki: World Institute for Development Economics, Research, United Nations University.

17. Economics and spatial aspects of ecosystem use: land sharing versus land sparing – wildlife conservation and forestry

17.1 INTRODUCTION

Spatial dimensions of ecosystem use have a major influence on the supply of ecosystem services. They affect the composition of this supply and the volumes of the supply of these services. Considerable attention has been given by ecologists to the effects on the conservation of species of changing the spatial dimensions of ecosystem use. Moreover, it is clear that changes in the patterns and mosaics of land use have other major environmental consequences. Nevertheless, at the same time, the biophysical relationships involved are often complex and frequently cannot be generalized. While much attention has been given to the biophysical consequences of the spatial use of ecosystems, much less attention has been given to the economic assessment of these patterns.

The purpose of this chapter is to show the relevance of economics to assessing the value of alternative spatial forms of ecosystem use. To do this, its focus is mainly on the economics of alternative patterns of logging and their consequences for the conservation of forest wildlife and other ecosystem services. In order to do so, it concentrates primarily on a concrete case, namely the economics of conserving the orangutan by means of alternative patterns of logging forests, for example, intensive forms of logging versus less intensive selective forms. Therefore, it extends the discussion in Chapter 12.

This chapter is relevant to the debate about the virtues of land sharing versus land sparing as alternative approaches to supplying ecosystem services. Aspects of the debate about this are discussed in the next section of this chapter. Then the subject of the economics of selective (reduced) impact logging as a means to conserve forest-dependent species and desired ecosystem services is introduced and some background is provided on the case modelled here, namely the conflict between orangutan conservation and forestry. Detailed modelling and a discussion follow.

17.2 THE DEBATE ABOUT THE DESIRABILITY OF LAND SHARING VERSUS LAND SPARING AS A STRATEGY FOR PROVIDING ECOSYSTEM SERVICES

There has been considerable debate about whether land sharing or land sparing is the most desirable means of providing ecosystem services. The dominant paradigm, at present, is that intensive use (a land-saving strategy) for the supply of food from agriculture and from aquaculture and timber from forests, is likely to be superior, that is 'superior' in providing conservation services (and most other ecosystem services) to a land-sharing strategy involving less intense land use but requiring increased extension of land use to obtain the same level of production (Brussaard et al., 2010; Phalan et al., 2011; and Chapters 5 and 8 of this book). While this may appear to call for greater intensification along the same lines as those evident in the Green Revolution, advocates of this second wave of intensification hope that it can be achieved by the use of more environmentally friendly means, such as greater reliance on ecologically-friendly biological mechanisms to achieve the desired results (Brussaard et al., 2010, p. 36; see also Chapters 5 and 8 in this book). One needs, however, to be wary about generalizing about this subject. Whether land sparing or land sharing is to be preferred is likely to vary with its geographical context or location, and certainly the nature of the intensification needs to be closely examined.

Some methods of land sharing for the supply of food may require only a small amount of agricultural food production to be forgone but can have a large positive effect on the supply of other wanted ecosystem services, for example the retention of natural vegetation around watercourses. The amount of extension of land use to compensate for food loss on land already used for agriculture may be quite low in this case, and consequently, the adoption of policy measures to ensure the retention of this vegetation may be justified. Similarly, banning the filling of ponds to increase the ease of the use of land by farm machinery may be justified.

Furthermore, increasing intensity of the agricultural use of land on steep slopes or on land subject to desertification (which is already used for agriculture) may not only result in lack of sustainability of food production, but can have serious adverse environmental consequences, unless special methods of culture are available. China, for example, has adopted policies to reduce the intensity of the use of these areas for agriculture (see the previous chapter of this book and Tisdell and Xue, 2013). In regions subject to desertification, light grazing by livestock combined with their nomadic movement would possibly be the most desirable policy for reducing desertification and for the long-term supply of livestock products, that is, extensive use of the land rather than its intensive use.

Phalan et al. (2011) compared crop yields and the densities of bird and tree species across gradients of agricultural intensity in south-west Ghana and northern India. They came to the conclusion that 'more species were negatively affected by agriculture than benefited from it, particularly for species with small global ranges. For both taxa in both countries, land sparing is the more promising strategy for minimising negative impacts of food production, at both current and anticipated levels of production' (Phalan et al., 2011, p. 1289). It is not apparent, however, that this general conclusion follows from this evidence.

Sometimes, changes can be made to land use on farms and elsewhere that result in little reduction in yields but that add considerably to the sustainable supply of ecosystem services. For example, on sloping lands, soil- and water-conserving barriers can be installed along the contours of their slopes. In some cases, it may also be desirable on sloping land to switch from annual crops to perennials. Occasionally, strategic retention of tree cover on agricultural land helps sustain long-term provision of environmental services as well as agricultural production. For instance, retaining tree cover on elevated areas can reduce the occurrence of dryland salting in regions where this can be a problem. Changes in land-use practices of this type do not fit neatly into the agricultural intensification versus agricultural extension dichotomy. They can add both to the sustainability of production on land already used for agriculture (and other purposes) and can help sustain the supply of wanted environmental services. This is not to say that such 'win–win' outcomes are always possible. Different changes in patterns of land use require individual assessment in order to determine their consequences for commercial production of commodities and the supply of environmental services. Furthermore, it is important to provide some valuation of changes in ecosystem services which occur as a result of alterations in land use. Many ecologists, including Phalan et al. (2011), do not do this. For example, ideally one should not merely count the number of species associated with alternative forms of land use but also take into account their (economic) value, which in turn is likely to be influenced by, among other things, the degree of their endangerment (Tisdell, 2014). I have discussed some additional issues relevant to this topic elsewhere. See, for example, 'Conserving and planting trees on farms: lessons from Australian cases', 'Integrating economics within the framework for the evaluation of sustainable land management (FESLM)' and 'Agricultural sustainability and conservation of biodiversity: competing policies and paradigms', Chapters 7, 8 and 9 respectively in Tisdell (2003).

The additional sections of this chapter (which examine the economics of conserving forest-dependent wildlife by altering logging practices) also suggest that whether or not mixed (shared) land use is desirable from an economics cum conservation point of view, depends on specific circumstances. Land-use practices which affect the conservation of the orangutan are examined to illustrate this point.

17.3 INTRODUCTION TO SELECTIVE LOGGING AS A MEANS TO CONSERVE FOREST DEPENDENT WILDLIFE AND WANTED ECOSYSTEM SERVICES

Comparatively little attention has been given in the literature to the economics of selective (reduced impact) logging as a means for conserving forest-dependent wildlife species and maintaining other ecological services. However, the subject of the spatial optimization of the use of ecosystems has received much attention by ecologists. Although Hof and Bevers (1998; 2002), for example, analyse in detail the ecological and environmental consequences of the spatial use of forest ecosystems and the effects of logging regimes on their ecological services, economic considerations are not central to their expositions. Their aim is to examine 'the use of optimization in the management of an ecosystem with the objective of directly capturing spatial ecosystem relationships and processes' (Hof and Bevers, 1998, p. 1). Because the problems posed by these authors are complex, they cannot always be solved by mathematical optimization methods. They usually have to rely on simulation approaches and approximations for solutions. Often linear approximations are needed to solve these problems and linear programming is a useful tool for obtaining solutions (see, for example, Hof and Bevers, 2002, Ch. 5).

The model presented in this chapter for considering the spatial use of a forest is simple and also involves linearity. Nevertheless, it is able to identify circumstances in which multiple use of a part of the forest is optimal, (land sharing) and others in which spatial use of the forest should be partitioned into its dominant uses (land sparing for conservation purposes). It is assumed that the spatial use of the forest should be such as to maximize the profit from logging subject to the condition that a focal wildlife species continues to survive in the forest at a particular level of population. To some extent, therefore, the valuation system adopted is lexicographic because the focal wildlife species is required to continue to exist at a targeted level of population, for example, a minimum viable population.

The rest of this chapter unfolds as follows: first, background material, such as the pre-existing literature, is considered and then the general nature of the chosen forest-use problem is outlined. Subsequently, a linear model is introduced to simplify the problem and solve it. It is argued that the model is not as restrictive as it may seem to be at first sight. Ways of extending the model and applying it to the assessment of other services (such as water quality) are suggested, and some of the complications raised by dynamics are highlighted.

Although the modelling used is not all-encompassing, its simplicity highlights important issues that otherwise can be easily overlooked in valuing ecosystem services and in determining optimal ecosystem use.

17.4 BACKGROUND INFORMATION RELEVANT TO THE MODELLING

Forests continue to disappear as forested land is converted to non-forest uses (such as agriculture) and many natural forests are still subject to heavy logging, often as a prelude to the conversion of the land to agriculture, thereby threatening the conservation of forest species. For example, the survival of the endangered orangutan is threatened by deforestation in Borneo and Sumatra and by the conversion of the forested land to agricultural use, mostly for the cultivation of oil palm (Swarna Nantha and Tisdell, 2009). However, orangutans (and several other forest species) can survive in selectively logged land or on land that is logged but not so heavily (Husson et al., 2008). In view of this relationship, some conservationists support non-intensive logging as a land use. For example, Payne and Prudente (2008, p. 145) state: 'Orangutans can survive and breed happily in natural forests that are well managed for timber production. This has already been proved since the 1990s in the 55 000 hectare Deramakot Forest in Sabah [Malaysia] certified under Forest Stewardship Council principles and criteria', and they claim that this 'forest sustains one of the highest recorded populations of densities of orangutans in logged lowland dipterocarp forest'. They advocate the adoption of similar selective logging procedures for managing forests sustainably in other areas containing orangutans. Selective or low-intensity forestry enables some forest species to survive, whereas virtually all will perish if forested land is converted to agricultural use.

The environmental requirements for the survival of the orangutan (*Pongo* spp.) are of particular interest because they require a very large continuous forested area for their survival *in situ*. Therefore, the economic opportunity cost of conserving the orangutan by completely protecting forested areas is high. Nevertheless, because of the human-like features of the orangutan (see Plous, 1993; Tisdell et al., 2005), residents of higher income countries probably strongly support its conservation, whereas in those developing countries where it is present, support for conserving the orangutan by fully protecting forest areas is likely to be limited because of the high economic sacrifice involved.

Rijksen and Meijaard (1999) estimated that a minimum viable metapopulation of 5000 adult orangutans is needed for the survival of this species and typically this requires about 10000 km^2 of suitable habitat because orangutans are sparsely distributed in their habitats: densities in Borneo range from 0.1 to 3.6 per km^2 (Soehartono et al., 2007, p. 8). On the other hand, Singleton et al. (2004) state that a single population of at least 500 orangutans is needed for its long-term viability. If the carrying capacity of the conserved habitat is on average one orangutan per km^2, 500 km^2 (and if it is two on average, 250 km^2) of suitable habitat is needed to reach the target of Singleton et al. (2004). In any case, the amount of forested area required to conserve the orangutan is considerable and the opportunity cost of protecting it can be high. For example, if all the land involved is suitable for oil palm production, the pre-tax profit from oil palm production forgone is likely on average to be US$528–790 annually per ha (Swarna Nantha and Tisdell, 2009, p. 490).

While ecological aspects of selective logging have been explored for particular species, for example, by Lindenmayer and Possingham (1994) for Leadbeater's possum, little attention has been given to the economics of selective or low-intensity logging (sometimes called reduced impact logging, RIL) as a strategy for conserving forest species but considerable attention has been given to the relationship between spatial ecology and the optimization of managed ecosystems (Hof and Bevers, 1998; 2002).

Nevertheless, this problem has been entirely neglected by economists. For example, Bowes and Krutilla (1989, pp. 32–4 and pp. 36–7) compare the dominant-use management of forested areas with their multiple-use management and suggest that the latter type of management is usually superior. Later, Vincent and Binkley (1993) argued that efficient multiple-use forestry may require land-use specialization. Consequently, in contrast to Bowes and Krutilla, they believe dominant use of forests is likely to be more efficient than their multiple use. They argue that when a fixed level of management effort can be allocated between identical forest stands, economies in their management favour specialization in their purpose. 'Multiple products are produced at the forest level, but management at the stand level tends towards dominant use' (Vincent and Binkley, 1993, p. 373).

Boscolo and Vincent (2003) develop this theme further, and argue that non-convexities in timber production, biodiversity provision and carbon sequestration are important and favour specialization in land use. They find that joint production of biodiversity and timber is likely to be less economic than the allocation of the forested area by dominant use. Using Malaysian data, they conclude that dominant use allocation of forested land 'is often likely to be superior to uniform management for the production of biodiversity and timber' (Boscolo and Vincent, 2003, p. 266). Despite this finding, it is important to examine more specifically the trade-offs involved in mixed land-use compared to partitioned use of land (in this case, a forest). The simple linear model presented below enables conditions to be found for which multiple use is optimal and others for which dominant use is optimal.

17.5 INTRODUCING THE MODELLING USED

The type of logging considered here is logging modified to assist the conservation of a focal wildlife species. It is a form of RIL (reduced impact logging), but it is reduced impact in relation to the conservation of a particular species (or set of species). It is a form of selective logging. While this type of selective logging modifies conventional logging, it is unclear whether it would be as financially demanding on loggers as RIL or sustainable forest management (SFM), the economics of which is reviewed by Putz et al. (2008) and by García-Fernández et al. (2008, p. 1468).

The objective of the modelling here is to determine whether multiple use or dominant use of a forest is the most economic means to conserve a targeted population level of a forest-dependent species, for example, a minimum viable level of population of the species. It considers whether it is more economic to conserve a focal forest species by:

1. Setting aside a fully protected area that enables a minimum viable population (or some other target level of population) of the species to survive and then allow heavy logging of the remainder of the land.
2. Or, not to set aside a protected area and allow light (selective) logging (multiple use) over a sufficient area to maintain a targeted viable population of the focal species.

Option 1 involves specialized partitioned use of the forested area by dominant purposes, whereas Option 2 involves mixed (multiple) land-use of a portion of the forested area with the remainder of this area (if any) being used for a single purpose, namely heavy logging. Note that, in practice, the level of population of a species constituting a minimum viable population is not certain. Usually the probability of its survival increases with the level of its population able to be supported by its environment (Hohl and Tisdell, 1993). Depending on the probability of survival of a focal species wanted by a policy-maker, the target level of the population of the focal species can be adjusted in the model detailed below. If a higher probability of survival is desired, its population threshold is adjusted upwards.

It might be objected that the model presented here concentrates on the conservation of a single desired species in a forested area and that it is linear. However, these objections can be exaggerated. First, in practice, a single wildlife species or a small set of such species is often the object of conservation in a particular geographical location.[1] Second, the model outlined can be extended to include the conservation of a focal set of wildlife species by specifying population targets for each of the species in the set. However, the larger the set, the greater are likely to be the conservation restraints on selective logging. Third, in some cases, the species to be conserved is an umbrella species, and consequently, the habitats preserved for its conservation will also enable some other valued species to survive. Fourth, the geographical dimensions of biodiversity conservation need to be considered. Boscolo and Vincent (2003) concentrate on the conservation of local biodiversity but the conservation of global biodiversity may be more important. If an endangered wildlife species is conserved in a locality at the expense of other common wildlife species not in danger of extinction, this helps to maintain global biodiversity even though local biodiversity is reduced. Consequently, the approach taken here can support global biodiversity conservation.

The fact that the modelling here assumes linear relationships is another possible criticism. However, linear relationships can be regarded as approximations to non-linear ones. They are also useful in that their coefficients can as a rule be readily approximated empirically. Furthermore, linear models can be regarded as an initial step towards developing non-linear ones. Hof and Bevers (1998, Ch. 5) make it clear that important insights into spatial optimization of the use of ecosystems can be obtained from linear models.

The analysis proceeds as follows: the general relationships between the density of a focal species and the intensity or selectivity of logging in its range (as well as the profitability of logging in relation to logging intensity) are explored. In this exposition, logging intensity is treated as a continuous variable, and the decision-making problem is then simplified by focusing on just two alternative intensities of logging, namely heavy logging and light (selective) logging. Choices based on these options are modelled and it is found that the economic value of a dominant land-use policy (a land-sparing policy) compared to a mixed land-use policy (a land-sharing policy) cannot be determined a priori. The specific factors that determine this optimal choice are identified, and consequently the main problem posed in this chapter is solved. The concluding discussion identifies important limitations of the model.

17.6 RELATIONSHIPS BETWEEN THE SELECTIVITY OF LOGGING, THE DENSITY OF A FOREST SPECIES AND THE PROFITABILITY OF LOGGING

The density of the population of forest wildlife species depends on the intensity (or the selectivity) of logging and so does the profitability of logging. A problem arises in specifying the relationships precisely, because logging processes can be diverse and can have multiple attributes or characteristics. For simplicity, however, suppose that logging intensity (or selectivity) can be represented by a variable, x, which falls in the range $0 \leq x \leq 1$, where zero means that logging is absent and 1 implies that it is at its maximum intensity. An alternative interpretation is that zero corresponds to an extremely high degree of selectivity so the survival of the focal species is not adversely affected by its changed habitat, whereas 1 corresponds to a situation in which no selectivity is practised in removing trees.

Suppose that y represents the density per km^2 of the focal forest-dependent species (for example, orangutans). Then the long-term density of this species as a function of the intensity of logging might be like that shown by relationship ABCD in Figure 17.1. The long-term density of the species declines as logging intensity approaches x_1 and then falls to zero.

However, for some forest species, this density may at first rise and then fall because they benefit from some forest disturbance. Also for some species, a precipitous decline at a threshold like that of x_1 in Figure 17.1 may not occur – the decline may be more gradual. Furthermore, the relationship involved may be non-linear. For instance, the density of the species may decline at an increasing rate as the intensity of logging increases.

Alternative types of logging are diverse. Even light logging can be disadvantageous to some species if it targets the harvest of trees on which these species depend. It is assumed that at a low intensity of logging, the loss of forest resources on which the focal species depends is low but increases with the intensity of logging. Husson et al. (2008, p. 92) found that orangutan 'densities' are lower in moderately to heavily logged forest

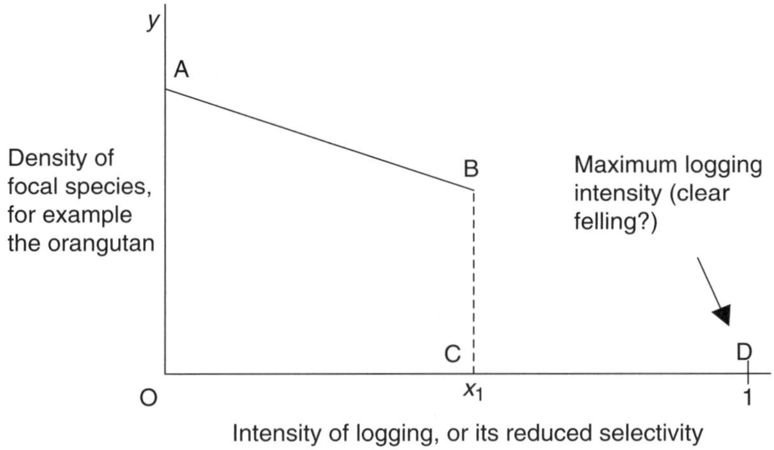

Figure 17.1 A hypothetical relationship between the density of a focal forest-dependent species and the intensity of logging (or its reduced selectivity) in its abode

than in unlogged areas of comparable habitat, in accordance with the majority of studies already published on this subject. In general, they find that logged sites have a lower density of orangutans than unlogged sites (Husson et al., 2008, p. 93). Nevertheless, little difference seems to occur between the density of orangutans in unlogged areas and those in 'lightly' logged areas. This is because when light logging takes place in the habitats of the orangutan, it is usually trees of the Diptercarpaceae family that are logged. These trees are not an important food source for primates, including orangutans (Husson et al., 2008, p. 83). Husson et al. come to the qualified conclusion that 'a well-managed, selective-logging operation that only removes those species of trees and does minimal damage to the surrounding forest may not significantly alter the forest structure and food availability from an orangutan's perspective' (Husson et al., 2008, p. 83).

The profit per km^2 from logging also depends on the intensity of logging or how selective it is. Profit is usually higher for conventional (higher intensities of logging) than for low intensity logging. One possible relationship between the profitability of logging and its intensity is shown in Figure 17.2 by curve ODEFG. In this case, logging at an intensity of less than x_0 is unprofitable. Profit is positive for intensities of logging greater than x_0 and is assumed to rise as the intensity (less selectivity) of logging increases.

In Figure 17.2, x_1 is the maximum intensity of logging potentially compatible with the long-term survival of the focal species (see Figure 17.1).

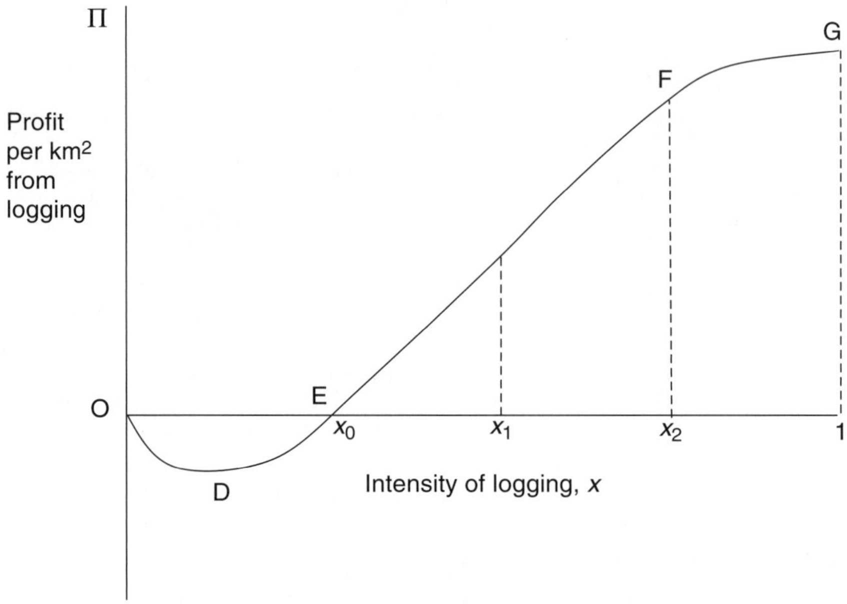

Figure 17.2 A possible relationship between the profitability per km² from logging and its intensity (or reduced selectivity)

However, in reality, this threshold is uncertain. Therefore, to be on the safe side if the survival of the species is at stake, it may be wise to opt for a lower intensity of logging than this.

Given a compact forested area of size H km^2, the envisaged spatial optimization problem is to organize the spatial management of the forest so that profit from logging is maximized subject to the condition that a focal wildlife species continues to survive at a population level K, where K is the targeted viable population of this species within this forested area. For simplicity, two discrete alternative spatial strategies for the allocation of the forested area (H km^2) for logging and for the conservation of the focal species[2] are examined. These are:

1. Completely protect a portion of the total available forested area so as to conserve just K of the focal species and allow the remaining portion to be used for high intensity logging.[3]
2. Do not completely protect any part of the total forested area but allow light logging in a portion of it sufficient to conserve a population K of the focal species and permit heavy logging in the remainder. The level of light logging and heavy logging are assumed to be pre-specified.

Note that heavy logging is assumed to be sufficiently heavy to extinguish the focal species. Further, light logging (which might also be designated moderate logging) is assumed to reduce the density of the focal species compared to no logging. Given the two alternative strategies listed above, which is the more profitable alternative?

17.7 THE SOLUTION TO THIS PROBLEM

In order to easily visualize the problem, consider Figure 17.3. The total available forested area for allocation (H km^2) is equal to the area of the rectangle marked A plus that marked B and is shown in the top portion of Figure 17.3. The area is assumed (on average) to be uniform in the quality of its forest and in the density of the focal species it can carry.[4] Option 1 is illustrated by the top large rectangle and Option 2 by the lower one. In case 1, the area of the rectangle identified by A is fully protected so as to conserve K of the focal species and the remainder is used for heavy logging (or it could be used for plantations or other forms of agriculture). In case 2, the protected area is made available for light logging and an area indicated by the rectangle identified by C is withdrawn from heavy logging and

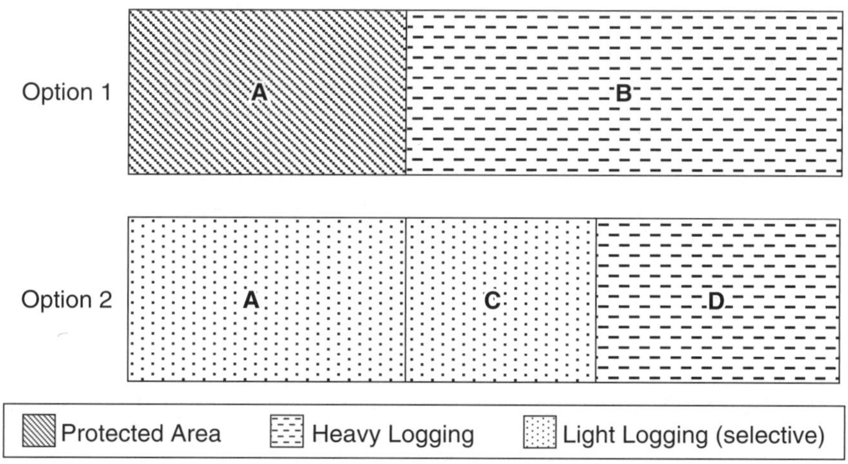

Figure 17.3 *An illustration of the alternatives of (1) a combination of a protected area (A) plus a heavily logged area (B); and (2) a lightly logged (selective) area (A + C) plus a smaller heavily logged area (D) (no fully protected area exists) as ways of conserving a targeted level of the population of a forest-dependent focal species*

used for light logging so as to ensure that the lightly logged area conserves K of the focal species.[5] The rectangular shapes of the forested land area and its uses are assumed for ease of exposition.

The change in total profit from logging when Option 2 rather than Option 1 is adopted equals the increase in profit from being able to log area A, less the reduction in profit from area C as a result of altering its use from heavy to light logging. If the former amount exceeds the latter amount, profit from logging rises as a result of the changed strategy for conserving the focal wildlife species. On the other hand, if the latter amount exceeds the former amount, profit from logging falls as a result of the change in strategy.

To examine this matter further, consider some mathematical analysis. Let K represent the target population of the focal wildlife species, and let λ be its density per km^2 on protected land and $\theta\lambda$ be that on lightly logged land where $0 < \theta < 1$. The species is assumed to disappear in the long term on heavily logged land. In this case, if Option 1 is adopted, the required size of protected area A in Figure 17.3 is K/λ km^2. If Option 2 is adopted and light logging occurs, the area needed to conserve K of the species (marked A plus C) is $K/\theta\lambda$ km^2. [It is assumed here that this, $K/\theta\lambda$ km^2, is not greater than H km^2. If it is, light logging is incompatible with the survival of the targeted level, K, of the focal species.] Therefore, the area marked C equals $(K/\theta\lambda - K/\lambda)$ km^2.

Suppose that the profit from heavy logging is Π per km^2 and that from light logging is $\varepsilon\Pi$ where $0 < \varepsilon < 1$. The total change in returns from logging when Option 2 is adopted rather than Option 1 can be expressed as:

$$\Delta R = \frac{K}{\lambda}\varepsilon\,\Pi - \left(\frac{K}{\theta\lambda} - \frac{K}{\lambda}\right)(\Pi - \varepsilon\Pi) \qquad (17.1)$$

$$= \frac{K}{\lambda}\varepsilon\Pi - \left(\frac{K}{\theta\lambda} - \frac{K}{\lambda}\right)(1 - \varepsilon\Pi) \qquad (17.2)$$

The first term on the right-hand side of these equations is the increased profit from being able to lightly log the previously protected area and the second term is the reduction in profit from having to forgo heavy logging in an area corresponding to area C in Figure 17.3.

Other things being held constant, the likelihood that ΔR is negative (that is that light logging is less financially rewarding than the alternative) increases as ε becomes smaller (that is, the larger is the reduction in profit per km^2 from light logging compared to that from heavy logging). Secondly, other things held constant, the smaller is θ

(that is, the larger the reduction in the density of the focal species when light logging occurs compared to no logging), the more likely is the light logging strategy to give lower returns from logging than Option 1. This is because the coefficient $(\frac{K}{\theta\lambda} - \frac{K}{\lambda})$ in equation (17.2) increases as θ becomes smaller.

17.8 THE SENSITIVITY OF THE OPTIMAL LAND USE TO CHANGED PARAMETERS

The above analysis identifies conditions for which it is economically optimal to follow a fully protected land policy for a portion of forested land compared to a light logging type of policy to ensure the survival of a focal forest-dependent wildlife species, and vice versa. Estimates of only a few parameters are needed to make an optimal economic choice. This is a practical advantage. Note that (given the wildlife conservation constraint) it cannot be decided *a priori* whether a protected area policy combined with heavy logging or a light logging type of policy is the more profitable land-use policy from the point of view of loggers. This is illustrated by the following simple examples. If both θ and ε equal 0.5, Expression (17.2) reduces to zero. In this case, light logging results in both a halving of the density of the focal species per km^2 and a halving of profit per km^2 compared to heavy logging. In Figure 17.3, it implies that the area marked by C equals the area identified by A.

Now consider the example in which $\varepsilon > 0.5$, (that is, the profit per km^2 from light logging is less than half that from heavy logging) and $\theta = 0.5$ (the density of the focal species is halved compared to that in the absence of logging). The light logging option raises returns from logging compared to Option 1 in this case, other things being unchanged. If $\varepsilon < 0.5$ and $\theta = 0.5$, then the opposite result follows.

Consider now variations in θ (the density of the focal species) when ε (the light logging profitability coefficient) is set at 0.5. If $\theta > 0.5$, the density of the focal species falls by less than a half when light logging occurs compared to no logging, and the returns from Option 2 exceed those from Option 1. On the other hand, if $\theta < 0.5$, the returns from Option 2 are less than those for Option 1.

In general, the likelihood that the light logging option (Option 2) gives greater total returns from forestry than the protected area option (Option 1) increases the smaller is the reduction in the density of a focal species under conditions of light logging compared to its density when no logging occurs, and the smaller is the reduction in profit from logging per km^2 when light rather than heavy logging is practised. This assumes that a given targeted level of the focal species is to be conserved. The opposite relationship also holds.

The model presented here is most relevant to conserving a single species, such as the orangutan, at minimum economic cost in terms of profit forgone. Therefore, it concentrates on maintaining a single ecosystem service. If a wider range of ecosystem services is desired, this may strengthen the case for setting aside fully protected areas. On the other hand, circumstances can arise when it is socially defensible to support the light logging option even though it is not the least cost one in terms of profit forgone. The light logging option may, for example, be accepted politically because it ensures greater employment spread over a wider geographical area than setting aside a fully protected area and allowing intensive forest use outside of this area.

17.9 FURTHER DISCUSSION

The previous modelling assumes static spatial relationships and corresponds to models outlined by Hof and Bevers (1998) in Part I of their *Spatial Optimization for Managed Ecosystems*. While such models have relevant applications (as pointed out by Hof and Bevers, 2002, Ch. 5), dynamic changes in habitat availability can be important in assessing the conservation of many wildlife species, and the analysis should be extended to increase its generality. However, such extensions can be complex. For instance, the relationship (illustrated in Figure 17.1) showing the density of a forest species as a function of the intensity of logging should be extended to take account of the passage of time and to allow for additional spatial dimensions of environmental change. For example, the relationship ABCD shown in Figure 17.1 (replicated by ABCD in Figure 17.4) might apply to the density of orangutans soon after an area has been logged. However, after a longer period of time, the forest regenerates and a relationship like that shown in Figure 17.4 by AECD might apply if the population of the focal species in the area is not subject to migration. If immigration of members of the focal forest species, for example, orangutans, from other areas is possible, then after enough time elapses, a relationship like that shown by AEFG in Figure 17.4 might apply on previously logged land.

Ideally, spatial and dynamic factors should be incorporated into the above analysis. For example, the conservation of selected small fully protected reserves (refuges) in logged landscapes can be valuable in facilitating recolonization of a logged area once it begins to regenerate. While these refuges may not (on their own) support minimum viable populations of focal forest species in the long term, they may support the populations for

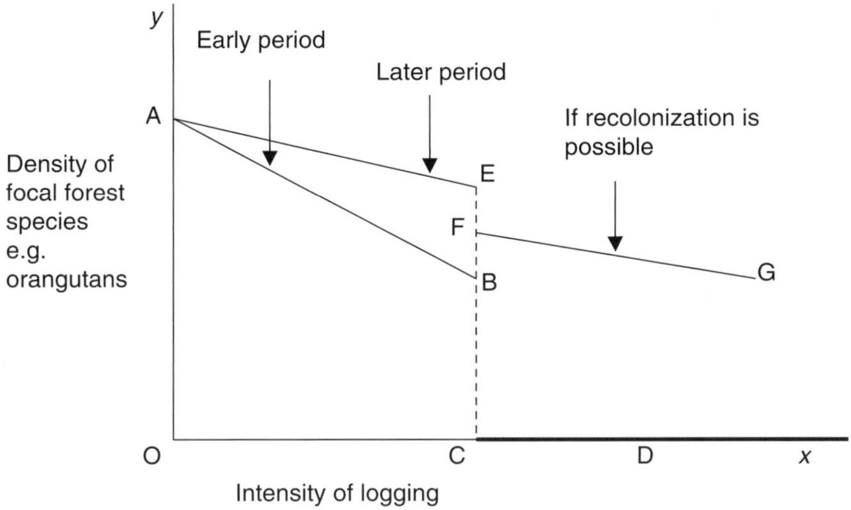

Figure 17.4 *Diagram used to highlight the fact that the density of a forest species following logging depends on the passage of time and the scope for a species to recolonize a logged area*

sufficient time to enable recolonization of nearby logged areas to occur. Consequently, they increase the chances of survival of the species locally. The provision of such refuges can be regarded as another form of selective logging and is also amenable to economic analysis. The appropriate selective spatial conservation patterns will, however, vary with the ecology of the forest species to be conserved. Therefore, extension of the analysis is probably best done by specifying the requirements for the conservation of specific forest species or groups of these in logging environments rather than by listing the myriad of general abstract possibilities.

Available evidence indicates that light-to-moderate selective logging does not endanger the orangutan. However, logging which is intense, rapid and present over a large area is a serious threat to its survival, as is the geographical spread of agriculture (Tisdell and Swarna Nantha, 2008; 2009). There are also forest species other than the orangutan for which selective logging is compatible with their conservation. As the number of different forest species to be conserved at a minimum viable population (or a higher target level) increases when selective logging is practised, the economic returns from selective logging are likely to decrease, because as the number of forest species to be conserved increases, the number of the habitats that need to be conserved usually also increases. Consequently, loggers will be required to be ever more selective in logging and to take increasing care of the environment. For example, more restrictions may be placed on the type of trees which can be harvested and on landscape patterns to be preserved. This can be expected to reduce the profitability of selective logging. A stage is therefore likely to be reached where selective logging is no longer a viable economic option for conserving the set of focal species and dominant use at the landscape level is likely to be a more economic solution than selective logging.

Note that the modelling used here does not provide a complete total social cost–benefit analysis of the economics of the use of forest ecosystems. Most forests have uses other than for logging and wildlife conservation and the economic value of these will vary with the intensity and selectivity of logging. For example, forests provide ecoservices (such as hydrological services, carbon sequestration), may cater for ecotourism, in some cases have visual value (see, for example, Haight et al., 1992), and can provide subsistence products to local villagers in developing countries. While intense logging is likely to reduce the economic value of these services substantially, this may not happen if light selective logging is practised.

The agency costs of enforcing environmental regulations should also be considered in deciding whether the conservation of forest wildlife species should be combined with logging or whether these ought to be spatially separated. The agency costs of monitoring and enforcing environmental regulations are likely to be higher when forests are jointly used for multiple purposes than when the forested area is partitioned and each portion is only used for a single purpose. Furthermore, when the available funds for the supervision of forest use and the enforcement of relevant environmental regulations are limited, the prevalence of illegal logging is likely to increase if the use of the forest is not partitioned.

An additional factor to consider when forestry is combined with wildlife conservation is its effect on the resilience and robustness of habitats required to conserve a focal species. If their disturbed habitats are less resilient and less robust than in undisturbed forests, this can be relevant to deciding whether multiple or single-purpose use of portions of the forest is desirable. For example, it has been argued that populations of many wildlife species show less persistence in fragmented habitats (Hof and Bevers, 1998, p. 3). Where multiple use is practised, a margin of safety may have to be built into the logging regime as a precaution to address this problem. This places extra constraints on selective logging, adds to its costs and reduces the returns from logging.

The basic model introduced in this chapter can be applied to other similar environmental problems. For example, suppose a forested watershed is to be spatially managed to ensure that the quality of water flowing from this area is of a particular standard. This quality is assumed to depend on the nature of the tree cover in the watershed. Different combinations of heavy and selective logging on the forested land may result in the water quality standard being satisfied. The problem then is to determine the spatial combination of logging (selective and otherwise) that supplies the environmental service required (a specified level of water quality) and that also maximizes the return from logging. A more detailed analysis of this type of problem is available in Hof and Bevers (2002, Chs 2 and 3). The solution cannot be determined a priori. In particular, multi-purpose (multi-functional) use of all the forested land may not be the most economical way to achieve the valued environmental service. This is also the case for other types of land use, for example, the use of land for agriculture (see, for example, Tisdell, 2009).

17.10 CONCLUSIONS

The United Nations Earth Summit held in Rio de Janeiro in 1992 stated in Chapter XI of Agenda 21 'the need for sustaining the multiple roles and functions of forests, and for enhancing their protection, sustainable management and conservation'. This need was further reaffirmed in 2000 in the United Nations Millennium Declaration (Chapter IV) and at the Second Earth Summit held in Johannesburg in 2002 (Sist et al., 2008, p. vii). While the development of methods for RIL pays attention to these objectives, this development pays little or no attention to the conservation of species. Sist et al. (2008, p. vii) observe:

> Most RIL operations are still based, as are all other selective logging systems operating in the tropics, on a very simple rule: the minimum diameter cutting, which is applied to all commercial species. These cutting limits are set to accommodate processing technologies and market demands, rather than the biology and conservation of the harvested species (Sist et al., 2003).

In addition, RIL methods fail to take account of the conservation of non-harvested species, such as forest wildlife species, some of which may, nevertheless, indirectly benefit from RIL. Therefore, an even more holistic approach to managing tropical forests than that envisaged by Sist et al. (2008) is called for.

The desirability of selective logging can be assessed from different points of view. Of particular interest to foresters is the ability of RIL to maximize the (discounted) economic yield from forests when only a forest's value for commercial logging is considered. However, this ignores the value of other commodities produced by forest ecosystems, including their provision of many environmental services. Nevertheless, the above-mentioned objective of RIL can be an appropriate goal on forested land allocated exclusively for logging if the use of the forested land has been optimally partitioned in a manner that maximizes the overall net value of the forest ecosystem in satisfying environmental constraints and in producing logs.

To conclude: economic optimization is unlikely to require the supply of *all* ecosystem services in all places where they are capable of being supplied. In fact, this is likely to be inconsistent with maximizing the total economic value obtained from an ecosystem as a whole.[6] It can be optimal to use different portions of the ecosystem to supply different sets of services. In some portions of an ecosystem, it may be optimal to supply sets of ecosystem services jointly, but in other portions it can be best to focus on supplying a single service. In some cases, it can even be economically optimal to suitably partition an area to ensure that each of its portions optimally supplies a single ecosystem service. In this case, different single ecosystem services are provided by different portions of the ecosystem. There is also an interdependence issue to consider. This is because the economic value of ecosystem services supplied in any portion of an ecosystem depends upon the quantity supplied of these services by other portions of the ecosystem as well as on the demand for these services. It is essential that such factors be taken into account when valuing the economic supply of ecosystem services. More importantly, the above analysis demonstrates that it is dangerous to assume a priori that land sparing is superior to land sharing (or vice versa) in ensuring the optimal supply of wanted ecosystem services. Optimal patterns of land use depend on the circumstances.[7]

NOTES

1. In this type of analysis, it is not unusual to concentrate on spatial management of the use of forests to conserve particular wildlife species. For example, Hof and Bevers (2002) do this in considering the conservation of the northern spotted owl and the western prairie fringed orchid.
2. Although the consequences of only two alternative forestry regimes are examined here, a larger number of alternatives can be considered by means of iteration and search procedures using computation.
3. The model outlined here assumes that a particular level of the population of a focal wildlife species should be conserved or, more generally, that a desired environmental service provided by an ecosystem should be of a specified value. However, this requirement (a standard one in many cases) may also be expressed as an inequality. For example, there may be an overarching requirement that the population of a selected wildlife species should equal or exceed a particular level. Or in relation to an example introduced later in this chapter, the quality of run-off water from a catchment should equal or exceed a specified standard. In the situations considered here, these inequalities become binding. This is because it is supposed that as logging becomes less selective, the economic returns from logging increase but the supply of the focal environmental service is reduced.

4. This assumption can be consistent with some heterogeneity within subdivisions (cells) of the forest.
5. The possibility set for land use shown in Figure 17.3 assumes that the whole of the forested area does not have to be fully protected to conserve the focal wildlife species. For example, if λ is the average density per km^2 of the focal wildlife species within a completely protected forested area, it is assumed that $K/\lambda < H$. If K/λ equals H, there is no scope for logging in this model. Where $\theta\lambda$ is the density of the focal wildlife species on selectively logged land and $0 < \theta < 1$, it is also supposed that $K/\lambda \leq H$. If this expression exceeds H, there is no scope for conserving the focal species at its targeted level of population by engaging jointly in selective logging. In the case where this expression equals H, set D in Figure 17.3 is empty, because all the forested area must be used to serve multiple purposes if the focal wildlife species is to attain its targeted level of population. In any particular case, these boundary possibilities should be checked.
6. The presence of infra-marginal externalities contributes to such results (see, for example, Tisdell, 1970; 2005, Ch. 5; Tisdell, 2009; Walsh and Tisdell, 1973). Note that if the supply of a favourable environmental service is infra-marginal, there is no case, from a Paretian efficiency point of view, to pay for this benefit. Payment for an environmental service is misplaced in this case. Furthermore, in this case, the *marginal value* of the supply of such an environmental service beyond some threshold value is *zero*.
7. Sections 17.3–17.10 of this chapter contain (with minor changes) material which was previously published in 2014. The details are C.A. Tisdell (2014), 'Conserving forest wildlife and other ecosystem services: opportunity costs and the valuation of alternative logging regimes' in K.N. Ninan (ed.), *Valuing Ecosystem Services: Methodological Issues and Case Studies*, Cheltenham, UK and Northampton, MA, USA: Edward Elgar Publishing.

REFERENCES

Boscolo, M. and J.R. Vincent (2003), 'Nonconvexities in the production of timber, biodiversity and carbon sequestration', *Journal of Environmental Economics and Management*, **46**, 251–63.

Bowes, M.D. and J.V. Krutilla (1989), *Multiple-use Management: The Economics of Public Forestlands*, Washington, DC: Resources for the Future.

Brussaard, L., P. Caron, B. Campbell, L.M. Lipper, S. Manika, R. Rabbinge, D. Babin and M. Pullerman (2010), 'Reconciling biodiversity consideration and food security: Scientific challenges for a new agriculture', *Current Opinion in Environmental Sustainability*, **2**, 34–42.

García-Fernández, C., M. Ruiz-Pérez and S. Wunder (2008), 'Is multiple-use forest management widely implementable in the tropics?', *Forest Ecology and Management*, **256**, 1468–76.

Haight, R.G., R.A. Monserud and J.P. Chew (1992), 'Optimal harvesting with stand density targets in managing Rocky-Mountain conifer stands for multiple forest outputs', *Forest Science*, **38**, 554–76.

Hof, J.G. and M. Bevers (1998), *Spatial Optimization for Managed Ecosystems*, New York: Columbia University Press.

Hof, J.G. and M. Bevers (2002), *Spatial Optimization in Ecological Applications*, New York: Columbia University Press.

Hohl, A. and C.A. Tisdell (1993), 'How useful are environmental safety standards in economics? The example of safe minimum standards for protection of species', *Biodiversity Conservation*, **2** (2), 168–81. Reprinted in C.A. Tisdell

(2002), *The Economics of Conserving Wildlife and Natural Areas*, Cheltenham, UK and Northampton MA, USA: Edward Elgar Publishing.

Husson, S.J., S.A. Wich, A.J. Marshall, R.A. Dennis, M. Acrenaz, R. Brassey, M. Gumal, A.J. Hearn, E. Meijaard, T. Simorangkir and I. Singleton (2008), 'Orangutan distribution, density, abundance and impacts of disturbance', in S.A. Wick, S.S.U. Atmoko, T.M. Setia and C.P. Scheik (eds), *Orangutans: Geographical Variation in Behavioural Ecology and Conservation*, Oxford: Oxford University Press, pp. 77–96.

Lindenmayer, D.B. and H.P. Possingham (1994), *The Risk of Extinction: Ranking Management Options for Leadbeater's Possum Using Population Viability Analysis*, Canberra, ACT: Centre for Resource and Environmental Studies, Australian National University.

Payne, J. and C. Prudente (2008), *Orangutans: Behaviour, Ecology and Conservation*, London: New Holland Publishers.

Phalan, B., M. Orial, A. Balmford and R.E. Green (2011), 'Reconciling food production and biodiversity conservation: Land sharing and land sparing compared', *Science*, **333**, 1289–91.

Plous, S. (1993), 'Psychological mechanisms in the human use of animals', *Journal of Social Issues*, **49**, 11–52.

Putz, F.E., P. Sist, T. Fredericksen and D. Dykstra (2008), 'Reduced-impact logging: Challenges and opportunities', *Forest Ecology and Management*, **256**, 1427–33.

Rijksen, H.D. and E. Meijaard (1999), *Our Vanishing Relative: The Status of Wild Orangutants at the Close of the Twentieth Century*, Dordrecht: Kluwer.

Singleton, I., S.A. Wich, S.J. Husson, S.S.U. Atmoko, M. Leighton, N. Rosen, K. Traylor-Holzer, R. Lacy and O. Byers (2004), *Orangutan Population and Habitat Viability Assessment: Final Report*, Apple Valley, MN, USA: IUCN/SSC Conservation Breeding Specialist Group.

Sist, P., C. García-Fernández and T. Fredericksen (2008), 'Moving beyond reduced-impact logging towards a more holistic management of tropical forests', *Forest Ecology and Management*, **256**, vii–ix.

Sist, P., R. Fimbel, R. Nasi, D. Sheil and M.-H. Chevallier (2003), 'Towards sustainable management of mixed diptorcorp forests of South East Asia: Moving beyond minimum diameter cutting limits', *Environmental Conservation*, **30**, 364–74.

Soehartono, T., H. Djoko Susilo, N. Andayani, S.S. Utami Atmoko, J. Sihite, C. Saleh and A. Sutrisno (2007), *Strategi dan Rencana aksi Konservasi Orangutan Indonesia 2007–2017*, Departemen Kehutanan. Jakarta: Direktorat Jenderal Perlindungan Hutan dan Konservasi Alam. Available at: http://www.dephut.go.id/files/Orangutan%20Action%20Plan%202007-2017_0.pdf. (Accessed 5 June, 2008).

Swarna Nantha, H. and C.A. Tisdell (2009), 'The orangutan–oil palm conflict: Economic constraints and opportunities for conservation', *Biodiversity Conservation*, **18**, 487–502.

Tisdell, C.A. (1970), 'On the theory of externalities', *The Economic Record*, **46** (113), 14–25.

Tisdell, C.A. (2003), *Economics and Ecology in Agriculture and Marine Production: Bioeconomics and Resource Use*, Cheltenham, UK and Northampton, MA, USA: Edward Elgar Publishing.

Tisdell, C.A. (2005), *Economics of Environmental Conservation*, 2nd edn, Cheltenham, UK and Northampton, MA, USA: Edward Elgar Publishing.

Tisdell, C.A. (2009), 'Complex policy choices regarding agricultural externalities: Efficiency, equity and acceptability', in V. Beckmann and M. Padmanabhan (eds), *Institutions and Sustainability: Political Economy of Agriculture and the Environment – Essays in Honour of Konrad Hagedorn*, Dordrecht: Springer, pp. 83–106.

Tisdell, C.A. (2014), *Human Values and Biodiversity Conservation: The Survival of Wild Species*, Cheltenham, UK and Northampton, MA, USA: Edward Elgar Publishing.

Tisdell, C.A. and H. Swarna Nantha (2008), 'Conservation of the proboscis monkey and the orangutan in Borneo: Comparative issues and economic considerations', in V.K. Gupta and A.K. Verma (eds), *Perspectives in Animal Ecology & Reproduction*, 5, Delhi: Daya Publishing House, pp. 225–50.

Tisdell, C.A. and H. Swarna Nantha (2009), 'Supply-side policies to conserve biodiversity and save the orangutan from oil palm expansion: An economic assessment', in J.B. Aranoff (ed.), *Handbook of Nature Conservation – Global, Environmental and Economic Issues*, New York: Nova Science Publications, pp. 171–80.

Tisdell, C.A. and D. Xue (2013), 'Managing ecosystem services for human benefit: Economic and environmental policy challenges', in E. Creighton and P. Davovich (eds), *Environmental Policy: Management, Legal Issues and Health Aspects*, New York: Nova Science Publishers, pp. 87–106.

Tisdell, C.A., C. Wilson and H. Swarna Nantha (2005), 'Association of public support for survival of wildlife species with their likeability', *Anthrozoos*, **18** (2), 160–74.

Vincent, J.R. and C.S. Binkley (1993), 'Efficient multiple-use forestry may require land-use specialization', *Land Economics*, **69**, 370–76.

Walsh, C. and C. Tisdell (1973), 'Non-marginal externalities: As relevant and as not', *The Economic Record*, **49** (3), 447–55.

Index